Nanomaterials for Environmental Applications

Emerging Materials and Technologies

Series Editor:
Boris I. Kharissov

Functional Nanomaterials for Regenerative Tissue Medicines
Mariappan Rajan

Uncertainty Quantification of Stochastic Defects in Materials
Liu Chu

Recycling of Plastics, Metals, and Their Composites
R.A. Ilyas, S.M. Sapuan, and Emin Bayraktar

Viral and Antiviral Nanomaterials
Synthesis, Properties, Characterization, and Application
Devarajan Thangadurai, Saher Islam, and Charles Oluwaseun Adetunji

Drug Delivery using Nanomaterials
Yasser Shahzad, Syed A.A. Rizvi, Abid Mehmood Yousaf, and Talib Hussain

Nanomaterials for Environmental Applications
Mohamed Abou El-Fetouh Barakat and Rajeev Kumar

Nanotechnology for Smart Concrete
Ghasan Fahim Huseien, Nur Hafizah A. Khalid, and Jahangir Mirza

Nanomaterials in the Battle Against Pathogens and Disease Vectors
Kaushik Pal and Tean Zaheer

MXene-Based Photocatalysts: Fabrication and Applications
Zuzeng Qin, Tongming Su, and Hongbing Ji

For more information about this series, please visit: https://www.routledge.com/Emerging-Materials-and-Technologies/book-series/CRCEMT

Nanomaterials for Environmental Applications

Edited by
Mohamed Abou El-Fetouh Barakat and
Rajeev Kumar

CRC Press is an imprint of the
Taylor & Francis Group, an **informa** business

First edition published 2022
by CRC Press
6000 Broken Sound Parkway NW, Suite 300, Boca Raton, FL 33487-2742

and by CRC Press
2 Park Square, Milton Park, Abingdon, Oxon, OX14 4RN

CRC Press is an imprint of Taylor & Francis Group, LLC

Library of Congress Cataloging-in-Publication Data
Names: Barakat, Mohamed Abou El-Fetouh, editor. | Kumar, Rajeev, editor.
Title: Nanomaterials for environmental applications / Mohamed A. Barakat
Barakat and Rajeev Kumar.
Description: First edition. | Boca Raton: CRC Press, [2022] |
Series: Emerging materials and technologies | Includes bibliographical references and index. |
Summary: "The book offers a comprehensive review of the latest advances in nanomaterials-based technologies for the treatment of emerging contaminants in wastewater. It describes the latest developments in synthesis protocols, including synthesis of different kinds of nano-structure materials using various physical and chemical methods. Aimed at researchers and industry professionals, this work will be of interest to chemical, environmental, and materials engineered concerned with the application of advanced materials for environmental and water remediation"—Provided by publisher.
Identifiers: LCCN 2021047660 (print) | LCCN 2021047661 (ebook) | ISBN 9780367653385 (hbk) | ISBN 9780367653484 (pbk) | ISBN 9781003129042 (ebk)
Subjects: LCSH: Environmental protection—Materials. | Water—Purification—Materials. | Air—Purification—Matierials. | Nanostructured materials.
Classification: LCC TD170.2 .N355 2022 (print) | LCC TD170.2 (ebook) |
DDC 363.7—dc23/eng/20211115 LC record available at
https://lccn.loc.gov/2021047660 LC ebook record available at
https://lccn.loc.gov/2021047661

ISBN: 978-0-367-65338-5 (hbk)
ISBN: 978-0-367-65348-4 (pbk)
ISBN: 978-1-003-12904-2 (ebk)

DOI: 10.1201/9781003129042

Typeset in Times
by codeMantra

Contents

Preface

The release of toxic substances is not only causing pollution, but also responsible for the global warming and climate change. Emission and disposal of the pollutants into the natural environment cause adverse effects on the ecosystem. Several studies have reported that environmental pollution (soil, air and water pollution) is directly related to the climate change. The issue of environmental pollution becomes more challenging due to the existence of persistent and emerging contaminants such as pharmaceuticals and hormones, along with the conventional pollutants. Traditional methods and technologies are not much efficient to deal with the emerging contaminants. Nanotechnology is emerging as a potential technology for the identification, quantification and removal of environmental pollutants. Nanomaterials are playing a vital role in the development of innovative technology for the detection and decontamination of microbes, gases, and organic and inorganic pollutants from the air, water and soil.

This book describes the key features of a variety of nanomaterials used in the separation and purification of the contaminants from air, water and soil. This book briefly describes the fundamentals of some of the main environmental remediation techniques in which nanomaterials are used nowadays, together with some examples. A brief overview of the features and fabrication techniques of nanomaterials along with their physicochemical properties, environmental decontamination behaviour, reusability and mechanisms is given. A detailed account of the parameters governing the removal efficiency, such as role of morphology, facets exposed and its implication on the dynamics of surface species, is elucidated.

Furthermore, this book provides an insight into the materials commonly used for air, water and soil remediation. Challenges in the applications, such as performance stability in various conditions, scaling up, production efficiency and the possibility of secondary pollution, are highlighted. Applications of nanotechnology to combating airborne diseases, particularly COVID-19, are discussed in keeping up with the latest events. Moreover, this book also provides a comprehensive review of the major aspects of nanomaterials in the environment, elaborating on the sources, fate, transport, risk assessment and ecotoxicity. It also examines the current state of nanomaterials research to identify the knowledge gaps and justify the need for additional research.

Overall, this book is meant to be a reference book for students, researchers and scientists who are looking for advanced knowledge of the nanomaterials and their applications. The contributors are well-known researchers and scientists. We are very thankful to the authors of all chapters for their hard work in making of this book. In the end, all thanks to CRC Press/Taylor & Francis Group for publishing the book.

Editors

Mohamed Abou El-Fetouh Barakat is a Professor of Environmental Sciences at King Abdulaziz University, Saudi Arabia. He is an experienced academician and prolific researcher in the field of environmental sciences, including industrial waste management and pollution control, catalysis and nanotechnology, modified adsorbents, membrane technology, phytoremediation of pollutants in soil, recycling of industrial solid wastes, wastes to energy production and environmental impact assessment (EIA) studies.

Prof. Barakat has published more than 140 papers in ISI journals and two US patents. He has served as an editorial board member of many reputable journals, and he is a reviewer on Elsevier Editorial Board System. According to Scopus, his h-index is 47, and the number of citations totals to 10,340. He was awarded 'State Award for Excellence in Advanced Technology Sciences', Academy of Sciences, Egypt (2021), Jeddah Prize of Creativeness in Saudi Arabia (2018), Award of Excellences in Scientific Publications at KAU in Saudi Arabia (2011–2016), Environmental Research Award from the Academy of Scientific Research and Technology in Egypt (2003 and 2013) and King Hassan II Prize for the Environment in Morocco (2008).

Prof. Barakat is currently a professor in the Faculty of Meteorology, Environment and Arid Land Agriculture, King Abdulaziz University (KAU), Saudi Arabia. He is also a Professor of Mineral Technology at Central Metallurgical R& D Institute (CMRDI), Cairo, Egypt. His experience further includes academic research in Japan, the United States and Germany, as well as initiating and leading industrial research projects in Egypt and Saudi Arabia in collaboration with research institutions in the United States.

Rajeev Kumar received his M.Phil. (2008) and Ph.D. degrees in Applied Chemistry (2011) from Aligarh Muslim University, India. He is currently working as an associate professor at the Environmental Science Department, King Abdulaziz University, Jeddah, Saudi Arabia. His research activities are in the areas of wastewater treatment and materials science. Rajeev studies the adsorption and photocatalytic properties of nanomaterials for the removal of contaminants from wastewater. He has completed 14 research projects and published two books and 84 research articles. He is actively involved in the M.Sc. and Ph.D. thesis supervision. Recently, he received the Kingdom of Saudi Arabia Award for Environmental Management in the Islamic World, Second Prize, Best researcher, Edition 2018–2019.

Contributors

Mohammed Alsawat
Department of Chemistry, College of Science
Taif University
Taif, Kingdom of Saudi Arabia

Tariq Altalhi
Department of Chemistry, College of Science
Taif University
Taif, Kingdom of Saudi Arabia

Muzammil Anjum
Department of Environmental Sciences
PMAS Arid Agriculture University
Rawalpindi, Pakistan

Muhammad Mufti Azis
Department of Chemical Engineering, Faculty of Engineering
Universitas Gadjah Mada
Sleman, Indonesia

Mohamed Abou El-Fetouh Barakat
Department of Environmental Science, Faculty of Meteorology, Environment and Arid Land Agriculture
King Abdulaziz University
Jeddah, Saudi Arabia

Moisés Canle
Chemical Reactivity and Photoreactivity Group (React!)
Department of Chemistry, Faculty of Sciences & CICA
University of A Coruña
A Coruña, Spain

Archana Charanpahari
Chemistry Division
Berkeley Careers
Ghaziabad, India
and
Galgotias University
Greater Noida, India

Thabitha P. Dasari Shareena
RCMI Center for Environmental Health
Jackson State University
Jackson, Mississippi

Asok K Dasmahapatra
RCMI Center for Environmental Health
Jackson State University
Jackson, Mississippi

Hariprasad E
Department of Chemistry
Vasavi College of Engineering
Hyderabad, India

M. Isabel Fernández
Chemical Reactivity and Photoreactivity Group (React!)
Department of Chemistry, Faculty of Sciences & CICA
University of A Coruña
A Coruña, Spain

Sachin Ghugal
School of Chemical Sciences
Hyderabad Central University
Hyderabad, India

E Lokesh Goud
Department of Agronomy, School of Agriculture
Lovely Professional University
Jalandhar, India

Prerna Higgins
Department of Chemistry
Sam Higginbottom University of Agriculture Technology & Sciences
Allahabad, India

Antonius Indarto
Department of Bioenergy Engineering and Chemurgy, Faculty of Industrial Technology
Institut Teknologi Bandung
Bandung, Indonesia

Runit Isaac
Department of Chemistry
Sam Higginbottom University of Agriculture Technology & Sciences
Allahabad, India

Azeem Khalid
Department of Environmental Sciences
PMAS Arid Agriculture University
Rawalpindi, Pakistan

Bhupendra Koul
Department of Biotechnology, School of Agriculture
Lovely Professional University
Jalandhar, India

Prasann Kumar
Department of Agronomy, School of Agriculture
Lovely Professional University
Jalandhar, India

Rajeev Kumar
Department of Environmental Science, Faculty of Meteorology, Environment and Arid Land Agriculture
King Abdulaziz University
Jeddah, Saudi Arabia

Chandra Shekhar Kushwaha
Department of Polymer Science, Bhaskaracharya College of Applied Sciences
University of Delhi
New Delhi, India
and
Research Technology Develop Centre
Sharda University
Greater Noida, India

Mohamed H. H. Mahmoud
Department of Chemistry, College of Science
Taif University
Taif, Kingdom of Saudi Arabia
and
Central Metallurgical R & D Institute
Cairo, Egypt

Rashid Miandad
Department of Environmental Sciences
University of Peshawar
Peshawar, Pakistan

Mohamed Mobarak
Physics Department, Faculty of Science
Beni-Suef University
Beni-Suef, Egypt

Essam A. Mohamed
Faculty of Earth Science
Beni-Suef University
Beni-Suef, Egypt

Suhas Mukherjee
Propellant Engineering Division
Vikram Sarabhai Space Centre
Trivandrum, India

Daniel Pramudita
Department of Chemical Engineering, Faculty of Industrial Technology
Institut Teknologi Bandung
Bandung, Indonesia

Samia Qadeer
Department of Environmental Sciences
University of Narowal
Narowal, Pakistan

J. Arturo Santaballa
Chemical Reactivity and Photoreactivity Group (React!)
Department of Chemistry, Faculty of Sciences & CICA
University of A Coruña
A Coruña, Spain

Moaaz K. Seliem
Faculty of Earth Science
Beni-Suef University
Beni-Suef, Egypt

Ali Q. Selim
Faculty of Earth Science
Beni-Suef University
Beni-Suef, Egypt

Dimple Sharma
School of Chemical Engineering and Physical Sciences
Lovely Professional University
Phagwara, India

S. K. Shukla
Department of Polymer Science, Bhaskaracharya College of Applied Sciences
University of Delhi
New Delhi, India
and
Research Technology Develop Centre, Sharda University
Greater Noida, India

Bhatt Tushar Shriram
Vikram Sarabhai Space Centre
Trivandrum, India

Shaziya H. Siddiqui
Department of Chemistry
Sam Higginbottom University of Agriculture Technology & Sciences
Allahabad, India

Harminder Singh
School of Chemical Engineering and Physical Sciences
Lovely Professional University
Phagwara, India

Jandeep Singh
School of Chemical Engineering and Physical Sciences
Lovely Professional University
Phagwara, India

NB Singh
Department of Polymer Science, Bhaskaracharya College of Applied Sciences
University of Delhi
New Delhi, India
and
Research Technology Develop Centre
Sharda University
Greater Noida, India

Sonika Singh
Guru Amar Dass Public School
Jalandhar, India

Veinardi Suendo
Department of Chemistry, Faculty of Mathematics and Natural Science
Institut Teknologi Bandung
Bandung, Indonesia

Md. Abu Taleb
Department of Environmental Science, Faculty of Meteorology, Environment and Arid Land Agriculture
King Abdulaziz University
Jeddah, Saudi Arabia

Paul B Tchounwou
RCMI Center for Environmental Health
Jackson State University
Jackson, Mississippi

Muhammad Waqas
Department of Botanical and Environmental Science
Kohat University of Science and Technology
Kohat, Pakistan

1 Applications of Nanomaterials in Environmental Remediation

Moisés Canle, M. Isabel Fernández, and J. Arturo Santaballa
University of A Coruña

CONTENTS

1.1 INTRODUCTION

Nanomaterials show large advantages due to their morphology and to the fundamental changes in the magnitude of many physicochemical properties as the size of particles reduces towards the nanometric scale. A good way to visualize this is to think about a glass of water and the properties of the liquid, which we all know. As we reduce the number of water molecules to a very small one, say ten molecules, the properties change and we cannot tell any more for sure what are the properties of this set of molecules. Very small objects behave very differently than large macroscopic ones. This is related, on the one hand, to the ratio of surface atoms to bulk atoms in nanoparticles, which may lead to a situation where surface contributions predominate over bulk contributions to properties. On the other hand, as particles become smaller and approach their *de Broglie* wavelength ($\lambda = h/(m \cdot v)$), size effects become more evident because of quantum confinement. If the spacing between quantum levels (ΔE) is higher than thermal excitation energy, *i.e.* $\Delta E > k_B \cdot T$, quantum effects predominate and size effects become very important.

DOI: 10.1201/9781003129042-1

Different types of materials: metal and metal oxide/sulphide allomorphs, polymers, carbon allotropes, etc. are available nowadays and may be transformed, doped or prepared in different morphologies (sizes and shapes), with a variety of surface characteristics. The active surface and the number of reactive sites on the surface or the porosity, all related to surface phenomena, are some key properties in environmental applications that benefit from nanometric size. Also, electrical, thermal, optical and magnetic properties may undergo favourable modifications on reducing the scale [1].

The rapidly growing world population and subsequent welfare needs have led to an unprecedented level of transformation of raw materials and production of all sorts of goods and, consequently, to a large pressure on the different environmental compartments. The last few years have seen an enormous increase in the variety and concentration of pollutants in the environment [2,3]. Many emerging organic pollutants are characterized by their persistence. Some, such as persistent and mobile organic compounds (PMOCs) and, particularly, the so-called 'forever chemicals', may pose serious risks to health and the environment, affecting also in some cases the climate, the quality of our food and life in general. Pharmaceuticals and personal care products (PPCPs), agrochemicals and dyes are widely consumed, in most cases in very large amounts. Trace levels of these are found in the atmosphere, in water bodies (wastewaters, ground- and surface waters, and even drinking waters) and in soils, due to inappropriate or inexistent treatment of these substances in water treatment plants [4,5]. For example, studies carried out in different countries have detected more than 80 drugs of different therapeutic families in amounts as high as ppm in water [6]. The mobility of chemicals among the different environmental compartments depends a lot on their chemical structure and derived physicochemical properties: vapour pressure, solubility, the presence of acidic hydrogens, ionization sites, etc. Depending on these properties, the different compounds will be mobilized or not between compartments and, if they are, will travel shorter or longer distances in the atmosphere and hydrosphere. Also, depending on their chemical structure and derived properties, they will be more refractory or more reactive and, therefore, show longer or shorter lifetimes. The continuous input of this type of pollutants to the environment may lead to a chronic low-level exposure, as well as to environmental accumulation, potential bioaccumulation and biomagnification [7].

It has been shown that the removal/elimination of a variety of the mentioned pollutants from emissions, effluents and discharges, and from the different environmental compartments using the currently available methods, often fails or is not effective enough [8], and ecotoxicological effects have been observed [9]. Hence, the development and optimization of new technologies for the elimination of PMOCs have become essential. Among these, nanoscience-based technologies arise as highly promising due to the enhanced properties of nanomaterials for adsorption and redox processes that may contribute to reducing/eliminating PMOCs [10]. Although applications are available for remediation of gaseous effluents and also soils, most research on the field has concentrated on water pollution, so this chapter will focus mainly on the remediation of water bodies. However,

TABLE 1.1
Main Nanomaterial-Based Technologies Used Nowadays in Water Treatment

Technology	Pros	Cons	Degree of Development
Nanofiltration	Very effective.	Expensive. Energy consuming. Not (reasonably) suitable for large volumes or flows.	Commercially available.
Nanoadsorbents	Effective/very effective. May be cheap.	Not a general solution. Many PMOCs are not adsorbed. Adsorption depends on environmental conditions.	Commercially available.
Nanostructured materials-based heterogeneous photocatalysis	Very effective. Cheap.	Need of artificial illumination. Needs separation of photocatalyst (sometimes difficult). Not suitable for large volumes or flows.	Commercial and pre-commercial for small volumes or flows.
Nanoparticle-based Fenton and photo-Fenton processes	Effective depending on environmental conditions. Cheap. May use sunlight.	Needs low pHs. Presence of Fe^{2+}/Fe^{3+} in solution. Formation of precipitates. Not suitable for drinking water treatment.	Commercially available for small-scale processes.
Metal-based (zero-valent) processes	May allow treatment of substances that cannot be treated using other techniques.	Expensive. May introduce metal pollution. Not suitable for large volumes or flows. Not suitable for drinking water.	Under development.
Nanostructured materials-based sensors	Highly sensitive. High reproducibility.	Expensive (so far).	Under development.

the fundamentals of the technology are exactly the same, and the main difference when dealing with gases or solids is the way to access or extract PMOCs.

Table 1.1 briefly compiles different nanomaterials-based technologies used nowadays in water treatment, comparing some of their characteristics: pros, cons and degree of development of the technology.

Nanomaterials may also contribute to pollutants detection through the development of new sensors and monitoring devices that may be especially useful to generate early alerts that help control accidental or illegal disposal of chemicals [11–13].

1.2 ADSORPTION

Adsorption is a surface phenomenon in which a substance called 'adsorbate' that is immersed in a medium (normally, a liquid, but may also be a gas) diffuses towards the surface of the 'adsorbent' and interacts with it, being retained. The extent to which this interaction/retention takes place depends a lot on the physicochemical characteristics of both adsorbent and adsorbate. It may be of a physical nature, called 'physisorption', corresponding normally to weak interactions that are easy to overcome (i.e. by soft heating), or of a chemical nature, called 'chemisorption', corresponding to stronger interactions (though weaker than a chemical bond) that are difficult to break. The adsorbate–adsorbent interaction affects dramatically the electronic structure of the former, which controls its interaction with the surface during the adsorption process, and may enhance its reactivity in heterogeneous catalytic processes [14]. Accordingly, the whole adsorption process depends on the available surface and its porosity, two parameters that increase largely on going to the nanoscopic scale. Not all points on the surface are available for interaction with the adsorbate; the points of interaction, binding sites, must fulfil some conditions related to electron density, acidity and even morphology depending on the chemical characteristics of the adsorbate. According to this, the commonest and simplest approach to adsorption is Langmuir's model that includes a number of assumptions: (1) the adsorption–desorption process approaches equilibrium, (2) the surface of the catalyst is homogeneous, (3) the different binding adsorption sites on the surface are energetically equivalent, (4) the adsorbent–adsorbate interaction takes place through the same kind of functional groups, (5) each binding site interacts with only one adsorbate species, (6) a single layer of adsorbate is formed on the surface, and (7) there is no lateral interaction between adsorbed species once on the surface [15–17]. Although these assumptions imply an oversimplification of the problem, they lead to the well-known Langmuir adsorption isotherm. Despite its obviousness, the very concept of 'isotherm' refers to a given temperature, so any adsorption study must be carried out under temperature-controlled conditions, and these must be similar to those under which the environmental remediation process is going to be applied.

When A is the adsorbate, S is the adsorbent, and A_{ads} is the adsorbed species, and when the equilibrium

$$A + S \rightleftarrows A_{ads}$$

is achieved, the rate of adsorption equals the rate of desorption:

$$r_{ads} = k_{ads} \cdot [A] \cdot [S] = k_{des} \cdot [A_{ads}] \tag{1.1}$$

where k_{ads} and k_{des} are the rate constants for adsorption and desorption, respectively, and [S] means the number of free binding sites per unit of surface on the adsorbate. Thus:

$$K_{eq\ a-d} = \frac{k_{ads}}{k_{des}} = \frac{[A_{ads}]}{[A] \cdot [S]} \tag{1.2}$$

Taking into account that the total number of binding sites (occupied and unoccupied) is $[S]_{tot} = [S] + [A_{ads}]$, and defining the fraction of occupied binding sites as $\Theta = [A_{ads}]/[S]_{tot}$, the Langmuir isotherm can be written in its common form:

$$\Theta = \frac{K_{\text{eq } a-d} \cdot [A]}{1 + K_{\text{eq } a-d} \cdot [A]} \tag{1.3}$$

which may be linearized by simply taking the reciprocal on both sides of the equation.

The typical shape of this isotherm is shown in Figure 1.1 for the adsorption of dimethyl aniline on TiO_2 P25. When $1 >> K_{\text{eq } a-d}$, a linear dependency is observed, but when $K_{\text{eq } a-d} >> 1$, the system reaches a plateau [18].

Consideration of the different failures of assumptions (1)–(7) (see above) has led to the development of different models that take into account the inhomogeneity of the surface, the possibility of multiple adsorption, interactions between adsorbate molecules on the surface or the potential formation of multilayers of the adsorbate [20]. Thus, the Freundlich isotherm (Eq. 1.4), where x is the mass of adsorbate, m is the mass of adsorbent, and K and n are characteristic constants for each adsorbate and adsorbent at a temperature T,

$$\frac{x}{m} = K \cdot C^{1/n} \tag{1.4}$$

allows to account for the heterogeneity of the surface, as the heat of adsorption, ΔH_{ads}, changes as a function of the number of binding sites that are occupied,

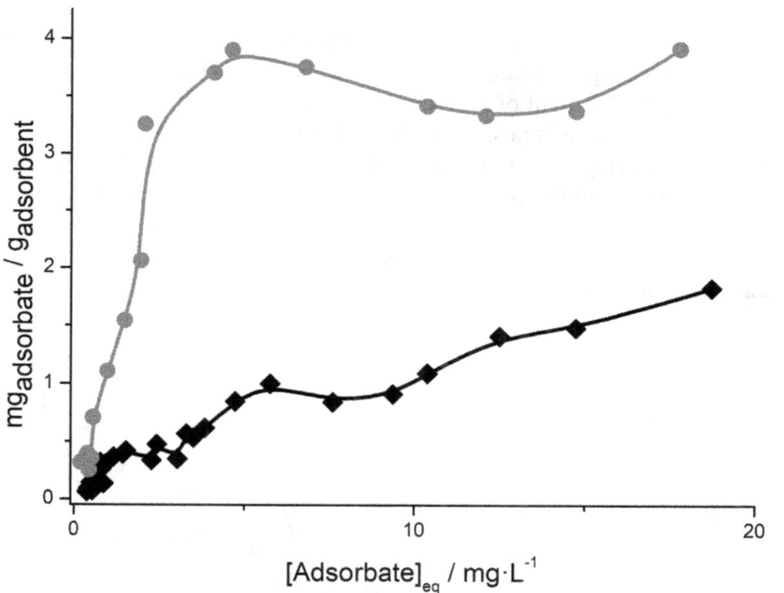

FIGURE 1.1 Adsorption isotherms for two different adsorbates: (◆) aniline and (●) dimethyl aniline, both on the same adsorbent, TiO_2 P25. T = 298.0 K, natural pH. Tendency lines are not mathematical fits.

sites with the most favourable interactions being occupied first, while homogeneous surfaces with equivalent active sites show a constant ΔH_{ads} [19].

The Temkin adsorption isotherm takes into account the interactions between adsorbate molecules once on the surface, assuming these interactions reduce the heat of adsorption and using this approach to correct the equilibrium rate constant in the Langmuir model [20].

It is common to observe a behaviour such as the one shown for the adsorption of aniline on TiO_2 P25 in Figure 1.1, *i.e.* a series of successive increases in the adsorption, resembling the addition of successive Langmuirian behaviours. The BET isotherm (which stands for 'Brunauer, Emmett and Teller') is used in such cases, based on a model that takes into account the formation of a multilayer adsorption isotherm with a random distribution of covered binding sites: some empty, some forming with a monolayer and some forming a multilayer, as in Figure 1.2, but with no lateral interaction between them. The BET theory assumes that each layer can be described according to the Langmuir behaviour (see above).

In line with this hypothesis, Figure 1.1 shows the formation of three layers and the incipient formation of a fourth one. The first formed layer is the one where the contact between adsorbent and adsorbate takes place, and the interaction might be a chemisorption, with a higher enthalpy of adsorption. The successive layers are formed by weak intermolecular interactions between adsorbate species that accumulate in layers and are explained by physisorption processes. Thus, a mathematical expression can be derived for the BET isotherm, which is usually applied to the adsorption of gases on solid surfaces [21].

The surface of an adsorbent may have ionizable functional groups, which may be protonated or deprotonated depending on the pH of the medium. At the pH corresponding to the point of zero charge (pH_{PZC}), the surface is neutral. When $pH < pH_{PZC}$, the surface is positively charged, and if $pH > pH_{PZC}$, it is negatively charged. Considering the pH of the medium and the pK_a of the different ionization sites on the molecule, it is possible to estimate whether the interaction between adsorbate and binding sites will be favourable or not [18]. Though frequently omitted, the study of adsorption under controlled pH is fundamental for the success of adsorption as a remediation technology.

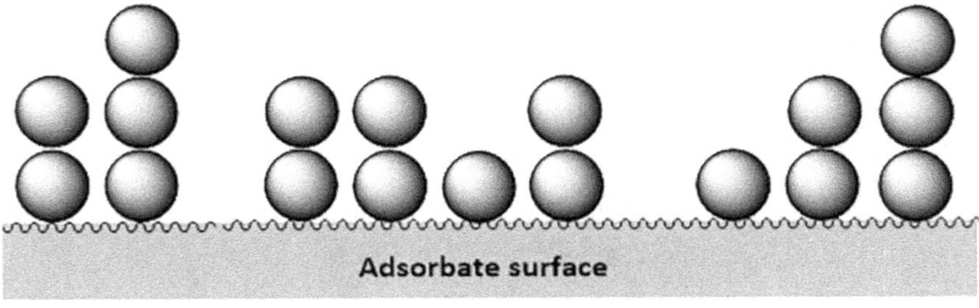

FIGURE 1.2 Surface of an adsorbate covered with a random distribution of non-interacting adsorbate entities, with free binding sites, monolayers, double layers and multilayers.

The same as most materials used for environmental remediation, adsorbents should, ideally, be abundant, cheap, easy to use, chemically and mechanically stable, harmless to both health and the environment, reusable and recyclable. It is most relevant that once they have reached the end of their useful life, the generated residue can be easily inertized or valorized to avoid accumulation of residues. Also, from the point of view of process, adsorbents must be efficient (i.e. show high yields of adsorption), allow easy on-demand desorption of adsorbates, whenever possible be selective, and be easy to separate from the medium. It is absolutely paramount that materials used in environmental remediation do not constitute a new source of pollution themselves, for which biodegradable nanomaterials are being intensively investigated [22]. Nanosized adsorbents are usually difficult to separate, unless nanofiltration is applied, a very expensive separation method. For this reason, paramagnetic particles can be incorporated into adsorbent nanomaterials to allow removal using simple magnets [23].

Nanoadsorbents can be prepared basically through two methods: top-down (starting from macroparticles and reducing their size, trying to control their morphology at the same time, through physical and chemical methods; see Figure 1.3a) and bottom-up (building the particles from their elementary or molecular constituents to reach the desired size and morphology; see Figure 1.3b).

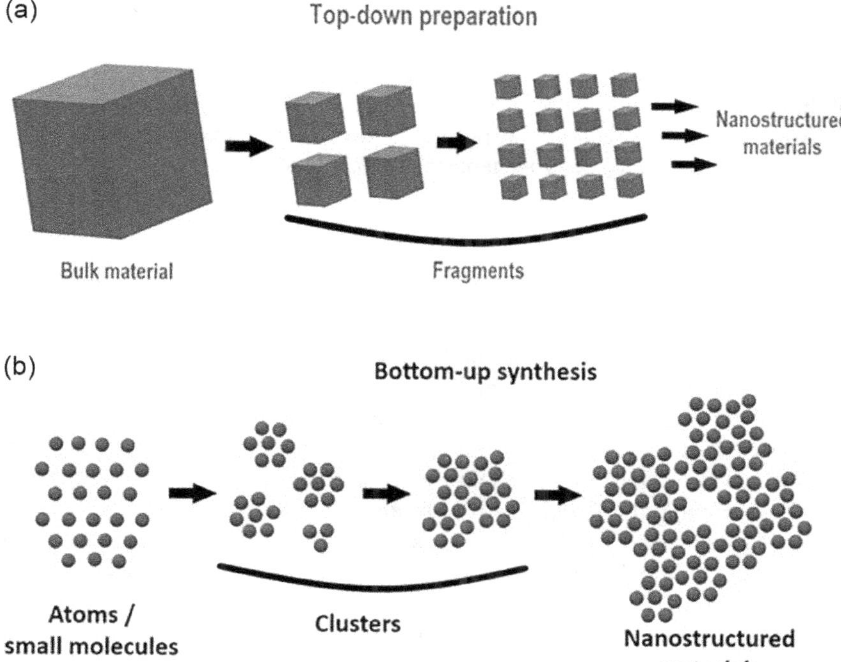

FIGURE 1.3 Pictorial concept of (a) top-down preparation of nanostructured materials and (b) bottom-up synthesis of nanostructured materials.

Nowadays, the preferred method of preparation of nanomaterials is the bottom-up approach, as it allows a better control of the process and of the morphology and properties of the products [24]. Attention is being paid also to the biosynthesis of nanomaterials, assuming these may be more biodegradable or environmentally friendly [25].

Adsorption processes are highly suitable for removal of a variety of PMOCs from different types of gaseous and liquid effluents. Usually, adsorption processes are easy and economical to apply [26–28]. Different types of materials are currently used as adsorbents: activated carbon, polymers, zeolites, etc., and many have been prepared from the valorization of different types of residues [29,30]. Developments in the field of nanomaterials have led to a very important increase in the efficiency of adsorbents and rate of adsorption, mainly due to the largely increased active surface and porosity and to the presence of a very larger number of active sites [31,32].

Nanosized/nanostructured adsorbents are nowadays used mainly for elimination/reduction of heavy metals, oil and dyes from water [33], but are promising, for example, for remediation of perfluoroalkyl substances (PFASs), the so-called 'forever chemicals' [34].

The possibilities of adsorption in environmental remediation have been reviewed by different authors [10,11,22,24,35]. Although adsorption has been used for years for this purpose, its application in the form of nanoparticles is still on its early years. There is a lot of fundamental research to be done in order to control the process and its associated variables appropriately. Very attractive possibilities are open, such as the potential surface modification (functionalization) to carry out selective adsorption processes, or the effective film deposition for efficient flow chemistry heterogeneous catalytic processes.

1.3 PHOTOCATALYSIS

Advanced oxidation processes (AOPs) and advanced reduction processes (ARPs) are highly promising alternatives for environmental remediation, especially for media that allow easy transmission of light, i.e. atmospheres and waters [36]. These processes are characterized by the formation of highly reactive oxygen species (ROS), mainly hydroxyl radicals ($HO^•$) and superoxide radical anions ($O_2^{•-}$). $HO^•$ is the strongest possible oxidant in aqueous medium, $E^0(HO^•, H^+/H_2O) = 2.73\,V$ *vs.* NHE [37], while positive vacancies on the surface of the photocatalyst (commonly denoted as h^+) are also strongly oxidant ($E^0 = 2.53\,V$ *vs.* SHE) [38]. The presence of ROS ensures high reactivity (i.e. fast processes) and a low selectivity (no discrimination between oxidizable/reducible species). AOPs are more commonly used nowadays, very specially in environmental remediation, so we will focus on them. Among AOPs, heterogeneous photocatalysis has widely been employed, achieving in many cases complete mineralization of both organic and inorganic pollutants [39]. AOPs make use of different types of near-UV/visible (NUV–Vis) lamps (recently LED devices) and, ideally, of sunlight, commonly under environmental conditions of P and T [40]. The higher the overlap between sunlight irradiance spectrum and the absorption of the photocatalyst, the higher the efficiency of photocatalysis processes. In fact, this objective is one of the main challenges in the field and has been attempted through a number of approaches, including doping of the chemical structure of the photocatalyst (mainly with metals) and introduction of defects in the crystalline network [38].

Heterogeneous photocatalysis is based on the use of semiconductors, commonly in the form of nanoparticles, or as films supported onto surfaces. When nanoparticles are used, they tend to agglomerate, with the result of loss of some surface and, correspondingly, of reactivity. When nanoparticles are deposited as films, assuming nanoparticles as spherical, the reactive surface of the hemisphere in contact with the surface would not available, with the result of a massive loss of active surface in the material. The energy gap (E_G) between the valence band (VB) and the conduction band (CB) must be equal to or lower than the energy of the incoming photons, *i.e.* $E_G \leq h\upsilon$ [38]. Under these conditions, the photon is adsorbed and an e^- is promoted from VB to CB, leaving a positive vacancy on the VB, h^+, and generating an electron–hole pair (e^-–h^+), as in Figure 1.4.

The e^-–h^+ pair is highly unstable and usually undergoes annihilation in just some nanoseconds, returning to the initial situation before photon absorption. Depending on the conditions of the experiment and, mainly, on the material, a very small percentage of h^+–e^- pairs undergo separation [41]. The size of the VB–CB gap, *i.e.* the value of E_G, is key for the success of the photocatalytic material. Ideally, materials with E_G values within the photon energies provided by sunlight are sought, to avoid unnecessary energy consumption with irradiation lamps.

When the h^+–e^- separation takes place in an aqueous medium, the most important reactions that occur are the reduction of O_2 to $O_2^{\bullet-}$ and oxidation of adsorbed

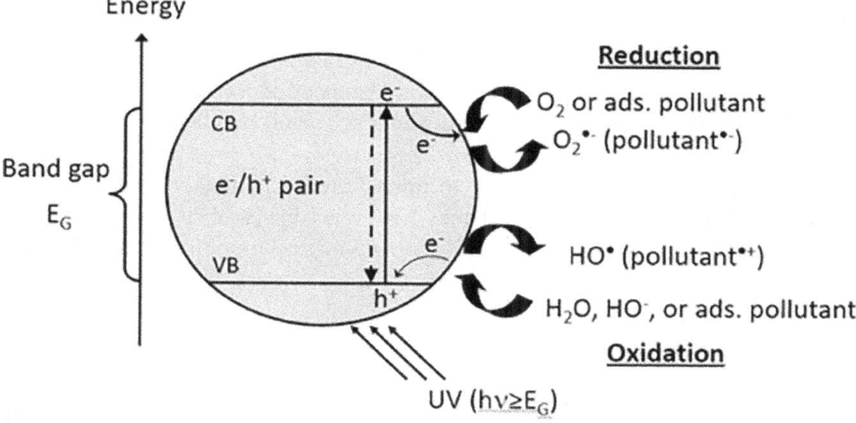

FIGURE 1.4 Working principle of heterogeneous photocatalysis processes over a semiconductor nanoparticle.

H_2O to $HO^{\bullet}$ (Figure 1.3). The most important reactions taking place are the following (Eqs. 1.5–1.8) [41]:

$$H_2O_{ads} + h^+ \rightarrow H_2O^{\bullet +} \rightarrow H^+ + HO^{\bullet} \quad (1.5)$$

$$h^+ + HO^-{}_{ads} \rightarrow HO^{\bullet} \quad (1.6)$$

$$O_{2ads} + e^- \ O_{2ads}^{\bullet -} \qquad E^0 = -0.33 \text{ V } vs. \text{ NHE} \quad [42] \quad (1.7)$$

$$O_2^{\bullet -} + H^+ \rightleftarrows HO_2^{\bullet} \qquad pK_a(HO_2^{\bullet}) = 4.8 \quad [43] \quad (1.8)$$

If other species are adsorbed on the surface, they may also undergo reduction (Eq. 1.9) or oxidation (Eq. 1.10). For example, for a given pollutant:

$$\text{Pollutant} + e^- \rightarrow \text{Pollutant}^{\bullet -} \quad (1.9)$$

$$\text{Pollutant} + h^+ \rightarrow \text{Pollutant}^{\bullet +} \quad (1.10)$$

Reaction of adsorbed species may increase the efficiency of disappearance of the pollutant, but may also open new reaction channels that lead to products that are different from those of reaction with ROS and may have also a different stability and/or toxicity.

Further ROS may also be generated during heterogeneous photocatalysis. For example, H_2O_2 is formed by dismutation of $O_2^{\bullet -}$ and $HO_2^{\bullet}$ (Eq. 1.11) [44] and may then be reduced to $HO^{\bullet}$, as in Eq. (1.12):

$$O_2^{\bullet -} + HO_2^{\bullet} + H^+ \rightarrow H_2O_2 + O_2 \quad (1.11)$$

$$H_2O_2 + H^+ + e^- \rightarrow HO^{\bullet} + H_2O \qquad E^o = -0.870 \text{ V } vs. \text{ NHE} \quad [45] \quad (1.12)$$

$HO^{\bullet}$ is regarded as mainly responsible for degradation of organic pollutants, as supported by experimental observation of differences of several orders of magnitude difference between the reactivity dependences with $[HO^{\bullet}]$ and $[O_2^{\bullet -}]$ or other ROS [46–48].

$HO^{\bullet}$ reacts typically by addition to unsaturated functional groups, yielding HO-adducts (Eq. 1.13), although it may also react by $H^{\bullet}$ abstraction to yield a H-deficient radical species (Eq. 1.14), or by one-e^- oxidation, forming HO^- and a radical cation (Eq. 1.15).

$$\text{Pollutant} + HO^{\bullet} \rightarrow (\text{Pollutant} - OH)^{\bullet} \quad (1.13)$$

$$\text{Pollutant} + HO^{\bullet} \rightarrow \text{Pollutant}(-H)^{\bullet} + H_2O \quad (1.14)$$

$$\text{Pollutant} + HO^{\bullet} \rightarrow \text{Pollutant}^{\bullet +} + HO^- \quad (1.15)$$

Each of these short-lived intermediates, (Pollutant-OH)$^•$, Pollutant(-H)$^•$ and Pollutant$^{•+}$, will undergo chemical reaction through different reaction pathways, generating a variety of products, most of which will be hydroxylated and, therefore, highly polar, soluble in water and easily adsorbed on mineral surfaces such as clays, which may increase their time of residence in the environment. If enough time is allowed for the reaction to take place, it might finally lead to mineralization, i.e. transformation into the highest possible oxidized ionic species (CO_3^{2-}, NO_3^-, PO_4^{3-}, SO_4^{2-}, etc.). However, a number of refractory compounds are known, such as triazines, that form an intermediate that is stable and does not react further [49,50].

In all cases, $HO^•$ reaction takes place very fast, typically with bimolecular rate constants $k > 10^9$ $M^{-1}{\cdot}s^{-1}$ or higher, *i.e.* close to or at the diffusion control limit in water, and in an unselective manner [51,52]. $HO^•$ and h^+ show differences in regioselectivity during the oxidation of aromatic compounds and may lead to different products [53]. Recently, competing pathways between reaction with $HO^•$ and oxidation of chemisorbed substances by h^+ have been proved for the same process [54].

It is particularly important to remark that heterogeneous photocatalysis is a surface process, *i.e.* the reactive species need to diffuse from the bulk of the solution to the surface, get adsorbed (possibly with subsequent structural modifications), undergo reaction, desorb and diffuse again into the bulk of the solution [55]. The possibility that ROS are generated on the surface and migrate into the solution has been studied, and it has been shown that $HO^•$ only diffuses some hundred Å away from the surface into the bulk of the solution [56]. ESR studies have also established that ROS migrate only a few atomic distances from the surface [57]. Therefore, as frequently observed experimentally, if a compound is not adsorbed, it will not react by heterogeneous photocatalysis. Just a very small proportion of initial adsorption is needed, as the adsorption/desorption equilibrium will be displaced as the reaction proceeds.

Titanium dioxide (TiO_2) is the most broadly used photocatalytic nanosized material for environmental remediation, for atmospheres [58] and waters [59], and both in suspension and supported in films. It is easy to manipulate under standard environmental conditions [60], cheap, abundant and readily available, easy to recover, of low toxicity and of high photochemical stability. Its main drawback is the low degree of spectral overlap with sunlight irradiance, only about 4% at ground level, which reduces the effectiveness and possibilities of application of TiO_2-based heterogeneous photocatalysis under sunlight [41,61]. Therefore, strong efforts are put on the structural and/or surface modification of TiO_2, with introduction of defects on the crystalline network [62], decoration of the surface [63], production of 2D (nanosheets and films) [64] and 3D (nanotubes and nanowires) structures [65], as well as composites with other materials [66–70] to improve the photoefficiency of the process.

Nanosized TiO_2 tends to aggregate, with subsequent loss of active surface and photoactivity. To avoid aggregation, simultaneous sonication can be applied before or during the process. The obtained disaggregation is not definitive, the equilibrium being favourable towards reaggregation, but the kinetics of the aggregation process may be slow. Longer sonication times do not necessarily help disaggregation [71].

A very relevant problem with the use of nanosized particles, especially in suspension, is to be able to eliminate them from the medium before disposal of the photoproducts [72]. Common procedures for this are nanofiltration and ultracentrifugation, but both have a high cost in terms of energy and should be avoided. It has recently been shown that the presence of TiO_2 nanoparticles in soil affects microbial communities in the soil, and therefore soil ecosystem functions [73].

Taking into account that the reaction occurs between species that are adsorbed on the surface of the photocatalyst, the photocatalytic transformation of a pollutant can be adequately described using a modified Langmuir–Hinshelwood kinetic model [66], as in Eq. (1.16):

$$r = k_{LH} \quad \Theta = k_{LH} \cdot \frac{K_{LH} \cdot C}{1 + K_{LH} \cdot C} \tag{1.16}$$

where C is the concentration of pollutant once the adsorption equilibrium has been established, k_{LH} (mol/s cm^2) is an apparent kinetic rate constant per unit of surface area, and Θ (cm^2) accounts for the coverage of TiO_2 surface. K_{LH} is the Langmuir–Hinshelwood adsorption constant, which would be identical to the Langmuir adsorption constant, $K_{\mathrm{eq}\,a-d}$ (2), if the Langmuir model were strictly followed (see above). Usually, the concentration of pollutants is very low so that the product $K_{LH} \cdot C \ll 1$, and then the rate of the process reduces to pseudo-first-order kinetics (Eq. 1.17):

$$r = k_{LH} \cdot K_{LH} \cdot C = k_{\mathrm{app}} \cdot C \tag{1.17}$$

The observed kinetics are usually fitted nicely by the first-order mathematical model, as in Figure 1.5. In this way, it becomes easier to determine rate constants and half-lifetimes of the different processes and to compare rates of disappearance of different pollutants. From Figure 1.5, it becomes evident that the size effect and the change in active surface have a dramatic effect both on the rate and on the efficiency of the process.

Much research is ongoing on this field, mainly with the aim of improving the photoefficiency of the processes with sunlight, for which a variety of approaches are taken [74].

1.4 CHEMICAL REDOX PROCESSES

Chemical redox processes are a group of catalytic processes that take place, in general, through exchange of ions or electrons with the medium, reaching the reagents there. In general, a solid, which may be nanosized or not, releases ions into the bulk of the medium, or is oxidized releasing e^-. Therefore, in both cases the process may be considered as a homogeneous one, taking place in the

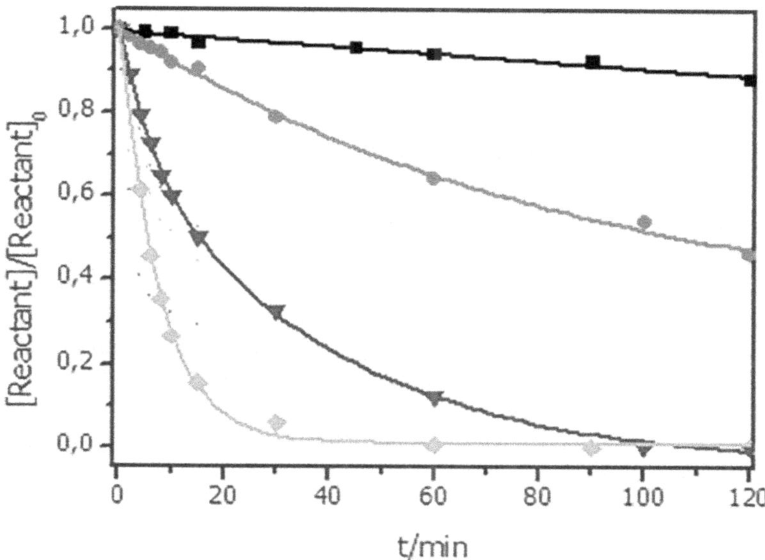

FIGURE 1.5 Typical fit of pseudo-first-order decay curves ([Reactant]/[Reactant]$_0$) *vs.* time fit using a modified Langmuir–Hinshelwood kinetic model. The photocatalysts used in these experiments were (◆) nanosized P25-TiO_2, (▼) TiO_2 nanorods, (●) TiO_2 nanotubes, (■) no catalyst.

bulk of the medium and not on the surface. The fact that nanoparticles are used helps, though, because, as discussed earlier in this chapter, the ratio surface/volume favours the processes taking place with respect to the use of macroparticles or aggregates of nanoparticles. The two main processes in this group are described in what follows.

1.4.1 Fenton and Photo-Fenton Processes

H_2O_2/Fe^{2+} mixtures show very powerful oxidizing properties in acidic medium, which is rationalized in terms of a set of processes generally known as *Fenton* reactions [75], in which ROS are generated, mainly $HO^{\bullet}$ radicals (Eqs. 1.18 and 1.19):

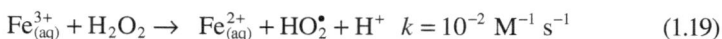

$$Fe^{2+}_{(aq)} + H_2O_2 \rightarrow Fe^{3+}_{(aq)} + HO^{\bullet} + HO^{-} \quad k = 76\ M^{-1}\ s^{-1} \tag{1.18}$$

$$Fe^{3+}_{(aq)} + H_2O_2 \rightarrow Fe^{2+}_{(aq)} + HO_2^{\bullet} + H^{+} \quad k = 10^{-2}\ M^{-1}\ s^{-1} \tag{1.19}$$

Under appropriate acidity conditions (pH *ca.* 2.7–2.8), Fe^{3+} is regenerated (Eqs. 1.20–1.23) and some other ROS are formed:

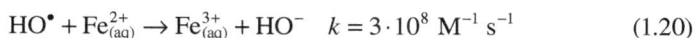

$$HO^{\bullet} + Fe^{2+}_{(aq)} \rightarrow Fe^{3+}_{(aq)} + HO^{-} \quad k = 3 \cdot 10^{8}\ M^{-1}\ s^{-1} \tag{1.20}$$

$$Fe^{3+}_{(aq)} + H_2O_2 \rightleftharpoons FeOOH^{2+} + H^+ \quad k = 3.1 \cdot 10^{-3} M^{-1}\ s^{-1} \tag{1.21}$$

$$FeOOH^{2+} \rightarrow HO_2^{\bullet} + Fe^{2+}_{(aq)} \quad k = 2.7 \cdot 10^{-2} M^{-1}\ s^{-1} \tag{1.22}$$

$$HO_2^{\bullet} + Fe^{2+}_{(aq)} \rightarrow HO_2^{-} + Fe^{3+}_{(aq)} \tag{1.23}$$

Fe^{2+} can also be regenerated by $HO_2^{\bullet}$ (Eq. 1.24):

$$Fe^{3+}_{(aq)} + HO_2^{\bullet} \rightarrow Fe^{2+}_{(aq)} + O_2 + H^+ \quad k < 2 \cdot 10^{3}\ M^{-1}\ s^{-1} \tag{1.24}$$

Sunlight or lamps emulating sunlight irradiance at ground level allow pollutant degradation in the presence of a source of iron and H_2O_2. This reaction is known as heterogeneous *photo-Fenton* process and may show remarkable performance in the environmental remediation. Under irradiation ($\lambda \geq 300\,nm$), Fe^{3+} complexes can regenerate Fe^{2+} (Eq. 1.25):

$$Fe(OH)^{2+} + h\nu \rightarrow Fe^{2+}_{(aq)} + HO^{\bullet} \tag{1.25}$$

Finally, in the *electro-Fenton* reaction, H_2O_2 is generated in situ through Eq. (1.26):

$$O_2 + 2H_3O^+ + 2e^- \rightarrow H_2O_2 + 2H_2O \tag{1.26}$$

and Fe^{2+} is regenerated through Eqs. (1.21 and 1.22).

The different ROS generated are highly reactive and able to efficiently transform pollutants in the medium, leading to their eventual mineralization. Fenton processes have a number of advantages: iron is cheap, readily available and non-toxic, and H_2O_2 is easy to manipulate. Fenton processes have been shown to degrade a variety of pollutants, including polycyclic aromatic hydrocarbons (PAHs), polychlorinated biphenyls (PCBs), phenols and drugs [76–78]. A serious drawback is that iron salts precipitate and the addition of Fe^{2+} and H_2O_2 imposes a high economical cost. Electro-Fenton processes show even increased costs due to energy consumption. In turn, photo-Fenton is applied in many cases using sunlight, which largely reduces associated costs.

1.4.2 Metal-Based Processes

Nanosized metal-based processes make use of zero-valence ($Metal^0$) properties, provided that the reduction potential of the metal itself is appropriate. The process taking place is a one-e^- reduction, or a series of successive one e^- reductions (Eqs. 1.27–1.29):

$$M^0 + Pollutant \rightarrow M^+ + Pollutant^{\bullet -} \tag{1.27}$$

$$M^0 + Pollutant^{\bullet -} \rightarrow M^+ + Pollutant^{\ominus} \tag{1.28}$$

$$M^{+} + \text{Pollutant}^{\bullet -} \rightarrow M^{2+} + \text{Pollutant}^{\ominus} \quad (1.29)$$

where '•−' represents an excess of one e^- and '⊖' represents an excess of two e^-. Processes in Eqs. (1.27 and 1.28) usually lead to bond-breaking processes of C–X bonds, where X is an electronegative atom with a high electron density charge, such as a halogen. Following this process, the intermediate derived from the pollutant captures H^+ or $H^\bullet$ from the solvent (usually water) and reduction products are generated (Eq. 1.30).

$$\text{Pollutant}^{\bullet -} + H_2O \rightarrow \text{Reduction products} \quad (1.30)$$

In this way, reactions with nanosized zero-valence metals are useful for the degradation of halogenated solvents, or halogenated aromatic derivatives that usually have very long environmental lifetimes and are frequently refractory to other remediation processes. Fe^0 nanoparticles are frequently used for this purpose because of the very appropriate reduction potential of iron [79–82].

Zero-valence metals are usually oxidized on the surface, which facilitates reaction with other species that may be present with a higher reduction potential. This may be useful for the remediation of heavy-metal pollution [83,84].

Some zero-valence nanosized metals have also been shown to have bactericidal or virucidal properties, which make them suitable for disinfection of inner atmospheres (in buildings, for example) and remediation of water bodies infested with microbes or viruses [85–87].

The use of bimetallic nanoparticles has been explored, to take advantage from the presence of metals with different reduction potentials that may promote the degradation of pollutants [80,88]. This strategy also helps improve the problem already described of aggregation of nanoparticles (see above) and enhance the stability of zero-valence metals [89]. Along this strategy, noble metals have been used to build bimetallic nanoparticles, with enhanced rates of reduction of the pollutants, but a largely increased cost due to the price of the used metals (Pd, Au, etc.) [90].

1.5 FUTURE CHALLENGES

As usual in science, it is hard to imagine the potential developments the future will bring, which most of the times are a product of the creativity of researchers. Therefore, it is difficult to write down a list of the most important advances we expect to see in the field of the application of nanomaterials to environmental remediation. We will, therefore, limit our ambition and rather list a few issues that have still not been resolved, but that at least we know would be important to master in the short term.

The level of mobility of contamination in soils and sediments is the lowest possible among the different environmental compartments. For this reason, improving remediation methods for solid phases such as soils and sediments is of paramount importance to keep agricultural land and water basins free of pollution.

A proper control of the acid–base and density of charge properties of the surface of adsorbents is necessary to guarantee that environmental changes do not affect adsorption processes.

Understanding surface chemistry sufficiently to tailor adsorbents and photocatalysts to different needs of each case of environmental remediation is a major challenge. Efforts on the functionalization of surfaces are advanced enough to allow us to meet this challenge in the near future.

Chemically stable photocatalysts are abundantly available, even commercially, but their mechanical stability needs to be guaranteed so as to have them exposed to aggressive processes such as bubbling, stirring or strong turbulent flows without lixiviation of potentially toxic or undesired components into the medium.

A highly challenging issue in heterogeneous photocatalysis is currently the development of efficient heterogeneous photocatalysis systems that allow us to perform flow chemistry, as that would facilitate the transition from current treatment processes in batch regime, with large scale-up problems, to treatment of large volumes directly in tubing systems under illumination.

Nanoparticle-based sensorization is not a technology of remediation itself, but rather a set of technologies that will facilitate fast and selective remediation. Environmental monitoring is performed nowadays mainly through spot sampling, passive sampling and the use of indirect correlation measurements. Although the combinations of these methods can be very powerful, nanotechnology has opened the opportunity to perform monitoring with increased selectivity, sensibility and measurement stability, with a minimal amount of sample and in a much faster manner [91,92,93]. Development of fast sensors that allow monitoring of physical properties of pollutants, or at least of families of pollutants with similar structure, will be essential for a proper control of illegal or irregular contamination events and, therefore, facilitate environmental remediation. Having these sensors available would allow us to control pollution upstream and follow it to the points of origin, which will be, no doubt, the best strategy to control it.

ACKNOWLEDGEMENTS

The authors are indebted to the students and researchers who worked in the React! group through the last two decades, and also to the different agencies and companies that financed the group research activities. We specifically wish to acknowledge current financial support from the regional government of the *Xunta de Galicia* (Project GPC ED431B 2020/52).

REFERENCES

[1] Hochella, M.F., Mogk, D.W., Ranville, J., Allen, I.C., Luther, G.W., Marr, L.C., McGrail, B.P., Muruyama, M., Qafoku, N.P., Rosso, K.M., Sahai, N., Schroeder, P.A., Vikesland, P., Westerhoff, P., Yang, Y. Natural, incidental, and engineered nanomaterials and their impacts on the earth system. *Science* 2019, 363, eaau8299. doi: 10.1126/science.aau8299.

[2] Lapworth, D.J., Baran, N., Stuart, M. E., Ward, R.S. Emerging organic contaminants in groundwater—a review of sources, fate and occurrence. *Environ. Pollut.* 2012, 163, 287. doi: 10.1016/j.envpol.2011.12.034.
[3] Pal, A., Gin, K.Y.H., Lin, A.Y.C., Reinhard, M. Impacts of emerging organic contaminants on freshwater resources—review of recent occurrences, sources, fate and effects. *Sci. Tot. Environ.* 2010, 408, 6062. doi: 10.1016/j.scitotenv.2010.09.026.
[5] Santos, J.L., Aparicio, I., Callejon, M., Alonso, E. Occurrence of pharmaceutically active compounds during 1-year period in wastewaters from four wastewater treatment plants in Seville (Spain). *J. Hazard. Mater.* 2009, 164, 1509. doi: 10.1016/j.jhazmat.2008.09.073.
[6] Petrovic, M., Barceló, D. Liquid chromatography-mass spectrometry in the analysis of emerging environmental contaminants. *Anal. Bioanal. Chem.* 2006, 385, 422. doi: 10.1007/s00216-006-0450-1.
[7] Heberer, T. Occurrence, fate, and removal of pharmaceutical residues in the aquatic environment—a review of recent research data. *Toxicol. Lett.* 2002, 131, 5. doi: 10.1016/s0378-4274(02)00041-3.
[8] Zuccato, E., Calamari, D., Natangelo, M., Fanelli, R. Presence of therapeutic drugs in the environment. *Lancet* 2000, *355*, 1789. doi: 10.1016/S0140-6736(00)02270-4.
[9] Malato, S., Blanco, J., Cáceres, J., Fernández-Alba, A.R., Agüera, A., Rodríguez, A. Photocatalytic treatment of water-soluble pesticides by photo-Fenton and TiO_2 using solar energy. *Catal. Today* 2002, 76, 209. doi: 10.1016/S0920-5861(02)00220-1.
[10] Rioboo, C., Franqueira, D., Canle, M., Herrero, C., Cid, A. Microalgal bioassays as a test of pesticide photodegradation efficiency in water. *Bull. Environ. Contam. Toxicol.* 2001, 67, 233. doi: 10.1007/s00128-001-0115-x.
[11] Khin, M.M., Nair, A.S., Babu, V.J., Murugan, R., and Ramakrishna, S. A review on nanomaterials for environmental remediation. *Energy Environ. Sci.* 2012, 5, 8075. doi: 10.1039/c2ee21818f.
[12] Sun, H. Grand challenges in environmental nanotechnology. *Front. Nanotechnol.* 2019, 1. doi: 10.3389/fnano.2019.00002.
[13] Zheng, Y., Li, M., Wen, X., Ho, H.-P., Lu, H. Nanostructured ZnO/Ag film prepared by magnetron sputtering method for fast response of ammonia gas detection. *Molecules* 2020, *25*(8), 1899. doi: 10.3390/molecules25081899.
[14] Jarju, J.J., Lavender, A.M., Espiña, B., Romero, V., Salonen, L.M. Covalent organic framework composites—synthesis and analytical applications. *Molecules* 2020, 25(22), 5404. doi: 10.3390/molecules25225404.
[15] Siaj, M., Temprano, I., Dubuc, N., McBreen, P.H. Preparation and olefin-metathesis activity of cyclopentylidene-oxo initiator sites on a molybdenum carbide surface. *J. Organom. Chem.* 2006, 691(24–25), 5jik497. doi: 10.1016/j.jorganchem.2006.09.033.
[16] Hanaor, D.A.H., Ghadiri, M., Chrzanowski, W., Gan, Y. Scalable surface area characterization by electrokinetic analysis of complex anion adsorption. *Langmuir* 2014, 30(50), 15143. doi: 10.1021/la503581e.
[17] Masel, R. *Principles of Adsorption and Reaction on Solid Surfaces.* Wiley Interscience, New York, 1996. ISBN 978-0-471-30392-3.
[18] Kolasinski, K.W. *Surface Science: Foundations of Catalysis and Nanoscience.* John Wiley & Sons, Ltd., Chichester, 2012. ISBN 9781119990352.
[19] Canle, M., Santaballa, J.A., Vulliet, E. On the mechanism of TiO 2-photocatalyzed degradation of aniline derivatives. *J. Photochem. Photobiol. Chem.* 2005, 175, 192. doi: 10.1016/j.jphotochem.2005.05.001.
[20] Burke, G.M., Wurster, D.E., Buraphacheep, V., Berg, M.J., Veng-Pedersen, P., Schottelius, D.D. Model selection for the adsorption of phenobarbital by activated charcoal. *Pharm. Res.* 1991, 8(2), 228. doi: 10.1023/a:1015800322286.

[21] Al-Ghouti, M.A., Da'ana, D.A. Guidelines for the use and interpretation of adsorption isotherm models—a review. *J. Hazard. Mater.* 2020, 393, 122383. doi: 10.1016/j.jhazmat.2020.122383.

[22] Sing, K.S.W. Adsorption methods for the characterization of porous materials. *Adv. Collid. Interfaces Sci.* 1998, 76–77, 3. doi: 10.1016/S0001-8686(98)00038-4.

[23] Guerra, F.D., Attia, M.F., Whitehead, D.C., Alexis, F. Nanotechnology for environmental remediation—materials and applications. *Molecules* 2018, 23(7), 1760. doi: 10.3390/molecules23071760.

[24] Kharissova, O.V., Dias, H.V.R., Kharisov, B.I. Magnetic adsorbents based on micro- and nanostructured materials. *RSC Adv.* 2015, 5, 6695. doi: 10.1039/c4ra11423j.

[25] El-sayed, M.E.A. Nanoadsorbents for water and wastewater remediation. *Sci. Tot. Environ.* 2020, 739, 139903. doi: 10.1016/j.scitotenv.2020.139903.

[26] El-Gendy, N.S., Omran, B.A., Fosso-Kankeu, E. (Ed.), *Nano and Bio-based Technologies for Wastewater Treatment—Prediction and Control Tools for the Dispersion of Pollutants in the Environment.* Scrivener Publishing LLC, 2019, p. 205. doi: 10.1002/9781119577119.ch7.

[27] Zare, K., Gupta, V.K., Moradi, O., Makhlouf, A.S.H., Sillanpaää, M., Nadagouda, M.N., Sadegh, H., Shahryari-ghoshekandi, R., Pal, A., Wang, Z.-J. A comparative study on the basis of adsorption capacity between CNTs and activated carbon as adsorbents for removal of noxious synthetic dyes—a review. *J. Nanostruct. Chem.* 2015, 5(2), 227. doi: 10.1007/s40097-015-0158-x.

[28] Biškup, B., Subotić, B. Removal of heavy metal ions from solutions using zeolites. III. Influence of sodium ion concentration in the liquid phase on the kinetics of exchange processes between cadmium ions from solution and sodium ions from zeolite A. *Sep. Sci. Technol.* 2005, 39(4), 925. doi: 10.1081/SS-120028454.

[29] Gupta, V., Moradi, O., Tyagi, I., Agarwal, S., Sadegh, H., Shahryari-Ghoshekandi, R., Makhlouf, A., Goodarzi, M., Garshasbi, A. Study on the removal of heavy metal ions from industry waste by carbon nanotubes—effect of the surface modification—a review. *Crit. Rev. Environ. Sci. Technol.* 2016, 46(2), 93. doi: 10.1080/10643389.2015.1061874.

[30] Ekmekyapar, F., Aslan, A., Bayhan, Y.K., Cakici, A. Biosorption of copper (II) by nonliving lichen biomass of Cladonia rangiformis hoffm. *J. Hazard. Mater.* 2006, 137(1), 293. doi: 10.1016/j.jhazmat.2006.02.003.

[31] Gupta, V.K., Tyagi, I., Agarwal, S., Sadegh, H., Shahryarighoshekandi, R., Yari, M., Yousefi-nejat, O. Experimental study of surfaces of hydrogel polymers HEMA, HEMA–EEMA–MA, and PVA as adsorbent for removal of azo dyes from liquid phase. *J. Mol. Liq.* 2015, 206, 129. doi: 10.1016/j.molliq.2015.02.015.

[32] Sadegh, H., Ghoshekandi, R.S., Masjedi, A., Mahmoodi, Z., Kazemi, M. A review on carbon nanotubes adsorbents for the removal of pollutants from aqueous solutions. *Int. J. Nano Dimens.* 2016, 7(2), 109. https://www.sid.ir/en/Journal/ViewPaper.aspx?ID=506535.

[33] Theron, J., Walker, J., Cloete, T. Nanotechnology and water treatment—applications and emerging opportunities. *Crit. Rev. Microbiol.* 2008, 34(1), 43. https://doi.org/10.1080/10408410701710442.

[34] Sadegh, H., Ali, G.A.M., Gupta, V.K., Makhlouf, A.S.H., Shahryari-ghoshekandi, R., Nadagouda, M.N., Sillanpa, M., Megiel, E. The role of nanomaterials as effective adsorbents and their applications in wastewater treatment. *J. Nanostruct. Chem.* 2017, 7, 1. doi: 10.1007/s40097-017-0219-4.

[35] Gagliano, E., Sgroi, M., Falciglia, P.P., Vagliasindi, F.G.A., Roccaro, P. Removal of poly- and perfluoroalkyl substances (PFAS) from water by adsorption—role of PFAS chain length, effect of organic matter and challenges in adsorbent regeneration. *Wat. Res.* 2020, 171, 115381. doi: 10.1016/j.watres.2019.115381.

[36] Anjum, M., Miandad, R., Waqas, M., Gehany, F., Barakat, M.A. Remediation of wastewater using various nanomaterials. *Arabian J. Chem.* 2019, 12, 4897. doi: 10.1016/j.arabjc.2016.10.004.
[37] Legrini, O., Oliveros, E., Braun, A. M. Photochemical processes for water treatment. *Chem. Rev.* 1993, 93, 671. doi: 10.1021/cr00018a003.
[38] Armstrong, D.A., Huiea, R.E., Koppenol, W.H., Lymar, S.V., Merényi, G., Neta, P., Ruscica, B., Stanbury, D.M., Steenken, S., Wardman, P. Standard electrode potentials involving radicals in aqueous solution—inorganic radicals, *Pure Appl. Chem.* 2015, 87, 1139. doi: 10.1515/irm-2013-00.
[39] Fujishima, A., Rao, T., Tryk, D. Titanium dioxide photocatalysis. *J. Photochem. Photobiol. C Photochem. Rev.* 2000, 1, 1. doi: 10.1016/S1389-5567(00)00002-2.
[40] Klavarioti, M., Mantzavinos, D., Kassinos, D. Pharmaceuticals and other organic Removal of residual pharmaceuticals from chemicals in selected north-central and aqueous systems by advanced oxidation northwestern Arkansas streams. *Environ. Int.* 2009, 35, 402. doi: 10.1016/j.envint.2008.07.009.
[41] Gaya, U.I., Abdullah, A.H. *J. Photochem. Photobiol. C Photochem. Rev.* 2008, 9, 1. doi: 10.1016/j.jphotochemrev.2007.12.003.
[42] Canle, L.M., Fernández, M.I., Martínez, C., Santaballa, J.A. Greening photochemistry: using light for degrading persistent organic pollutants. *Rev. Environ. Sci. Bio-Technol.* 2012, 11, 213. doi: 10.1007/s11157-012-9275-x.
[43] Sawada, Y., Iyanagi, T., Yamazaki, I. Relation between redox potentials and rate constants in reactions coupled with the system oxygen-superoxide. *Biochem* 1975, 14, 3761. doi: 10.1021/bi00688a007.
[44] Bielski, B., Cabelli, D., Arudi, R., Ross, A. Reactivity of $HO_2/O_2^{\cdot-}$ radicals in aqueous solution. *J. Phys. Chem. Ref. Data* 1985, 14, 1041. doi: 10.1063/1.555739.
[45] Bielski, B. Reevaluation of the spectral and kinetic properties of $HO_2^{\bullet}$ and $O_2^{\bullet-}$ free radicals. *Photochem. Photobiol.* 1978, 28, 645. doi: 10.1111/j.1751-1097.1978.tb06986.x.
[46] Koppenol, W., Butler, J. Energetics of interconversion of reactions of oxyradicals. *Adv. Free Radical Biol. Med.* 1985, 1, 91. doi: 10.1016/8755-9668(85)90005-5.
[47] Nosaka, Y., Yamashita, Y., Fukuyama, H. Application of chemiluminescent probe to monitoring superoxide radicals and hydrogen peroxide in TiO_2 photocatalysis. *J. Phys. Chem. B* 1997, 101, 5822. doi: 10.1021/jp970400h.
[48] Schwarz P., Turro N., Bossmann S., Braun A., Wahab A., Dürr H. A new method to determine the generation of hydroxyl radicals in illuminated TiO_2 suspensions. *J. Phys. Chem. B* 1997, 101, 7127. doi: 10.1021/jp971315c.
[49] Romeiro, A., Freitas, D., Azenha, M.E., Canle, M., Burrows, H.D. Effect of the calcination temperature on the photocatalytic efficiency of acidic sol–gel synthesized TiO_2 nanoparticles in the degradation of alprazolam. *Photochem. Photobiol. Sci.* 2017, 16, 935. doi: 10.1039/C6PP00447D.
[50] Azenha, M.E.D.G., Burrows, H.D., Canle, M., Coimbra, R., Fernández, M.I., García, M.V., Peiteado, M.A., Santaballa, J.A. Kinetic and mechanistic aspects of the direct photodegradation of atrazine, atraton, ametryn and 2-hydroxyatrazine by 254 nm light in aqueous solution. *J. Phys. Org. Chem.* 2003, 16, 498. doi: 10.1002/poc.624.
[51] Canle, M., Fernández, M.I., Santaballa, J.A. Developments in the mechanism of photodegradation of triazine-based pesticides. *J. Phys. Org. Chem.* 2005, 18(2), 148. doi: 10.1002/poc.874.
[52] Ervens, B., Gligorovski, S., Herrmann, H. Temperature-dependent rate constants for hydroxyl radical reactions with organic compounds in aqueous solutions. *Phys. Chem. Chem. Phys.* 2003, *5*, 1811. doi: 10.1039/B300072A.
[53] Canle, M., Fernández, M.I., Rodríguez, S., Santaballa, J.A., Steenken, S., Vulliet, E. Mechanisms of direct and TiO_2-photocatalysed UV degradation of phenylurea herbicides. *ChemPhysChem* 2005, 6, 2064. doi: 10.1002/cphc.200500004.

[54] Richard, C. Regioselectivity of oxidation by positive holes (h^+) in photocatalytic aqueous transformations. *J. Photochem. Photobiol. A Chem.* 1993, 72, 179. doi: 10.1016/1010-6030(93)85026-5.
[55] Romeiro, A., Azenha, M.E., Canle, M., Rodrigues, V.H.N., da Silva, J.P., Burrows, H.D. Titanium dioxide nanoparticle photocatalysed degradation of ibuprofen and naproxen in water—competing hydroxyl radical attack and oxidative decarboxylation by semiconductor hole. *Chem. Select* 2018, 3, 10915. doi: 10.1002/slct.201801953.
[56] Canle, M., Santaballa, J.A., Vulliet, E. On the mechanism of TiO_2-photocatalyzed degradation of aniline derivatives. *J. Photochem. Photobiol. A Chem.* 2005, 175, 192. doi: 10.1016/j.jphotochem.2005.05.001.
[57] Turchi, C.S., Ollis, D.F. Photocatalytic degradation of organic water contaminants—Mechanisms involving hydroxyl radical attack. *J. Catal.* 1990, 122, 178. doi: 10.1016/0021-9517(90)90269-P.
[58] Sun, L., Schindler, K.M., Hoy, A.R., Bolton, J.R. Spin-trap EPR studies of intermediates involved in photodegradation reactions on TiO_2—Is the process heterogeneous or homogeneous? In Helz, G.R., Zepp, R.G., Crosby, D.G. (Eds.), *Aquatic and Surface Photochemistry*, Lewis Publishers, Boca Ratón, 1994, p. 409.
[59] Ansón-Casaos, A., Sampaio, M.J., Jarauta-Córdoba, C., Martínez, M.T., Silva, C.G., Faria, J.L., Silva, A.M.T. Evaluation of sol–gel TiO_2 photocatalysts modified with carbon or boron compounds and crystallized in nitrogen or air atmospheres. *Chem. Eng. J.* 2015, 277, 11. doi: 10.1016/j.cej.2015.04.136.
[60] Burrows, H.D., Canle, M., Santaballa, J.A., Steenken, S. Reaction pathways and mechanisms of photodegradation of pesticides. *J. Photochem. Photobiol. B Biol.* 2002, 67, 71. doi: 10.1016/s1011-1344(02)00277-4.
[61] Klavarioti, M., Mantzavinos, D., Kassinos, D. Removal of residual pharmaceuticals from aqueous systems by advanced oxidation processes. *Environ. Int.* 2009, 35, 402. doi: 10.1016/j.envint.2008.07.009.
[62] Canle, M., Fernández, M.I., Martínez, C., Santaballa, J.A. Photochemistry for pollution abatement. *Pure Appl. Chem.* 2013, 85(7), 1437. doi: 10.1351/PAC-CON-13-01-10.
[63] Boukhatem, H., Khalaf, H., Djouadi, L., González, F.V., Navarro, R.M., Santaballa, J.A., Canle, M. Photocatalytic activity of mont-La (6%)-$Cu_{0.6}Cd_{0.4}S$ catalyst for phenol degradation under near UV visible light irradiation. *Appl. Catal., B Environ.* 2017, 211, 114. doi: 10.1016/j.apcatb.2017.03.074.
[64] Belekbir, S., El Azzouzi, M., El Hamidi, A., Rodríguez-Lorenzo, L., Santaballa, J.A., Canle, M. Improved photocatalyzed degradation of phenol, as a model pollutant, over metal-impregnated nanosized TiO_2. *Nanomaterials* 2020, 10, 996. doi: 10.3390/nano10050996.
[65] Zhang, Y., Xia, T., Shang, M., Wallenmeyer, P., Katelyn, D., Peterson, A., Murowchick, J., Dong, L., Chen, X. Structural evolution from TiO_2 nanoparticles to nanosheets and their photocatalytic performance in hydrogen generation and environmental pollution removal. *RSC Adv.* 2014, 4, 16146. doi: 10.1039/C3RA48066F.
[66] Wang, M., Iocozzia, J., Sun, L., Lin, C., Lin, Z. Inorganic-modified semiconductor TiO_2 nanotube arrays for photocatalysis. *Energy Environ. Sci.* 2014, 7, 2182. doi: 10.1039/C4EE00147H.
[67] Martínez, C., Canle, M., Fernández, M.I., Santaballa, J.A., Faria, J. Kinetics and mechanism of aqueous degradation of carbamazepine by heterogeneous photocatalysis using nanocrystalline TiO_2, ZnO and multi-walled carbon nanotubes-anatase composites. *Appl. Catal., B Environ.* 2011, 102, 563. doi: 10.1016/j.apcatb.2010.12.039.

[68] Martínez, C., Canle, M., Fernández, M.I., Santaballa, J.A., Faria, J. Aqueous degradation of diclofenac by heterogeneous photocatalysis using nanostructured materials. *Appl. Catal., B Environ.* 2011, 107, 110. doi: 10.1016/j.apcatb.2011.07.003.

[69] Martínez, C.; Viñariño, S.; Fernández, M.I.; Faria, J.; Canle, M.; Santaballa, J.A. Mechanism of degradation of ketoprofen by heterogeneous photocatalysis in aqueous solution. *Appl. Catal., B Environ.* 2013, 142, 633. doi:10.1016/j.apcatb.2013.05.018.

[70] Bougarrani, S., Sharma, P.K., Hamilton, J.W., Singh, A., Canle, M., El Azzouzi, M., Byrne, J.A. Enhanced photocatalytic degradation of the imidazolinone herbicide imazapyr upon UV/Vis irradiation in the presence of CaxMnOy-TiO_2 hetero-nanostructures—degradation pathways and reaction intermediates. *Nanomaterials* 2020, 10, 896. doi: 10.3390/nano10050896.

[71] Djouadi, L., Khalaf, H., Boukhatem, H., Boutomi, H., Kezzime, A., Santaballa, J.A., Canle, M. Degradation of aqueous ketoprofen by heterogeneous photocatalysis using Bi_2S_3/TiO_2–Montmorillonite nanocomposites under simulated solar irradiation. *Appl. Clay Sci.* 2018, 166, 27. doi: 10.1016/j.clay.2018.09.008.

[72] Hooshyar, Z., Rezanejade Bardajee, G., Ghayeb, Y. Sonication enhanced removal of nickel and cobalt ions from polluted water using an iron based sorbent. *J. Chem.* 2013, 2013, 786954. https://doi.org/10.1155/2013/786954.

[73] Karn, B., Kuiken, T., Otto, M. Nanotechnology and in situ remediation—a review of the benefits and potential risks. *Environ. Health Perspect.* 2009, 117, 1813. doi: 10.1289/ehp.0900793.

[74] Simonin, M., Richaume, A., Guyonnet, J.P., Dubost, A., Martins, J.M.F., Pommier, T. Titanium dioxide nanoparticles strongly impact soil microbial function by affecting archaeal nitrifiers. *Sci. Rep.* 2016, 6, 33643. doi: 10.1038/srep33643.

[75] Rafique, M.S., Tahir, M.B., Rafique, M., Khan, M.I. Chap. 13—Recent advances in the development of photocatalysis and future perspectives. In (Tahir, M.B., Rafique, M., Rafique, M.S., Eds.), *Micro and Nano Technologies, Nanotechnology and Photocatalysis for Environmental Applications.* Elsevier, 2020, pp. 221–223. ISBN 9780128211922- https://doi.org/10.1016/B978-0-12-821192-2.00013-9.

[76] Barbusiński, K. Fenton reaction - controversy concerning the chemistry. *Ecol. Chem. Eng.* 2009, 16(3), 347. http://tchie.uni.opole.pl/freeECE/S_16_3/Barbusinski_16(3).pdf.

[77] Redouane-Salah, Z., Malouki, M.A., Khennaoui, B., Santaballa, J.A., Canle, M. Simulated sunlight photodegradation of 2-mercaptobenzothiazole by heterogeneous photo-Fenton using a natural clay powder. *J. Environ. Chem. Eng.* 2018, *6*, 1783. https://doi.org/10.1016/j.jece.2018.02.011.

[78] Khenaoui, B., Malouki, M.A., Canle, M., Zehani, F., Boutanoi, N., Redouane-Salah, Z., Zertal, A. Heterogeneous photo-Fenton process for degradation of azo dye—methyl orange using a local cheap material as a photocatalyst under solar light irradiation. *Optik* 2017, 137, 6. doi: 10.1016/j.ijleo.2017.02.081.

[79] Khenaoui, B., Zehani, F., Malouki, M.A., Menacer, R., Canle, M. Chemical and physical characterization of a natural clay and its use as photocatalyst for the degradation of the methabenzthiazuron herbicide in water. *Optik* 2020, 219, 165024. doi: 10.1016/j.ijleo.2020.165024.

[80] Kharisov, B.I., Rasika Dias, H.V., Kharissova, O.V., Manuel Jiménez-Pérez, V., Olvera Pérez, B., Muñoz Flores, B. Iron-containing nanomaterials—Synthesis, properties, and environmental applications. *RSC Adv.* 2012, 2, 9325. doi: 10.1039/C2RA20812A.

[81] Khin, M.M., Nair, A.S., Babu, V.J., Murugan, R., Ramakrishna, S. A review on nanomaterials for environmental remediation. *Energy Environ. Sci.* 2012, 5, 8075. doi: 10.1039/C2EE21818F.

[82] Guo, M., Weng, X., Wang, T., Chen, Z. Biosynthesized iron-based nanoparticles used as a heterogeneous catalyst for the removal of 2,4-dichlorophenol. *Sep. Purif. Technol.* 2017, 175, 222. https://doi.org/10.1016/j.seppur.2016.11.042.

[83] Han, Y., Yan, W. Reductive dechlorination of trichloroethene by zero-valent iron nanoparticles—reactivity enhancement through sulfidation treatment. *Environ. Sci. Technol.* 2016, 50, 12992. https://doi.org/10.1021/acs.est.6b03997.

[84] Li, X.Q., Zhang, W.X. Iron nanoparticles—the core–shell structure and unique properties for Ni(II) sequestration. *Langmuir* 2006, 22, 4638. doi: 10.1021/la060057k.

[85] Ebrahim, S.E., Sulaymon, A.H., Saad Alhares, H. Competitive removal of Cu^{2+}, Cd^{2+}, Zn^{2+}, and Ni^{2+} ions onto iron oxide nanoparticles from wastewater. *Desalin. Water Treat.* 2016, 57, 20915. doi: 10.1080/19443994.2015.1112310.

[86] Gupta, A., Silver, S. Molecular genetics—silver as a biocide—will resistance become a problem? *Nat. Biotechnol.* 1998, 16, 888. doi: 10.1038/nbt1098-888.

[87] Bosetti, M., Masse, A., Tobin, E., Cannas, M. Silver coated materials for external fixation devices—in vitro biocompatibility and genotoxicity. *Biomaterials* 2002, 23, 887. doi: 10.1016/s0142-9612(01)00198-3.

[88] Gogoi, S.K., Gopinath, P., Paul, A., Ramesh, A., Ghosh, S.S., Chattopadhyay, A. Green fluorescent protein-expressing Escherichia coli as a model system for investigating the antimicrobial activities of silver nanoparticles. *Langmuir* 2006, 22, 9322. doi: 10.1021/la060661v.

[89] Tao, F. Synthesis, catalysis, surface chemistry and structure of bimetallic nanocatalysts. *Chem. Soc. Rev.* 2012, 41, 7977. doi: 10.1039/C2CS90093A.

[90] Wu, L., Ritchie, S.M.C. Removal of trichloroethylene from water by cellulose acetate supported bimetallic Ni/Fe nanoparticles. *Chemosphere* 2006, 63, 285. doi: 10.1016/j.chemosphere.2005.07.021.

[91] Wang, X., Chen, C., Chang, Y., Liu, H. Dechlorination of chlorinated methanes by Pd/Fe bimetallic nanoparticles. *J. Hazard. Mater.* 2009, 161, 815. doi: 10.1016/j.jhazmat.2008.04.027.

[92] Wang, L.; Ma, W.; Xu, L.; Chen, W.; Zhu, Y.; Xu, C.; Kotov, N.A. Nanoparticle-based environmental sensors. *Mater. Sci. Eng. R Rep.* 2010, 70, 265. doi: 10.1016/j.mser.2010.06.012.

[93] Wang, L.; Xu, L.; Kuang, H.; Xu, C.; Kotov, N.A. Dynamic nanoparticles assemblies. *Acc. Chem. Res.* 2012, *45*(11), 1916. doi: 10.1021/ar200305.

2 Multifunctional Nanomaterials for Environmental Remediation

Archana Charanpahari
Berkeley Careers
Galgotias University

Suhas Mukherjee
Vikram Sarabhai Space Centre

Sachin Ghugal
Hyderabad Central University

Hariprasad E
Vasavi College of Engineering

Bhatt Tushar Shriram
Vikram Sarabhai Space Centre

CONTENTS

DOI: 10.1201/9781003129042-2

2.1 INTRODUCTION

Humans and environment are dependent upon each other for their existence and survival. All the anthropogenic developmental activities have a deep impact on the environment. The introduction and usage of a wide variety of materials such as dyes, paints, lacquers, fertilizers, pesticides, personal care products, pharmaceuticals and heavy metals have made life comfortable for humans. However, every coin has two sides. The rapid industrialization coupled with urbanization has adversely affected the surrounding air, water and soil. The burgeoning demand of the huge populace has aggravated the overutilization of non-renewable natural resources, leading to their quick depletion. On the one hand, humans have amassed wealth to make their life luxurious and comfortable. On the other hand, they have interfered with the regenerative properties of the environment and caused pollution. The level of environmental deterioration can be foreseen from the continuous increase in global carbon emissions. Carbon dioxide emissions are estimated to increase to 43 Gt per year from current 29 Gt per year under new policy regime [1,2]. Thus, to prevent further deterioration of the environment and to preserve it for future generations, multifunctional materials can play a prominent role.

Multifunctional materials have multi-fold advantages. They are basically either material-based systems or hybrid materials equipped to perform multiple functions. Their operation may be performed in a sequential manner or concomitant mode [1]. Designing such materials is fundamentally challenging because it will mean exploring novel materials, identifying and establishing their properties and structure systems, or excavating new properties in existing structures and improving their performance. All these endeavours should culminate in obtaining and integrating myriad functions in one material that will perform consistently and excel even under adverse or harsh conditions. Nanomaterials have properties such as high surface area, good tunability and high performance as compared to bulk materials. Such multifunctional nanomaterials would be apt for real-world technological applications along with sustainable development in the next decade.

This chapter begins with the introduction to variety of pollutants present in the environment. Thereafter, an introduction to processes occurring during heterogeneous catalysis is presented. The subsequent section deals with the working principle and mechanism of photocatalysis/photoelectrocatalysis and adsorption, followed by a brief discussion about various nanomaterials used. This does not entail a detailed, exhaustive account of each and every material used for the removal of pollutants. It merely offers a bird's eye view of the general materials such as TiO_2 used for removing a variety of pollutants. On the similar lines, other materials also exhibit degradation of pollutants using photocatalysis and photoelectrocatalysis. Thereafter, multifunctional materials such as magnetic photocatalysts, round-the-clock photocatalysts exhibiting catalytic memory, upconversion photocatalysts and supported/immobilized photocatalysts are discussed. This chapter concludes with environmental remediation of air pollutants and actual application of the above technologies to real-life problems. However, these multifunctional nanomaterials are commonly used as catalysts, photocatalysts or adsorbents for environmental remediation of gaseous/liquid pollutants.

The interaction of catalysts with pollutants plays a kingpin role in deciding the efficiency of removal/conversion of pollutants. They must possess a moderate electronic interaction with pollutants, particularly on the surface. Hence, before discussing the type of pollutants it is necessary to present a section that specifically discuss the types of major electronic interactions that drive processes such as catalysis, photocatalysis, electrocatalysis and adsorption.

2.2 ELECTRONIC INTERACTION

Efficacy of any kind of reaction, viz. catalytic reactions, physical adsorption and chemical adsorption, for the separation of pollutants from gaseous or liquid stream exclusively depends on the structure of the surface and reactant molecules. The types of electronic interactions that play an important role in deciding the fate of these reactions or separation processes are often categorized as short-range and long-range interactions. Short-range interactions are mostly cohesive in nature. These kinds of interactions are mostly associated with the forces that arise due to chemical bonds in the molecules. On the other hand, the interactions involved in any type of chemical and surface reaction are long-range interactions. Unlike short-range interactions, the origin of these interactions is often classified as electrostatic and polarization forces [3]. Both these forces are responsible for propelling the processes such as adsorption and chemical reactions. Now let us discuss the scenarios/conditions under which these interactions evolve.

2.2.1 Charge–Charge Interactions

The interaction forces between two charges or two ions in any fluid or solid objects are the strongest form of interaction forces. The amplitude of these forces is larger than chemical binding forces. Due to the coulombic nature, the forces that arise due to charge–charge interactions in any fluid or ionic crystal strongly depend on the distance between the charges.

Figure 2.1 shows the interactions among charged molecules or groups present in the fluid stream and the charged surface. Jiang and co-workers [4] synthesized anionic metal–organic frameworks where the sulphonate group present in the adsorbate helps in the enhancement of the adsorption efficiency of amino groups present in the wastewater via charge–charge interactions.

2.2.2 Charge–Dipole Interactions

Charge–dipole interaction forces are one of the key driving forces that exclusively decide the efficiency of adsorption of gaseous pollutants from gaseous streams and conversion of these gases into less harmful products by ionic catalysts, where molecules with strong inherent dipole moments are adsorbed on ionic active metals/sites such as Pt, Cu, Rh and Ru substituted over catalytic substrates such as ceria, alumina and cobalt oxides. Subsequently, these adsorbed gases are converted into products by reactions over the surfaces.

Let us consider the case of carbon monoxide oxidation over ionic catalysts as depicted in Figure 2.3. Figure 2.3 renders the example of charge–dipole interaction of carbon monoxide and oxygen with ionic active sites A^+ and B^+ substituted

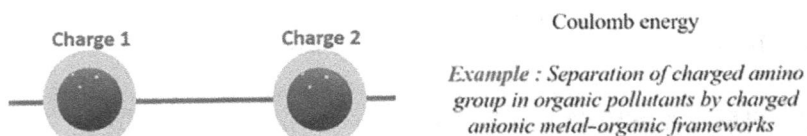

Coulomb energy

Example : Separation of charged amino group in organic pollutants by charged anionic metal-organic frameworks

FIGURE 2.1 Charge–charge interactions.

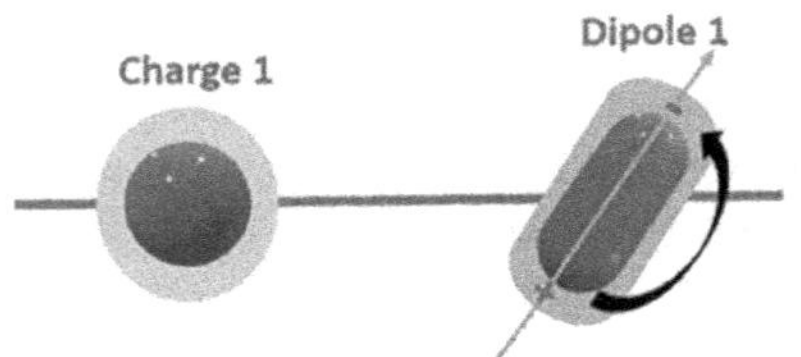

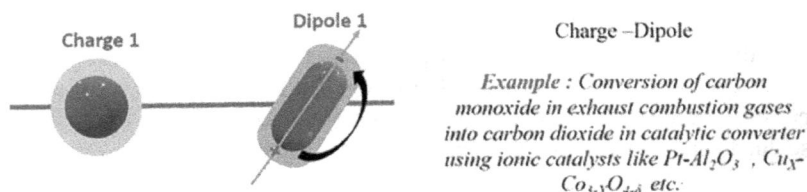

Charge –Dipole

Example : Conversion of carbon monoxide in exhaust combustion gases into carbon dioxide in catalytic converter using ionic catalysts like $Pt\text{-}Al_2O_3$, $Cu_X\text{-}Co_{3\text{-}X}O_{4\text{-}\delta}$ *etc.*

FIGURE 2.2 Charge–dipole interactions.

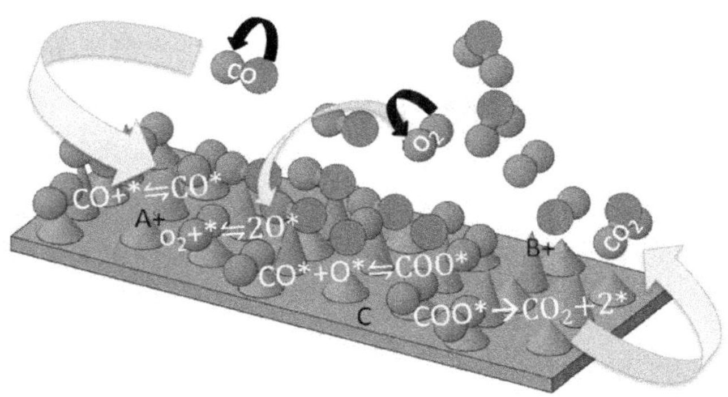

FIGURE 2.3 Charge–dipole interactions in CO oxidation over ionic catalyst.

in metallic substrate C. The process depicted in Figure 2.2 exemplifies the interaction of strong dipole in CO molecule and charged active sites. However, oxygen is attached to the surface via adsorption on the substrate and ionic active sites. Subsequently, both adsorbed intermediated species interact with each other via charge–charge interaction to form carbon dioxide. Finally, carbon dioxide is desorbed quickly from the catalyst surface due to the weak ion–dipole interaction of CO_2 with the surface.

The activity of Pt/alumina-based catalyst towards selective CO oxidation of a feed composed of 1% CO, 1% O_2, 65% H_2 and the remaining He for fuel cell applications was studied by Manasilp et al. [5]. Singh et al. [6] studied the selective carbon monoxide oxidation and CO_2 methanation over copper-substituted cobalt oxide. They reported strong interaction of CO and CO_2 with cobalt oxide and copper. Likewise, the removal of harmful components such as CO from the automotive exhaust is commonly achieved by the application of a catalytic converter installed at the exhaust of the IC engine of the automobile. In general, the catalytic converter consists of a monolith reactor containing parallel channels coated with a wash coat layer. These layers contain noble metals such as platinum, rhodium and palladium supported on alumina or ceria [7].

2.2.3 Dipole–Dipole Interactions

Dipole–dipole interaction is a weak form of interaction between two polar molecules where both the molecules have permanent dipoles. This kind of interaction energy is often called Keesom energy [3]. Figure 2.4 depicts the interaction between two permanent dipoles. Dipole–dipole interaction is one of the major interaction forces in the gas–liquid chromatographic separation technique. Separation techniques such as gas chromatography, liquid chromatography and high-performance liquid chromatography are frequently used for analysing the composition of pollutant components during wastewater treatment.

All of these separation techniques use a column with a stationary phase that is responsible for the separation of different kinds of molecules from the mobile

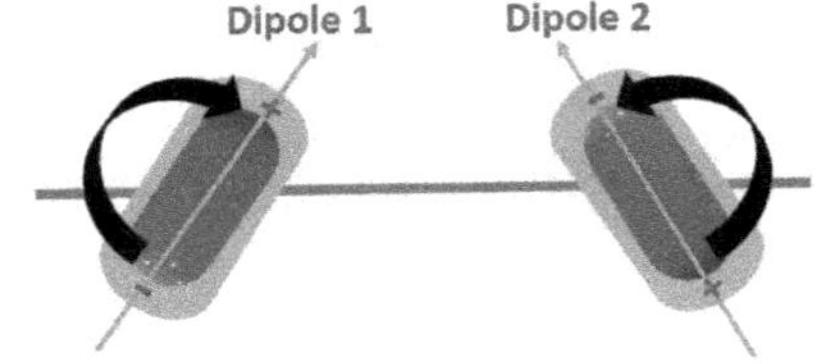

Keesom energy

Example : Separation of strong electronegative components like -NO_X, -SO_X, Halogens etc. from gas mixture in chromatographic separation technique using a charged polymer

FIGURE 2.4 Dipole–dipole interactions.

phase (gas or liquid) passing over it. The efficiency of this separation technique solely depends on the type of interactions between the stationary and mobile phases. Thus, the selectivity of adsorption to the stationary surface entirely depends on the interaction among permanent dipoles of molecules present in the mobile phase and dipoles present in the stationary phase. The most commonly used stationary phases in the chromatographic column are polydimethylsiloxane, methylsiloxane, polyethylene glycol, fluorosiloxane, etc. [7]. These columns are used for the separation of higher molecular weight organic compounds such as phenol, chlorophenol, benzene and butanol. However, for the separation of lower molecular weight compounds such as gases CO_2, CO, methane, NO_X and SO_X, molecular sieves and porous polymers are used. Among all types of interactions, dipole–dipole interactions are the strongest interaction that plays a major role in deciding the selectivity of the chromatographic column. In addition to dipole–dipole interaction, the gas separation chromatographic separation process also involves other interaction forces such as dipole–induced dipole and London dispersion/van der Waals' interactions [8].

2.2.4 Charge/Dipole–Induced Dipole Interactions

Charge/dipole–induced dipole interaction is a weak interaction that evolves due to the polarization of a nonpolar molecule by a charged molecule or surface having a strong dipole moment or charge (Figure 2.5).

While polarization of nonpolar molecules by the strong dipole, the positive side and negative side of the dipole induce negative charge and positive charge in nonpolar molecules, respectively. The adsorption of CO_2 on the catalytic surface involves the polarization of nonpolar molecules such as CO_2 by the surface.

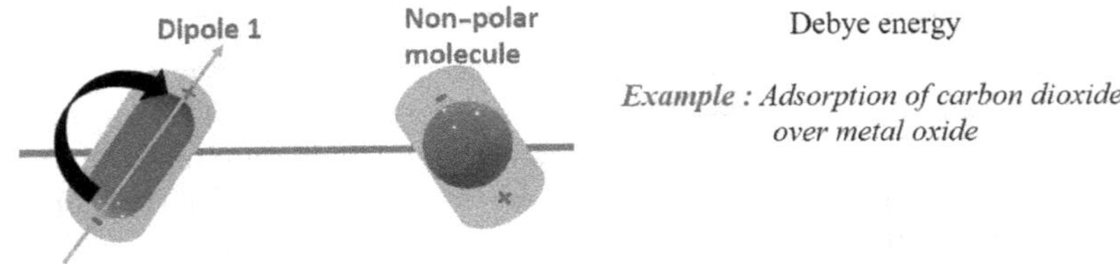

FIGURE 2.5 Dipole–induced dipole interactions.

Little and Amberg [9] described the linear adsorption of CO_2 on the metal and metal oxide surface through ion–induced dipole interaction between a metal ion and carbon dioxide as per the mechanism described below:

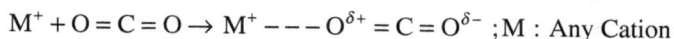

$$M^+ + O = C = O \rightarrow M^+ --- O^{\delta+} = C = O^{\delta-} \text{ ;M : Any Cation}$$

Dissociative water adsorption in the metal oxide surface is also instigated by the polarization of nonpolar metal oxide surface by the dipole present in the water molecule [10].

2.2.5 Interactions between Two Nonpolar Molecules

The weakest form of interaction between two nonpolar molecules exists due to the self-polarization of both the molecules due to their self-dipole creation. Often, this force is called London force that is responsible for condensing of liquid or freezing of solid at low temperature. According to the nature of these forces, they are purely attractive and emerge due to the induced dipole–induced dipole interactions among nonpolar molecules. The key force that results in the physical adsorption of gas molecules on activated charcoal or zeolite surface is the London dispersion force. London dispersion force is a form of van der Waals weak interaction among molecules. In addition to the physical adsorption of gases, other processes such as chromatographic separation [8] and adsorption of liquid organic molecules such as isooctane on iron surface [10] are also associated with the weak van der Waals interaction (Figure 2.6).

2.3 VARIETY OF POLLUTANTS

Any material causing undesirable physical, chemical or biological changes to the environment (air, water and soil) or any other organism is known as a pollutant. As mentioned above, the environment is replete with man-made chemicals and natural pollutants, which can't be cleansed by natural regenerative process, hence causing pollution. Depending upon the phase, they can be mainly classified as liquid- and gas-phase pollutants. In the liquid phase, the pollutants can briefly be classified into broad categories: dyes, phenols, petroleum products, pharmaceuticals, personal care products, insecticides and heavy metals. The adverse effects of each of them are briefly quoted.

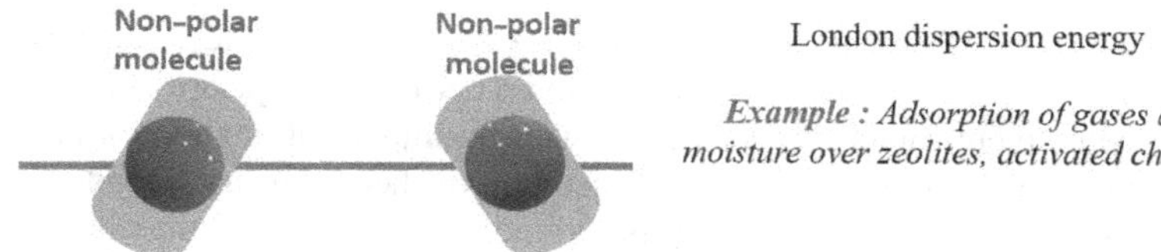

FIGURE 2.6 Interactions between two nonpolar molecules.

Phenols are aromatic compounds, are highly corrosive in nature and can affect the lungs and respiratory tract of humans. Thus, they pose risk to not only humans, but also their ecosystem. Even when present in small concentrations, they can cause dermatitis and burns. Bisphenol A has versatile industrial applications in plastic/resin industry, as a sealant and as lining of food and drinking bottles. However, it is an endocrine-disrupting compound posing threat to human health [11].

Dyes are ubiquitous materials, used in textile, paper, plastics, rubber, cosmetics, tanneries, paints and, leathers industries [12]. The dyes that are discharged as industrial wastewater into water bodies create havoc in the life of marine flora and fauna. Most of the dyes used in the textile industry are highly stable, soluble in water, resistant to reactions with chemical agents and of low biodegradability. Additionally, the use of non-controlled oxidative reactions can generate highly toxic by-products that cause secondary pollution. Some azo dyes have been related to cancer of human bladder in humans and chromosomal aberrations in mammalian cells [13]. The discharge into surface water leads to aesthetic problems and obstructs the light penetration and oxygen transfer into water bodies, thereby affecting aquatic life and sometimes leading to eutrophication [14]. The mutagenic potential of these compounds is directly related to the type and position of substituents such as an aromatic ring and amino nitrogen atom. Several aminoazobenzene dyes are mutagenic for Salmonella, and their potencies depend on the nature and position of substituents with respect to both the aromatic rings and the amino nitrogen atom. It has been observed that textile effluent contaminated with azo dyes induces chromosomal and nuclear aberrations in Allium cepa seeds, which were used as test system for the study [15].

Pesticides are highly toxic and carcinogenic in nature. Pesticide exposure leads to birth defects, foetal death and altered foetal growth. Pesticides cause asthma, allergies, hypersensitivity and hormone disruption. Pesticides such as methyl parathion, imidacloprid, triazophos, dichlorvos and alachlor are highly toxic in nature and have adverse effects on humans [16].

Heavy metals are discussed in a separate section altogether.

Pharmaceuticals in water streams have created a menace in the ecosystem, the causative factor being the resistance of pathogens to these compounds/chemicals due to dumping of unused pharmaceuticals into water streams. As a result, superbugs that are resistant to innumerable antibiotics are developed. This hinders the treatment of the infected host [17].

2.4 ENVIRONMENTAL REMEDIATION

For sustainable development, the development of non-polluting technologies for environmental restoration and energy conversion is one of the top priorities and challenges at the moment. Heterogeneous catalysis is an important method for achieving environmental remediation. Before proceeding further, let us try to understand the mechanism of heterogeneous catalysts and interactions involved therein.

Catalysis involves basically two components: one is the catalyst, and the other, the reactant/pollutant. Obviously, the catalyst is used because it alters the activation energy for a particular reaction. Catalysis occurs via the following mechanism [18].

i. diffusion of reactants to the surface,
ii. adsorption of reactants onto the surface,
iii. reaction on the surface,
iv. desorption of products off the surface,
v. diffusion of products from the surface.

However, the interaction between the catalyst and reactants differs in gas-phase and liquid-phase reactions. This is pivotal while considering the removal of liquid and air pollutants. For gas-phase reaction, the above mechanism is valid, whereas for liquid phase, the interaction of the catalyst with a variety of molecules plays a crucial role. They key reason is the coverage of the adsorption sites by not only reactants/products or intermediates, but also omnipresent solvent molecules, especially at the solid–liquid interface. Hence, thermodynamically, along with Gibbs free energy change due to adsorption of adsorbate for instance A. There is additional contribution from the desorption of previously adsorbed species B, solvent interactions, in solution and for material surface, as also sorbate–sorbate interactions for adsorbed A and B. The detailed discussion of these phenomena and interaction of solid and liquid surfaces is exhaustively presented in the review by Sievers et al. [19].

This section speaks briefly about the various techniques used for environmental remediation. Environmental remediation can be brought about by three major techniques, namely photocatalysis, photoelectrocatalysis and adsorption. The traditional techniques such as chemical precipitation, filtration, flocculation and adsorption are of high cost, require advanced processing or serve to transfer the pollutants from one medium to another causing secondary pollution. The other techniques are also fraught with limitations. For example, biological degradation requires exact pH and concentration, which are difficult to maintain while handling large quantity of effluents. Membrane filtration results in fouling of the membranes and is a costly alternative. The detailed limitations of these techniques are well described in the review by Chen et al. [20]. Advanced oxidation processes such as photocatalysis/photoelectrocatalysis are the best alternative because of manifold advantages. They are environment-friendly, do not cause secondary pollution and, most importantly, result in total mineralization of pollutants through generation of reactive oxygen species to harmless products such as CO_2 and water. Let us discuss briefly the working of each of these techniques.

2.5 WORKING PRINCIPLE OF PHOTOCATALYSIS, PHOTOELECTROCATALYSIS AND ADSORPTION

2.5.1 PHOTOCATALYSIS

Serpone et al. have given an exhaustive account of the origin of photocatalysis of the pre-1970 era. The first appearance of the terms **photokatalyse** and

photokatalytisch occurred in the 1910 textbook on photochemistry by the Russian scientist J. Plotnikow [21]. Photocatalysis can be defined as, 'the acceleration of the rate of chemical reactions (oxidation/reduction) brought about by the activation of a substrate by light (UV or visible)' [22]. Photocatalyst is the substance/material which is activated by light absorption. Photocatalysts alter the kinetics of a chemical reaction, but themselves remain unchanged. Heterogeneous photocatalysis is an advanced oxidation process, based on absorption of photons with energy higher than the band gap of any material, resulting in the excitation of semiconductor material and generation of excitons (electron–hole, e^-–h^+, pair) [23]. Upon light irradiation, an electron is excited from valence band to the conductive band (e_{CB}^-). The absence of electron results in the formation of a positive hole in the valence band (h_{VB}^+) (Eq. 2.1). The charge carriers hole (h_{VB}^+) and electron (e_{CB}^-) are powerful oxidizing and reducing agents, respectively. Both electrons and holes will migrate to the surface of the catalyst and participate in redox reactions. The oxidation of adsorbed H_2O is brought about by photogenerated holes, while dissolved oxygen is reduced by a photoexcited electron. It is noteworthy that oxidation and reduction processes occur simultaneously and at the same rate. Hence, extremely reactive oxygen species (ROS) ($OH^•$, $O_2^{•-}$, H_2O_2, O_3, etc.) are formed at the surface of the semiconductor, which will result in direct oxidation of the organic pollutant. Alternatively, valence band hole can also result in their mineralization of pollutant producing CO_2 and H_2O (Eq. 2.2). The h_{VB}^+ can react with water to generate $^•OH$ (oxidation potential 2.8 V) and oxidize pollutants (Eqs. 2.3 and 2.4).

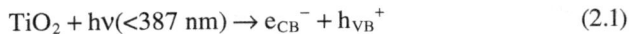

$$TiO_2 + h\nu(<387\ nm) \rightarrow e_{CB}^- + h_{VB}^+ \tag{2.1}$$

$$h_{VB}^+ + R \rightarrow \text{intermediates} \rightarrow CO_2 + H_2O \tag{2.2}$$

$$H_2O + h_{VB}^+ \rightarrow {}^•OH + H^+ \tag{2.3}$$

$${}^•OH + R \rightarrow \text{intermediates} \rightarrow CO_2 + H_2O \tag{2.4}$$

where R represents the organic compound. The formation of superoxide by reaction of conductive band electron is shown in Eq. (2.5) [24]. The mechanism of the electron–hole pair formation when the TiO_2 is irradiated is depicted in Figure 2.7 [25].

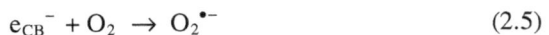

$$e_{CB}^- + O_2 \rightarrow O_2^{•-} \tag{2.5}$$

The presence of dissolved oxygen is an extremely important parameter. The reduction process of oxygen and the oxidation of pollutants proceed simultaneously to prevent the accumulation of electrons in the conduction band [23]. This will reduce the rate of recombination of e_{CB}^- and h_{VB}^+ [26,27].

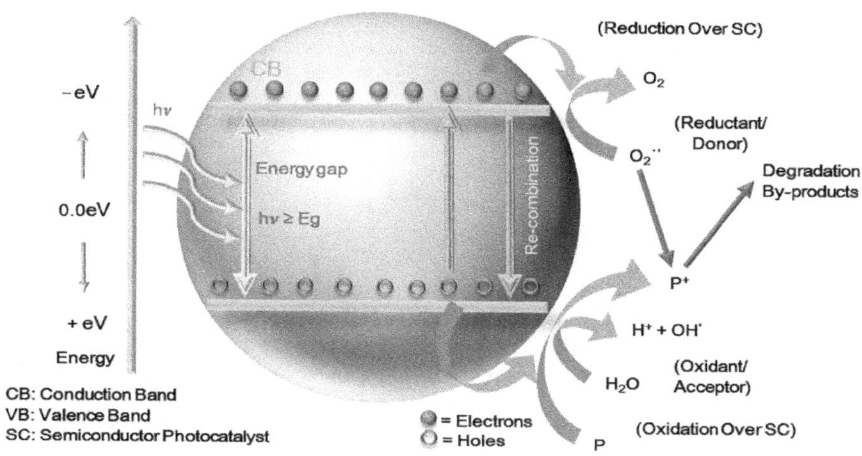

FIGURE 2.7 Mechanism of electron–hole formation in a TiO_2 particle in the presence of pollutant in the water.

2.5.2 Photoelectrocatalysis

Photoelectrocatalysis is a green technique that couples photocatalysis with electrochemical process. The basic process remains the same, but with a slight modification. For example, upon light illumination, electron–hole pairs are generated.

$$MOx \rightarrow MOx\text{-}e_{CB}^{-} + MOx\text{-}h_{VB}^{+}$$

where semiconductor is represented by MO_x [28], and VB and CB denote valence band and conduction band, respectively. Now, the low photonic efficiency due to recombination of electron–hole pairs is circumvented by facilitating their separation by gradient potential [29, 30]. This implies that the semiconductor is attached to the surface of a conductive substrate and used as a photoelectrode.

When a semiconductor is in contact with an electrolyte, the Schottky junction is formed. Any change in electrochemical potential of semiconductor is compensated for by band bending in semiconductor. Figure 2.8 (a-c) illustrates the possible band bending at the SC–electrolyte interface [31]. The region where bending occurs is termed as space charge layer or depletion layer. In the space charge layer, the majority of the carriers are depleted.

Consider the electrons in n-type TiO_2. It is this region where the applied bias potential facilitates improved charge separation [32]. Let us consider the example of TiO_2. When the bias potential is greater than flat band potential, then band bending occurs with depletion of electrons and holes moving to the surface. At the surface, they bring about oxidation of water molecules to yield OH· radicals responsible for degrading organics [31]. Electrons move through external circuit to counter electrode, where they bring about reduction reactions.

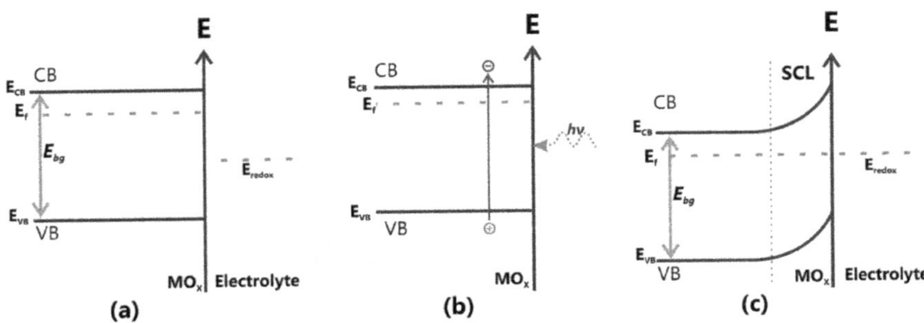

FIGURE 2.8 (a–c) Schematic representation of the energy band diagram in a semiconductor. (a) Semiconductor in contact with electrolyte. (b) Semiconductor under illumination, no depletion layer. (c) Semiconductor with applied bias greater than flat band potential.

2.5.3 Adsorption

Adsorption is a surface-based process in which molecules of a compound in gaseous or liquid state are accumulated on an adsorbent surface. The compound that is adsorbed is termed as adsorbate, while the material upon which adsorption occurs is called adsorbent. On the other hand, desorption is reverse of adsorption, which means that the adsorbed molecules leave the surface of the adsorbent. This process is schematically depicted in Figure 2.9.

Adsorption is of two types, depending on the nature of interaction between the adsorbate and the adsorbent. They are 'physical adsorption', or 'physisorption', and 'chemical adsorption', also called 'chemisorption'.

Physical adsorption, as the name suggests, involves weak van der Waals forces or electrostatic forces. Here, monolayer or multi-layer adsorption is possible and is reversible. Chemisorption involves the formation of strong chemical bonds such as covalent bonds between the surface and the adsorbed molecules. It is irreversible [33].

2.6 VARIOUS NANOMATERIALS USED FOR ENVIRONMENTAL DETOXIFICATION

Environmental remediation can be brought about by a variety of nanomaterials. Metal oxides were the forerunners of photocatalytic technology. The water splitting

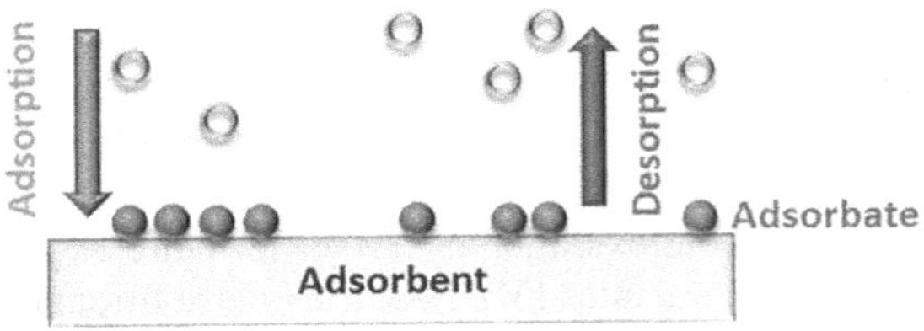

FIGURE 2.9 Schematic representation of adsorption and desorption processes.

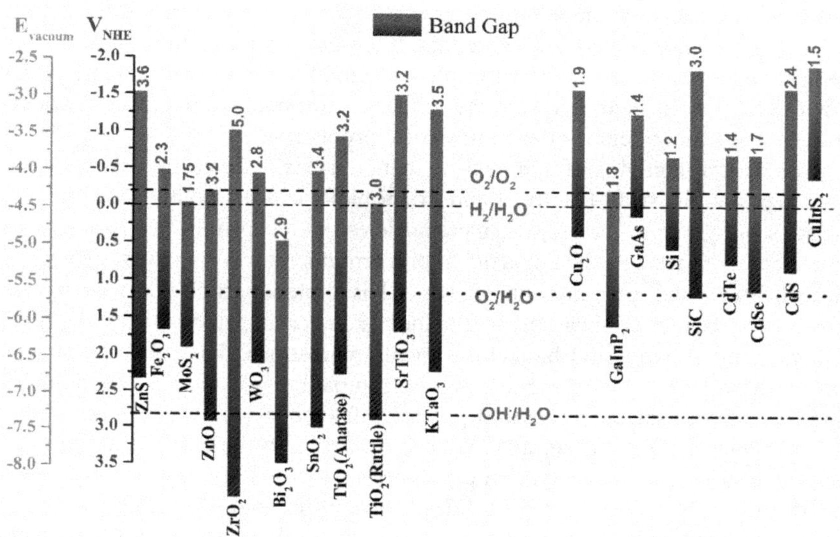

FIGURE 2.10 Valence band and conduction band positions for various materials along with their band gaps.

by Fujishima and Honda heralded an era of research in photocatalysis. Figure 2.10 depicts various materials used for photocatalysis, along with their band gaps.

In addition to metal oxides, metal chalcogenides and sulphides are also used for photocatalysis. But all these materials either require UV activation due to their wide band gap, or suffer from instability (as in case of chalcogenides or sulphides). Hence, scientists have sought for newer alternatives, for example carbon-based compounds such as graphene and carbon quantum dots (CQDs). Also, ternary compounds were investigated, such as $BiVO_4$, $ZnIn_2S_4$, Bi_2WO_6, $ZnWO_4$, $PbWO_4$ and Bi_2MoO_6.

2.7 SYNTHESIS METHODOLOGIES OF NANOMATERIALS

Synthesis method is crucial in determining the nature and properties of the nanomaterial. The methods of synthesis can broadly be classified as 'top-down' and 'bottom-up' approaches. In top-down method bulk materials are reduced to nanomaterials, while in case of bottom-up method, the nanomaterials are synthesized from the elementary level. The top-down method includes the physical processes such as crushing, milling and grinding methods. They will break the bulk materials into smaller particles. However, it is very difficult to obtain very small size distribution of nanoparticles. Hence, the top-down method has lacunae, namely inaptness in formulating evenly shaped nanomaterials and substantial crystallographic loss of the processed shapes.

In bottom-up method, materials are prepared atom-by-atom or molecule-by-molecule to make large quantity of materials. The advantages are uniform size, shape and well-distributed nanomaterials. It basically controls the chemical synthetic process in a precise manner to prevent undesirable particle growth, which provides better particle size distribution and morphology. It is environment-friendly and economical for the nanoparticle production.

Broadly, the methods of synthesis of nanomaterials are further divided into three types as physical, chemical and green biological methods. Depending on varieties of synthetic techniques, nanoparticles, nanocapsules and nanospheres have been synthesized. Nanoparticles preparation mainly involves two steps: emulsification and precipitation. Polymer-based nanoparticles are promising vehicles for various applications in conducting materials, electronics, photonics, sensors, drug delivery and biotechnological applications. The selection of non-toxic, non-antigenic, biodegradable and biocompatible polymers is the main criteria for polymer selection. Natural polymers for the synthesis of nanoparticles, such as albumin, sodium alginate, gelatin, starch, β-cyclodextrin and chitosan, are widely used [2]. Some synthetic polymers such as polyvinyl alcohol, polyanhydrides, polylactides, polyglycolides, polyacrylic acid, polyethylene glycol, poly-N-vinylpyrrolidone, polycyanoacrylates, polyorthoesters, poly(malic acid), polyglutamic acid, polycaprolactone and polymethyl methacrylate were also used. Others approaches for synthesizing nanomaterials include chemical vapour deposition method, thermal decomposition, hydrothermal synthesis, solvothermal method, pulsed laser ablation, templating method, combustion method, microwave synthesis, gas-phase method and conventional sol–gel method.

The uniform size distribution of nanoparticles (NPs) and the absence of solvent contamination in the synthesized thin films are the key advantages of physical methods over the chemical processes [34]. Chemical synthesis methods involve reduction reaction by organic and inorganic agents. Most commonly used reducing substrates are ascorbate, Tollens' reagent, sodium borohydride, N,N-dimethylformamide and elemental hydrogen.

In chemical reduction method, various polymeric components such as polyvinylpyrrolidone, polymethacrylate acid and polyvinyl alcohol are effectively used as stabilizers for the synthesis of nanoparticles. Sometimes, the Brownian motion of the formed droplets is responsible for particles collision with each other to form fused dimer [35].

Metallic nanoparticles are also synthesized by electrochemical and irradiation methods. Laser irradiation for metal nanoparticle synthesis has been used with varying pulse time duration [34]. The benefits of microwave-assisted synthesis of nanoparticles include narrow size and higher crystallinity without any agglomerates. Gamma irradiation and radiolysis of silver ions with ethylene glycol can be the important methods for successful adjustment of pH in nanoparticles preparation.

Sol–gel is a versatile chemical method with the involvement of two different phases, namely sol (solid particles in liquid) and gel (polymers in liquid). This leads to the formation of discrete particle network. Hydrothermal methods have been used for the synthesis of zirconium, strontium, cobalt, silver, iron and lanthanum nanoparticles. Polyol process of nanoparticle synthesis involves polyethylene glycol as solvent and varying concentration of nitrate ions as reactant to produce stable complexes with controlled size. Various types of polyols are a major determining factor for the size and texture of nanoparticles. Biological agent-based metallic nanoparticle synthesis is categorized into bio-reduction and biosorption processes. Several research groups have focused on non-green physicochemical methods of nanoparticle synthesis. However, the involvement of bacteria, fungi, plant extracts and plant bioactive components is responsible for a green, sustainable, resourceful and cheap method of nanoparticle preparation.

2.8 STRATEGIES FOR IMPROVING PHOTOCATALYSIS

Various modification strategies such as morphological modification, element doping, semiconductor coupling, co-catalysts deposition, defect formation (oxygen vacancy generation) and crystal facet control have been developed to achieve enhanced photocatalytic performance (Figure 2.11).

2.8.1 Interface and Defect Engineering

The existing photocatalysts suffer from the unsatisfactory charge separation and poor light absorption. Hence, it is generally desirable to optimize the band gap and band edge positions of the light-absorbing material for total light absorption and thereby for efficient photocatalytic activity. Therefore, it is important to

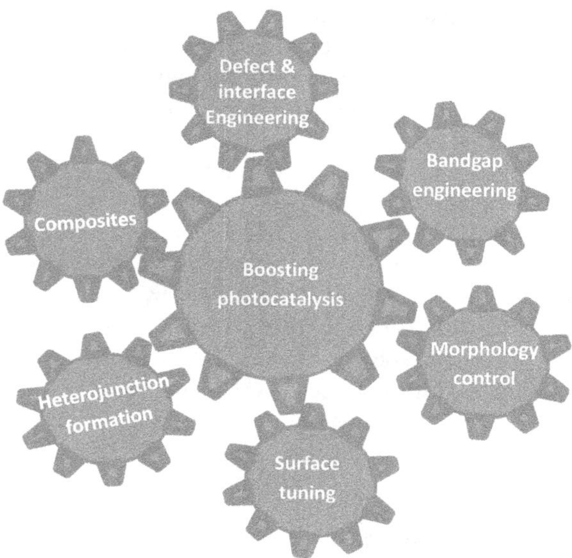

FIGURE 2.11 Strategies for boosting photocatalysis.

design, synthesize and modify existing photocatalysts to ensure high photocatalytic efficiency.

Introducing a heteroatom (doping) into the crystal lattice is a common strategy for band gap optimization. For the optimization of band gap of an oxide, both anion and cation doping is adopted. For example, nitrogen is commonly doped into the crystal lattice of an oxide, which in turn introduces vacancies in the lattice. Defect engineering is a versatile approach used in the modification of photocatalyst materials allowing the effect on the structure of the catalyst to alter the photocatalytic performance. Structural defects including zero-dimensional defects and higher-dimensional defects, interstitial defects such as vacancy defects, dislocations and grain boundaries will be effective and interesting [36].

Particularly, vacancy defects are thought to be the active sites and useful in capturing photogenerated electrons. Innumerable methods have been implemented, to introduce a defect, and among those few important strategies are thermal treatment, electrochemical treatment and photoreduction method. The low-temperature plasma-enhanced method is another versatile, green, time-saving, low-energy consumption approach to synthesize defective catalysts [37].

Photoinduced defect engineering is another novel strategy to fabricate defects. Ultrathin 2D black In_2O_{3-x} nanosheets synthesized using this approach exhibit enhanced photocatalytic activity. Under light irradiation, abundant oxygen vacancies were created in the nanosheet precursor, but the ultrathin layer structure was maintained while undergoing phase transformation to In_2O_{3-x}. The key point here is the abundant oxygen vacancies, which enhance the reaction rate and utilize majority of the absorbed solar energy. Under the light irradiation, these oxygen vacancies activate CO_2 gas molecules and stabilize the reaction intermediates, which leads to effective photothermal catalysis [38].

The crystal facet designing is another effective method for tuning the photocatalytic performance of the materials. Photocatalytic efficiency can be improved while modifying the facets of photoactive materials. This can be achieved by the synthetic methods based on surface and interface design. The charge separation and transfer of election-hole pairs in photocatalyst materials can be controlled by tuning the crystal facets.

The performance of a given photocatalyst varies according to the exposed crystal facets. The desired photocatalytic performance can be obtained by controlled growth of crystals which can be achieved by controlling the process of crystal synthesis. The wet chemistry routes including solvothermal and hydrothermal with the addition of solvent, capping agents or structure-directing agents are the most widely used methods for manipulating the nucleation and growth behaviour of crystals [39].

2.8.2 Heterojunction

Rapid electron–hole recombination and weak redox ability lead to low efficiency of photocatalytic reactions. Limited visible light absorption and the severe charge recombination are some of the key drawbacks that limit the practical applications of photocatalysts. A suitable construction of heterojunction is the most widely accepted strategy to solve these problems.

Generally, the relatively negative conduction band (CB) minimum and positive valence band (VB) maximum lead to a wide band gap and strong redox ability. It means that for strong redox ability, one has to compromise the light absorption in wide wavelength range, resulting in lower utilization of solar energy. To enhance the activity of a photocatalyst, construction of a semiconductor heterojunction has been recognized as one of the most promising ways among the known strategies because of its feasibility and effectiveness for the spatial separation of electron–hole pairs.

Now, the definition of heterojunction is expanded to semiconductor–semiconductor (S–S) heterojunction, semiconductor–metal (S–M) heterojunction and semiconductor–carbon group (S–C) heterojunction (carbon group: activated carbon, carbon nanotubes and graphene), which are different types of heterojunctions studied till now. The unequal Fermi level between two components at a heterojunction can cause a flow of electrons at the interface until their Fermi levels are aligned to the same level, inducing a built-in electric field and band bending. The separation and migration of photogenerated carriers can be promoted using the driving force of the internal field. The rational design of heterojunction delivers the possibility to possess wide light absorption range for materials, efficient separation of charge carriers and high redox ability, which again lead to the dramatic enhancement in photocatalytic efficiency.

Design of heterojunction photocatalysts composed of multiple semiconductors has proven to be a successful way to achieve the high photocatalytic performance. More often, the S–S heterojunctions can be classified into three types, namely conventional heterojunction, p–n heterojunction and direct Z-scheme heterojunction [40]. The conventional heterojunctions are of three types according to the different orientations/levels of conduction band and valence band of the two semiconductors. These are straddling band gap type (type-I), staggered band gap type (type-II) and broken band gap type (type-III). The most suitable heterojunction design for photocatalysis is type-II. In type-II heterojunction, the conduction band and valence band levels of first semiconductor are higher than the corresponding levels of second semiconductor. Under light irradiation, electrons in the conduction band of first semiconductor can be transferred to the conduction band of second semiconductor, while the holes in the valence band of second semiconductor can be transferred to the valence band of first semiconductor. As a result, the spatial separation of electron–hole pairs can be achieved (Figure 2.12).

2.9 TiO_2 FOR PHOTOCATALYTIC APPLICATIONS

Titanium dioxide (TiO_2) has been the most promising material due to the following characteristics: higher photoreactivity (usually up to ζ (photonic efficiency)$=10\%$), low cost, non-toxicity, chemical and biological inertness, and photostability. The reactivity of n-type TiO_2 can be attributed to the presence of oxygen vacancies that occur in these semiconductors. Let us comprehend the

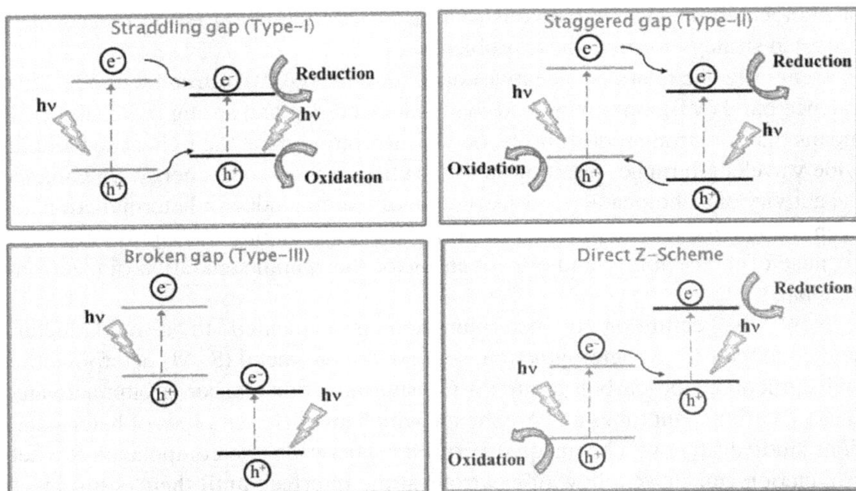

FIGURE 2.12 Various types of heterojunctions formed depending upon the relative positions of band edge of two semiconductors.

properties of TiO_2 and allied semiconductor, by delving into their electronic properties. Schneider [41] described the presence of two types of surface states: ionic states (Tamm states) and covalent surface states (Shockley states). The Shockley states are mainly distributed at the surfaces of covalently bound semiconductors, such as Ge and Si, and are known as dangling bonds. This is the key feature of nanomaterials distinguishing them from bulk materials. These surface states introduce additional energy levels in the middle of the band gap [41]. For electronic equilibrium between the surface and the bulk, a positively charged space charge layer is formed just beneath the surface of an n-type semiconductor because some of the mid-band gap electrons are transferred to its surface states. Hence, energy distance between the Fermi level and the conduction band increases at the surface, leading to a corresponding band bending. Thus, the accumulation of electrons and surface chemistry of TiO_2 determine the charge carrier dynamics of TiO_2.

2.10 ENVIRONMENTAL REMEDIATION THROUGH PHOTOCATALYSIS

The current chapter does not entail a detailed description of each and every photocatalytic work done by scientists all over the globe. It provides a bird's eye view of various processes used for detoxification with emphasis on the underlying mechanism and multi-utility of such catalysts.

TiO_2 is used in the degradation of various kinds of dyes, phenols and organics. Because of the wide band gap of TiO_2, it has been doped with various elements. Various types of doped TiO_2, namely cation-doped TiO_2, anion-doped TiO_2 and supported TiO_2, have been used for the degradation of dyes and phenolic compounds, as elaborately described in various papers and reviews [11]. The degradation of phenol involves attack of hydroxyl radicals in bulk liquid, which results in the formation of various intermediates. The $^{\bullet}OH$ radical attacks the phenyl ring of the phenol yielding alcohols such as resorcinol, catechol, 1,2,3-benzenetriol and hydroquinone. Subsequently, the phenyl rings disintegrate to give short-chain organic acids such as maleic, oxalic, acetic and substituted acids and finally CO_2 and H_2O [12,42].

Dye molecule is used as a quick probe for evaluating the efficiency of photocatalysts. Dyes have been classified according to various functional groups such as azo, triphenylmethane, anthraquinone, indigo, xanthene, fluorene and triazine [12].

Scientists have shown interest in evaluating the photocatalysts for the degradation of variety of functional groups. Ryu et al. [43] evaluated the commercial photocatalysts for a variety of groups such as dyes, acids, phenols and bases. Also, Vinu et al. [44] explored the versatility of their combustion-synthesized TiO_2 and commercial photocatalysts for all the dyes. They found difference in decolourization and mineralization rates. Also, they said that no photocatalyst can be termed as gold standard for photocatalysis. Thus, photocatalysts show substrate-specific activity.

Based on the work by Chong et al. [25] and Ryu et al. [43], some of the challenges are enumerated below for the photocatalytic system:

i. Catalyst improvement for a high photoefficiency that can utilize a wider spectrum of light with higher sensitivity.
ii. Improvement in the photocatalytic operation range (wide pH range and small amount of oxidant additives).
iii. Effective design of photocatalytic reactor system.
iv. Design of versatile, recyclable photocatalysts which can work for wide variety of substrates.

Pharmaceuticals and personal care products (PPCPs) are referred to as emerging contaminants. They comprise diverse collection of thousands of chemical substances, veterinary drugs, fragrances, lotions and cosmetics [45,46]. PPCPs are used worldwide and are categorized as potential endocrine-disrupting compounds (EDCs) [47]. These various PPCPs have been detected everywhere at the ng/L concentration levels in the effluents of wastewater treatment plants (WWTPs). Since conventional water treatment processes are insufficient for the removal of PPCPs from sewage water, heterogeneous photocatalytic degradation is a prospective way to suppress the contamination of emerging pollutants [48–51].

Antibiotic pollution is a major problem. Cephalexin (CEX) is a widely prescribed antibiotic for curing bacterial infection. NiO incorporated into clinoptilolite through ion exchange was evaluated for CEX degradation. Various parameters such as pH and catalysts loading under UV light were optimized.

The photocatalysts exhibited 89% degradation efficiency within 300 min and was found to be reusable up to three runs [52].

Carbamazepine (CBZ), 5H-dibenzo[b,f]azepine-5-carboxamide, is used for the treatment of epilepsy and psychiatric disorders. The photocatalytic degradation of 8 ppm CBZ was evaluated using TiO_2, ZnO and multi-walled carbon nanotubes under UV–Vis light. The type and loading of catalysts, use of H_2O_2 additive and addition of dissolved oxygen were varied. P25 (0.5 g/L) and P25 + 5mM of H_2O_2 or 50% O_2 (v/v) exhibited rate constants of ca. 0.3144 and 0.2005 min^{-1}, respectively. Martínez et al. observed various photoproducts, like 10,11-dihydro-CBZ-10, 11-epoxide after the degradation of carbamazepine [53].

Photodegradation of parabens using TiO_2 photocatalyst yielded short-chain carboxylic acids as mineralization products. Several emerging pollutants were tested over TiO_2 and ZnO photocatalysts for low to complete mineralization under solar light irradiation [54–59].

In summary, semiconductor-mediated photocatalysis is a sustainable, low-cost technique to tackle the contamination issue caused by heavy metals, pesticides, dyes, pathogens and other emerging organic pollutants.

2.11 ENVIRONMENTAL REMEDIATION THROUGH PHOTOELECTROCATALYSIS

Photoelectrocatalysis is a well-known technique used for environmental remediation. The comparison between the photocatalytic and photoelectrocatalytic and electrocatalytic processes is explained with the help of some salient examples.

Atrazine is an endocrine disruptor known to cause reproductive cancers in humans. Photoelectrocatalytic (PEC) degradation of atrazine herbicide was performed using WO_3 nanostructure in the nanosheets/nanorods as photoanode. Atrazine underwent complete degradation after 180 min, following pseudo-first-order kinetics, and 2-hydroxyatrazine was identified as the main intermediate species [60]. The s-triazine ring in cyanuric acid was partially broken, confirming the excellent PEC behaviour of this WO_3 nanostructure. The main degradation pathway was dechlorination-hydroxylation followed by dealkylation [60].

Amoxicillin (10 ppm) was completely degraded in just 30 min by the quaternary working electrode ZnO/ZnSe/CdSe/MoS_2 film with an applied bias of 0.5 V vs. Ag/AgCl. The reusability and stability of the quaternary electrode were demonstrated by three-run recycling experiments. The superior performance is attributed to better light absorption, improved charge separation due to the formation of p–n junction and reduced photocorrosion of CdSe due to protection offered by MoS_2 on the surface [61].

$BiVO_4$/BiOI/FTO was used as photoelectrocatalyst for the degradation of acetaminophen and ciprofloxacin [62]. The formation of p–n junction inhibited the recombination of charge carriers, proved by the improved photocurrent of 0.352 mA/cm^2 obtained at $BiVO_4$/BiOI photoanode than at their pristine counterparts, $BiVO_4$ and BiOI. The photoelectrocatalytic degradation using composite photoanode was found to be dependent upon the applied bias potential.

The degradation efficiencies of 68%, 62% and 85% were achieved within 120 min for acetaminophen, ciprofloxacin and Orange (II) dye, respectively. The catalyst was found to be recyclable.

TiO_2/Ti photoelectrocatalysis is an effective method for the degradation of sulfamethoxazole (SMZ) in aqueous solutions under UV irradiation [63]. The degradation rate by photoelectrocatalysis was found to be approximately 30 times higher than that by photocatalysis. SMZ degradation was enhanced with a higher applied anodic potential and NaCl concentration and lower solution pH and concentration of radical scavengers. The higher potential gradient reduces the recombination of holes and electrons, the lower pH favours photoadsorption through modulation of SMZ speciation and TiO_2 surface charge, and the higher Cl^- level encourages the formation of chlorine radicals.

Thus, in each case, photoelectrocatalysis is superior to photocatalysis as it enhances the degradation efficiency of photocatalyst considerably. Also, the reactive species remains the same as that in case of photocatalysis.

2.12 HEAVY METALS

Heavy metal poisoning by industrial waste is a serious threat to humans and aquatic environment. The most common heavy metals associated with adverse effects on human health are arsenic, chromium, mercury, zinc, nickel and cadmium.

Thus, the removal/recovery of heavy metals from water bodies has been important for both industry and human health. The oxidative removal of toxic heavy metals using semiconductor-mediated photocatalysis has extensively been reported and is discussed henceforth. The photocatalytic removal of toxic heavy metals is a cheap and sustainable method for technological development.

Contamination by arsenic (As) is a worldwide problem [64]. Inorganic trivalent arsenite [As(III)], pentavalent arsenate [As(V)] and As in methylated acidic form show adverse effects. As(III) is more toxic as compared to As(V) and is hard to remove from water by traditional water treatment processes. The photocatalytic oxidation of As(III) to As(V) can be achieved by strong reactive oxygen species. During the TiO_2-mediated removal of arsenic, the final mineralization product is As(V) [65]. Studies on the photocatalytic oxidation of As(III) to As(V) suggested that superoxide radical anion or valence band hole has a crucial role in the removal of As(III) [66–70]. Chromium exists in hexavalent [Cr(VI)] and trivalent [Cr(III)] states. Cr(VI) is highly toxic, while Cr(III) can be easily removed by precipitation as $Cr(OH)_3$ [71]. The photocatalytic reduction of Cr(VI) into Cr(III) is performed using TiO_2, P25, SnS_2, Bi_2O_3, WO_3, ZnS, CdS, CuInOS, FeOOH, $La_2Ti_2O_7$ and modified TiO_2 composites [72–80]. Reduction of chromium is pH dependent, and the presence of oxygen does not affect the chromium rate kinetics [81, 82]. Mercury poisoning was the main cause for Minamata tragedy, and methylmercury is deadly for all living organisms [83,84]. The photocatalytic removal of mercury and organic mercury has been reported by several studies [85–87]. The photoreduction of mercury is feasible over TiO_2 photocatalyst as the redox potential of mercury and methylmercury lies within the band edge potentials of TiO_2 [88–90]. The most common oxidation state of zinc, cadmium and nickel is +2. Photocatalytic reduction of zinc is thermodynamically unfavourable as the redox potential of zinc is lower than that of TiO_2. Thus, zinc can be removed only by adsorption technique. However, the reduction potential of nickel is less negative than that of TiO_2; therefore, reduction of Ni^{+2} to Ni is feasible [91]. The reports by Kabra et al. revealed that the hole scavenger has a prominent role in photocatalytic reduction of Ni(II) [81]. Contamination of cadmium occurred during World War II, and it is the main cause for 'Itai–itai' disease [92]. TiO_2-assisted photocatalytic reduction of cadmium is theoretically unfavourable as the redox potential of Cd^{+2}/Cd (−0.403 V) is below the TiO_2 conduction band (−0.5 V) [93].

2.13 FACTORS AFFECTING PHOTOCATALYSIS AND CATALYSIS

Photocatalytic activity is affected by intrinsic and extrinsic factors. Intrinsic parameters assumed to influence the dynamics of the photogenerated charge carriers include the crystallographic phase, the exposed crystal face, crystallite size, the presence of dopants, impurities, vacancies and different surface states originating from the presence of a variety lattice defects [29]. Some extrinsic parameters include the pH of the solution, structure of the pollutant, initial concentration, the presence of impurities, light intensity (photon flux), catalyst dosage and flow rate etc may affect the catalysis process. The plurality of variables affects the kinetics and mechanism of the photocatalysis. The detailed discussion about the parameters is given in various reviews [11]. Herein, we will confine our discussion to materials aspects.

Nanomaterials have smaller size as compared to bulk materials. This results in quadratic increase in surface area and reactive sites. However, smaller size does not necessarily mean high catalytic efficiency. This is because the generation and migration of charge carriers is in competition with their recombination. Surface properties and morphology play a crucial role in determining catalytic efficiency. For example, for the migration of charge carriers, smaller size results in shorter diffusion length, but it also requires potential or concentration gradient, which is governed by morphology, structure and surface properties.

A wonderful method for altering the surface area is tailoring the morphology. The morphology can be tuned by designing nanomaterials with core shell, yolk structure and hierarchical porous catalysts and through facet engineering of nanomaterials. We will discuss each of these structures separately and delve into the underlying mechanisms.

Photocatalytic activity is dependent on the structure and morphology of the photocatalytic material, especially at nanometre level. Hence, structural engineering or tailoring morphology has emerged as a major technique for boosting the photocatalytic efficiency. Hierarchical materials are referred to as nanostructured semiconductors having multidimensional domains at different levels or multimodal pore structures [94]. They are basically inspired by natural materials, for example trees, dendrites and urchin-shaped nanocatalysts. Such structures offer increased light harvesting, better charge carrier separation and improved mass transport and adsorption capacity. For example, porous materials allow rapid mass diffusion and accommodate active sites, while hollow particles offer short diffusion lengths and act as a reservoir to continuously replenish active sites with the monitored substance [95,96].

Hollow structures are generated by various chemical methods, with or without the use of templates [96]. The use of hard templates such as silica results in post-synthetic removal of these templates like etching and other methods. Nowadays, methods such as sol–gel, controlled hydrolysis, Ostwald ripening, Kirkendall effect and spray method are widely used. In addition to this, a wide variety of soft templating methods are also used, such as generating gas–liquid and liquid–liquid interface and the use of micelle for the generation of hollow nanomaterials. The detailed discussion about these methods can be found in various reviews [97]. These materials have versatile applications in antimicrobial, biomedical, in lithium-ion batteries, catalysis.

In catalysis, these hollow metal oxide nanomaterials are favourites as they provide high surface area, easy accessibility of pores, thermal resistance, toughness and environmental friendliness. In such materials, a large surface area and improved crystallinity would favour photocatalytic reactions. This is because the high surface area would facilitate the adsorption of pollutants and increase its rate. The high crystallinity would inhibit the recombination of electron–hole charge carriers and increase the photocatalytic reaction rate. Hollow materials perform better due to an additional factor of multiple light reflection [96,98].

Another important morphology is core–shell materials. These materials could be classified on the basis of the number of shells into single-shelled and multi-shelled materials [98]. The basic advantages of single-shelled hollow catalysts are the following: (1) a special hollow cavity structure and lower density; (2) a tunable surface-to-volume ratio and (3) short transport lengths for both mass and charge. Multi-shelled hollow micro-/nanostructured materials allow for more multiple reflections of the incident light, thereby enhancing the light harvesting efficiency and their catalytic performance.

2.14 FACETED NANOMATERIALS

The driving force for simple particle growth is the reduction in surface energy. Morphology evolution occurs due to the tendency of crystal system to minimizing the area of high surface energy faces to further reduce surface energy. Various solvents, impurities and additives in solution have a crucial role in influencing the ultimate shape of the crystals [99]. The solvent actually generates varied solvent–solute interactions along different orientations of a crystal. This is because surface atomic arrangement and thus surface affinity for the solvent to each crystal orientation is different, which obviously affects the growth rates to the crystal surfaces and hence the final shape of the crystal [99].

Faceted nanomaterials containing a facet with a high percentage of under-coordinated atoms possesses a superior reactivity to that with a low percentage of under-coordinated atoms [99]. Surface atomic arrangement and coordination changes with crystal facets in different orientations.

Photocatalysis requires the effective adsorption of reactant molecules/ions (i.e. water, O_2, CO_2 and organic compounds) on the surface of TiO_2 photocatalysts before surface electron transfer [100].

The substantial role of different facets in affecting the interaction can be demonstrated with the examples by comparing water adsorption, dye anchoring and lithium insertion on anatase (001) and (101) surfaces [101]. The anatase (001) surface allows the dissociative adsorption of water, while anatase (101) surface only allows the molecular adsorption of water. Even the interaction of dye molecules on photocatalyst surface is different for different facets. For example, the adsorption of C101 dye molecules on anatase (001) and (101) surfaces forms two anchoring geometries on TiO_2 pertaining to dye carboxylic groups to the TiO_2 surfaces (bridged bidentate vs monodentate). This phenomenon not only affects the photocatalytic efficiency and mechanism, but is also helpful in determining the efficiency of dye-sensitized solar cells.

2.15 MISCELLANEOUS CATALYSTS

2.15.1 Magnetic Catalysts

The basis of magnetic catalysts or photocatalysts/photoelectrocatalysts is imbibing magnetism in it so that they may be reused and recycled. Few salient examples are discussed herein to highlight the applications of these multifunctional materials.

A highly active Pd@Fe_3O_4 nanocatalyst was applied to US/EC degradation of ibuprofen [102]. The evaluation of total iron leaching and pH effect was used to determine the stability of the catalyst. The nanoparticles of Pd@Fe_3O_4 nanocatalyst exhibited outstanding catalysis performance for the degradation of IBP using US/EC reactions, recyclability and good reusability due to the good saturation magnetization of Pd@Fe_3O_4 owing to the high stability of the catalyst.

The photocatalytic efficiency of the synthesized Bi_2WO_6/Fe_3O_4/GSC composites was evaluated for the mineralization of oxytetracycline (OTC) and ampicillin (AMP) under solar light. The adsorption of both OTC and AMP on Bi_2WO_6/Fe_3O_4/GSC followed pseudo-second-order kinetics. Using Bi_2WO_6/Fe_3O_4/GSC catalytic system, the mineralization of AMP was obtained in 8h, and in case of OTC, 70% of antibiotic was mineralized to CO_2, H_2O and NO_3^- ions. The synergistic adsorption and photocatalysis (A+P) was the most efficient process for oxytetracycline and ampicillin removal and resulted in higher degradation of antibiotics. During A+P process, OTC and AMP were mineralized to CO_2, H_2O and NO_3^- ions. Due to magnetic separation, Bi_2WO_6/Fe_3O_4/GSC exhibited substantial recycle efficiency for ten catalytic processes [103].

The magnetic $ZnFe_2O_4$ nanoparticle-functionalized g-C_3N_4 sheets were synthesized through one-step solvothermal method [104]. CN-ZnFe composites exhibited superior visible light degradation of methyl orange. This can be ascribed to the effective electron–hole pair separation at the composite interfaces, to the smaller particle size and to being highly water dispersible. Intermediates studies indicated that $^{\bullet}OH$ radicals play a more crucial part than $^{\bullet}O_2^-$ in the degradation process. Thus, g-C_3N_4 sheets can function as good support to develop highly efficient g-C_3N_4-based magnetic photocatalysts in environmental pollution clean-up.

The naturally derived iron oxide nanowires from *Mariprofundus ferrooxydans* bacteria are a highly efficient visible light photocatalyst [105]. Annealing at 800°C converted the amorphous iron to oxide nanowires. α-Fe_2O_3 provided the best catalytic activity, degrading RhB (1.7×10^{-5} M) completely in less than an hour. The narrow band gap of the iron oxide coupled with high surface area of these bacterial nanowires led to enhanced optical absorption and generation of exciton pairs, providing unprecedented rhodamine B degradation efficiency. The catalyst was recyclable and had no loss in activity up to six cycles. Thus, it is a beautiful example of the use of a low-cost, natural material endowed with high surface area and magnetism for environmental clean-up.

Thus, the multifunctional, visible-light-active photocatalyst is a step towards practical realization of photocatalytic technology for addressing wastewater remediation.

2.15.2 Round-the-Clock Catalysts

Some of the nanocatalysts display a fascinating property known as catalytic memory. The beauty of these catalysts is that they exhibit catalytic activity not only under light illumination, but also after the activating light has been switched off. Such catalysts are known as round-the-clock catalysts. The underlying mechanisms and reasons for such a phenomenon have been elucidated through some examples.

Cu_2O nanospheres decorated with TiO_2 nanoislands have catalytic memory for the degradation of methyl orange dye and *E. coli*. This means that these catalysts work even after visible light has been cut off. Part of photoexcited electrons transferred from Cu_2O to TiO_2 were trapped by TiO_2 nanoislands under visible light illumination and then were released in the dark. and produced radicals even in the absence of light illumination. These catalysts are apt for wide range of environmental applications [106].

TiON/PdO fibres also demonstrated high disinfection efficiency on *E. coli* bacteria under visible light illumination. However, post-illumination catalytic disinfection capability after the visible light is shut off is fascinating and attributed to electrons trapped in PdO, which are transferred to nitrogen-doped TiO_2 [107].

Another excellent example is of Se nanorods that exhibit superior photocatalytic degradation of methylene blue dye as compared to commercial P-25 TiO_2. This is attributed to single crystallinity of nanorods, which suppresses the recombination of electron–hole pairs and maximizes their utilization. Also, typical photocatalysis occurs under visible light illumination, while memory photocatalytic effect occurs during dark [108].

The detailed and in-depth mechanistic study of such photocatalysts would help scientists reproduce and design such materials, which in turn would pave way for the development of technology for versatile, efficient, energy-conserving materials.

2.15.3 Upconversion and Downconversion Catalysts

The solar light consists of a wide spectrum of light, namely 5% UV, 47% visible and 47% infrared. Now, if the photocatalyst is only UV or visible active, the maximum portion of solar light would be unutilized. Hence, the design of photocatalysts utilizing the entire spectrum of solar light has been attracting attention recently.

An important strategy in this context is the use of upconversion or downconversion agents. In upconversion materials, the lower-energy photons are converted into higher-energy photons during emission. While in downconversion, the process is vice versa: high-energy photons are converted into lower-energy photons. The crux of this process is to tap the entire range of solar spectrum for catalytic activity. CQDs and lanthanide compounds are fascinating materials that are used as upconverting agents. Some examples are given to explain the utility of these compounds.

The photoactivity enhancement of g-C_3N_4 after Er^{3+} doping is monitored for the photodegradation of rhodamine B dye under visible light. The contribution of the upconversion agent is demonstrated by measurements using only a red laser. The Er^{3+} doping alters both the electronic and chemical properties of g-C_3N_4. The Er^{3+} doping reduces emission intensity and lifetime, indicating the formation of new non-radiative deactivation pathways, probably involving charge transfer processes [109].

CQDs are impressive upconversion agents used in composite and coupled photocatalysts. The primary reason for such utility is their ability to harness NIR light (700–1000 nm). The loading process and quantity of CQDs can affect the final photocatalytic activity of composited photocatalysts, because a higher quantity of CQDs does not mean better activity [110]. There should be an optimal value for composited photocatalysts built from a semiconductor and CQDs.

2.15.4 Supported Catalysts and Their Applications

Supported catalysts have been considered separately because of their versatility, multifunctional use and the different mechanism as compared to suspended photocatalysts/catalysts. A wide variety of materials, in various forms, such as microspheres, membranes, gel beads, fibres and powders, have been reported as supports. Innumerable materials have been reported and used as support, such as silica gel [111], mesoporous silica, zeolites, clays [112], ceramic membranes and monoliths, stainless steel, carbonaceous nanomaterials, glass plates and fibres, Vycor glass, hollow glass spheres and indium tin oxide (ITO)-coated glass, fluorine tin oxide-coated glass, quartz optical fibres, synthetic and natural polymers as films or membranes [113], cellulose membranes, natural and synthetic fabrics, various types of fly ash, Raschig rings, films and fabrics [112].

The major advantages of photocatalysts obtained by immobilization of semiconductors are (1) the reduction in electron–hole recombination resulting in high photocatalytic efficiency; (2) easy separation, recovery and reuse in reaction; (3) incorporation of chemical promoters alongside support materials; and (4) increase in catalytic efficiency due to combining the positive influence of immobilized materail as well as interaction of immobilized materail with semiconductor.

Photocatalytic efficiency in most of the cases increases on decreasing the particle size to nanometre scale. However, the actual use of nanomaterials in powder or slurry although provide high surface area, but the recovery of powered catalyst is not easy. Coupling of support and semiconductor is an added advantage for inhibiting the electron–hole recombination, making them more available for the generation of ROS. Thus, supports are used to increase the overall photocatalyst performance and to yield high activities by improving active species dispersion and accessibility of light and reactants. However, this is entirely dependent upon the nature, morphology, hydrophobicity of support, its porosity, interaction of semiconductor and support, and the immobilization technique used to obtain good dispersion of the supported nanomaterial. In addition, the efficient, selective photocatalyst system would entail good adsorption capacity for support liquid-phase pollutants.

The role of support in photocatalysis can be easily understood by this example. ZrO_2-, CeO_2-, Al_2O_3- and zeolite-supported TiO_2 were used for photocatalytic hydrogen generation from water–methanol system. Apart from crystallinity and surface, TiO_2-supported on zeolite and ZrO_2 exhibited the best activity, mainly due to the surface acidic properties of ZrO_2. Surface acidity prevented the recombination of electron–hole pairs, which was in accordance with the lifetime measurements done for all these supports [114]. Various compositions of ZrO_2-supported TiO_2 were used for the photocatalytic degradation of malachite green. They exhibited visible light photocatalytic degradation. In addition, bismuth doped TiO_2 showed 80% mineralization. This material also exhibited room temperature ferromagnetism due to oxygen vacancies [115].

N-doped TiO_2/NaY zeolite membrane electrode was evaluated for the degradation of phenol. As compared to electrocatalysis, the photocatalytic activity showed a twofold increase in photocatalytic degradation of phenol using this catalyst. Under the photoelectrocatalysis pattern, the N-doped TiO_2/NaY zeolite membrane for the degradation of phenol exhibited the best catalytic activity, degrading phenol by about 80% and 89% after 90 and 180 min of solar light irradiation, respectively [116].

A well-known instance of supported photocatalysts for multifunctional applications is the use of BiOBr/Ag copper mesh. It has a band gap of 2.52 eV, demonstrating water–oil separation up to 99.5%, rhodamine B photodegradation for three consecutive cycles up to 97% and anti-corrosive ability even after being dipped in 3.5% NaCl for 3 days [117].

The examples of multifunctional catalysts are given in Table 2.1.

TABLE 2.1
Various Multifunctional Nanomaterials and Their Applications

Catalyst	Type of Remediation	Pollutant	Application	Remarks	Ref.
Pectin/chitosan/zinc oxide (Pec/CS/ZnO) nanocomposite	Adsorbent photocatalytic	Carbamazepine (CBZ)	Degradation	Remediation of pharmaceutical compounds	[118]
Bi_2S_3/b-Bi_2O_3/$ZnIn_2S_4$	Photocatalyst	Tetracycline (TCN), Cr(VI), bacteria	Degradation reduction, antibacterial	Visible light assisted	[119]
MoS_2@$CoMoS_4$	Photoelectrocatalyst	Lomefloxacin, nitrophenols	Degradation, redox process	Full spectrum light photoelectrocatalytic	[120]
Ca/TiO_2/NH_2-MIL-125	Photocatalyst	Methyl orange (MO) and rhodamine B (RhB) dyes	Degradation	Visible light	[121]
Functionalized layered double hydroxides	Photocatalyst	Heavy metals, radionuclides, organic contaminants	Adsorption, degradation	UV light	[122]
MXene-based composites	Photocatalyst	CO_2, heavy metal ions, cationic dyes, bacteria	Reduction, adsorption, antibacterial	UV light	[123]
PM-CQD/TiO_2 nanocomposite	Photocatalyst	Methylene blue, erythromycin	Degradation	UV light, green synthesis of nanomaterials	[124]
Covalent organic frameworks (COFs)	Het. catalysts, electrocatalysts, photocatalyst	CO_2, organic pollutants	Cyclization, reduction, degradation	UV light	[125]

(Continued)

TABLE 2.1 (*Continued*)
Various Multifunctional Nanomaterials and Their Applications

Catalyst	Type of Remediation	Pollutant	Application	Remarks	Ref.
Polydopamine/ zirconium(IV) iodate	Adsorbent	Ampicillin	Adsorption	Removal of pollutants	[126]
CuFe and Ni-Fe nanomaterials	Electrocatalytic, photocatalytic	Methylene blue	Degradation	UV light	[127]
Molecularly imprinted polymers	Adsorbent	Heavy metals, dyes	Adsorption	Removal of pollutants	[128]
Ag(I)-Hived Fullerene Microcube	Catalyst, photocatalyst	4-Nitrophenol, Orange G dye	Reduction, degradation	Visible light radiation	[129]
Ag_3PO_4-rGO-coated textiles	Disinfecting photocatalyst	Organic dyes and pathogenic microbes	Degradation	Solar light	[130]
Pt/Al_2O_3 nanofibrous membranes	Catalyst membrane	Bisphenol A, CO to CO_2	Degradation	-	[131]

2.16 AIR POLLUTION

A variety of air pollutants are present in the atmosphere. These include carbon monoxide and volatile organic compounds such as benzene, acetone and NO_x [132]. According to EPA report, 80% of the air pollution is due to VOCs generated through various industrial and man-made activities.

Air pollution is caused by both natural and man-made activities. Air pollution occurs due to the presence of pollen, vacuum cleaners, hair dryers, air conditioners, dry cleaners and vehicle emissions [114]. Sometimes, although indoor pollutants are imperceptible or can't be smelt, they lead to nausea, dizziness and respiratory disorders. Also, they affect the psychological well-being of an individual, leading to depression, unhappiness and allied disorders [133]. The use of air filters is an important method for removing particulate matter. CO is another pollutant that is highly hazardous to health. This is because it combines with haemoglobin in the blood leading to poisoning and even death. Hence, CO conversion to methane is an important way to reduce pollution. The conversion of harmful gases to non-toxic counterparts and its utility are presented in the form of case study in the Annexure section.

2.17 FUTURE PROSPECTS

Thus, multifunctional materials with tantalizing properties constitute the backbone of development and research. Although laboratory-scale materials have achieved marginal success, the integration and actual functioning of the same material for diverse applications remains a far-fetched dream. There are prominent challenges to be addressed while scaling up and systematically integrating functions of these novel materials as smart materials under actual working conditions. The investigation of these materials for technological development is actively pursued by different organizations worldwide for societal welfare and utility. A peek into the real-world space applications is provided in the case study given in the Annexure.

ANNEXURE

Case Studies Utilizing Nanoporous Catalysts and Adsorbents for the Real-World Space Applications

Suhas Mukherjee and Bhatt Tushar Shriram

Vikram Sarabhai Space Centre

The case study: 'Effective and economic way of using catalytic gas-phase reaction and regenerable adsorption process for environmental remediation of carbon dioxide in human space missions'

Problem definition: For the human race, space is the next frontier to conquer. This is evident with the revived interest among various spacefaring nations to establish a permanent outpost in the low to medium earth orbits (in addition to the 'multi-national ISS(International Space Station)' and the twentieth-century marvels such as Skylab of the erstwhile USSR) and further take it to the next level by establishing human colonies on the moon and the Mars for long-term future. For humans to survive in the space, certain specific technologies are needed for creating an earth-like atmosphere (**Pressure: 1 atm abs; Temperature: 18°C–25°C; Relative humidity: 40%–80%; Partial pressure of oxygen: 0.210 atm abs; Partial pressure of nitrogen: 0.786 atm abs; Partial pressure of carbon dioxide: below 0.004 atm abs**) inside the spacecraft, as the space environment is devoid of the earth-like atmosphere with no trace of oxygen, to maintain the required air quality and earth atmospheric pressure inside the crew cabin. Thus, the spacecraft must be equipped with a suitable remediation technique to manage the humidity levels and CO_2 exhaled by the crew and other trace contaminants getting generated within the crew cabin. The system that continuously removes the carbon dioxide from the crew cabin and maintains an inhabitable environment inside the spacecraft is known as the Environmental Control and Life Support System (ECLSS). One of the critical aspects of the closed-loop ECLSS is the effective mitigation of carbon dioxide generated by the metabolic activity of the astronauts. Carbon dioxide mitigation/management involves selectively capturing/storing carbon dioxide gas from the atmosphere of the crew cabin and subsequently converting it into useful products in the most efficient way. The standard ECLSS system must be equipped with regenerable carbon dioxide adsorption and recycle system to facilitate continuous resupply of oxygen to the spacecraft by recovering oxygen from carbon dioxide. In a standard life support system, molecular sieve (MS)-based selective carbon dioxide adsorption system is used for the separation of carbon dioxide from the crew cabin air that is contaminated with a higher percentage of CO_2. Furthermore, the oxygen is recovered from the carbon dioxide by continuously converting it into water and methane via the 'Sabatier reaction' (reaction of CO_2 with hydrogen in the presence of metal oxide catalyst) route. Moreover, the water generated from the Sabatier process is further electrolyzed to generate oxygen and hydrogen. Oxygen is used for re-pressurization of the crew cabin for making up oxygen partial pressure. Recovered hydrogen from the electrolyser is recycled to the reaction chamber. By this process, the CO_2 removal system is designed to operate in a continuous closed-loop by minimizing the loss of oxygen. The system that operates in open-loop mode generally does not utilize CO_2 to recover oxygen to reduce the water demand of the crew. Thus, a continuous closed-loop system eventually reduces the amount of water needed to be carried along with the spacecraft for meeting 14 days of oxygen demand.

The case study deals with an integrated closed-loop ECLSS (Figure 2.13) for selectively adsorbing carbon dioxide from cabin air using molecular sieve-based adsorbents and converting it into useful products, viz. water and methane, using the Sabatier reaction. The Sabatier reaction involves the reduction of carbon dioxide with hydrogen as reactant over metal oxide catalysts such as ruthenium-γ-alumina and ruthenium-TiO_2 in the temperature range of 260°C–300°C and pressure range of 2–4 bar. Further, the products of the reaction would be separated

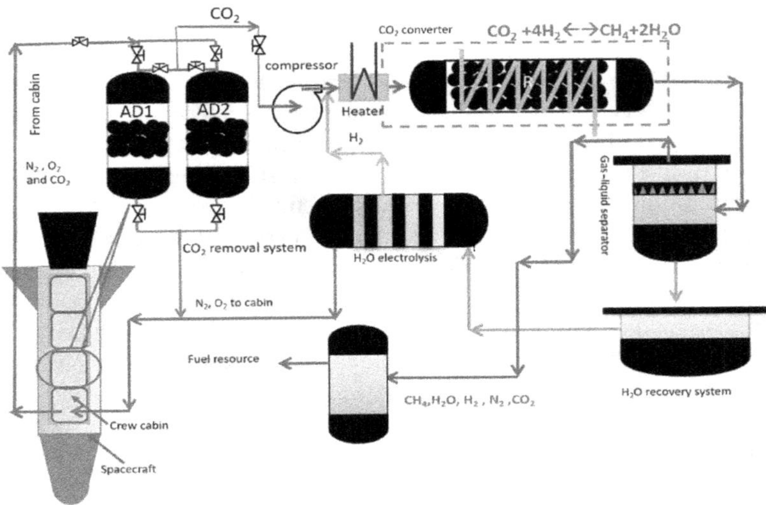

FIGURE 2.13 Scheme for carbon dioxide mitigation and conversion into useful products in a closed-loop ECLSS.

into their gas (methane, unused CO_2 and hydrogen) and liquid fraction (water). To make the system more economically viable, an electrolyser that splits the water into hydrogen and oxygen is also considered. The hydrogen gas evolved from the electrolysis process is recycled into the methanation reactor, and the oxygen is used to re-pressurize and maintain the cabin oxygen level. As a part of the case study, we have only considered the sizing of two major subsystems of this project, i.e. **'Molecular sieve-based carbon dioxide adsorption system'** and **'M_x (M: Rh, Ru, Re, etc.) $B_{N-x}O_{n-\delta}$ (B=cobalt, aluminium, titanium, etc.)'-based carbon dioxide methanation reactor.**

The role of adsorbents and catalysts with nanoporous distribution is of paramount importance in realizing the two subsystems for carbon dioxide mitigation through adsorption and the chemical reaction process. This case study discusses the indispensable role played by the catalysts and molecular adsorbents with multifunctional nanopores for selective adsorption and separation of carbon dioxide from the air and further converting it to water and other useful products through the application of processes such as gas adsorption and catalytic chemical reaction. Furthermore, at the end of each case, we will also briefly discuss the chemical structures, physical properties of the materials (adsorbents and catalysts) that are proposed to use in the system, and the related mechanisms of both the processes.

CASE STUDY I: EXPERIMENTS CONDUCTED FOR DESIGN OF MOLECULAR SIEVE-BASED CARBON DIOXIDE ADSORPTION SYSTEM

As part of the adsorption system development, experiments were conducted for selective adsorption of carbon dioxide from the simulated crew cabin air using

TABLE 2.2
Adsorption Parameters

Adsorption Parameters	Values
Adsorber system pressure	1 kgf/cm^2 abs
Inlet gas composition, v/v	76% Nitrogen, 4% carbon dioxide and 20% oxygen
Flow rate	600 sccm
Adsorption bed length	~300 mm
Loading	60 gm
Adsorbent type	Molecular sieve-4A
Size	4.3 mm (L)×4.3 mm (Φ)
Shape	Cylindrical: L/D = 1

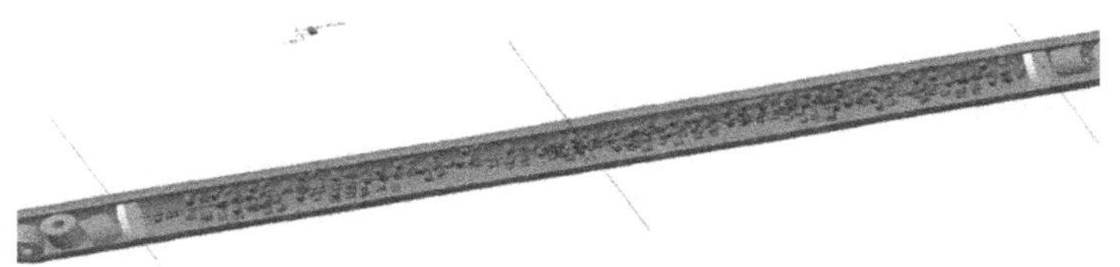

FIGURE 2.14 Adsorption bed packed with adsorbent and ceramic packing.

molecular sieves as an adsorbent. Carbon dioxide adsorption experiments were carried out in a laboratory-scale adsorber. The process parameters of the adsorption experiment are mentioned in Table 2.2. Adsorption experiments were carried out by passing the mixture of gases composed of 4 vol% CO_2, 76% N_2 and the rest oxygen through an adsorber packed with MS adsorbents. Figure 2.14 shows the packing arrangements of the adsorbents inside the bed supported with ceramic wool and packings. The quantity of adsorbed gas in the MS was estimated based on the signal obtained from gas chromatography (GC). The GC employs electronic interaction, mainly dipole–dipole interaction, among gaseous molecules and stationary column to separate and detect the concentration of gaseous molecules. The details of the interaction process have already been discussed in the interaction section of this chapter.

The gas outlet and inlet concentrations of the adsorbent bed are measured based on the timely basis capture of instantaneous carbon dioxide, nitrogen and oxygen peaks using the gas chromatography technique. The amount of carbon dioxide adsorbed on the material is computed based on the difference in the mathematical peak area between bed inlet and outlet chromatographic signals of carbon dioxide. On the other hand, a very minor change in nitrogen and oxygen concentrations is observed from the GC peak area.

The change in carbon dioxide peak area at the adsorption bed outlet at different time instants is depicted in Figure 2.15. It can be observed from the above-captured

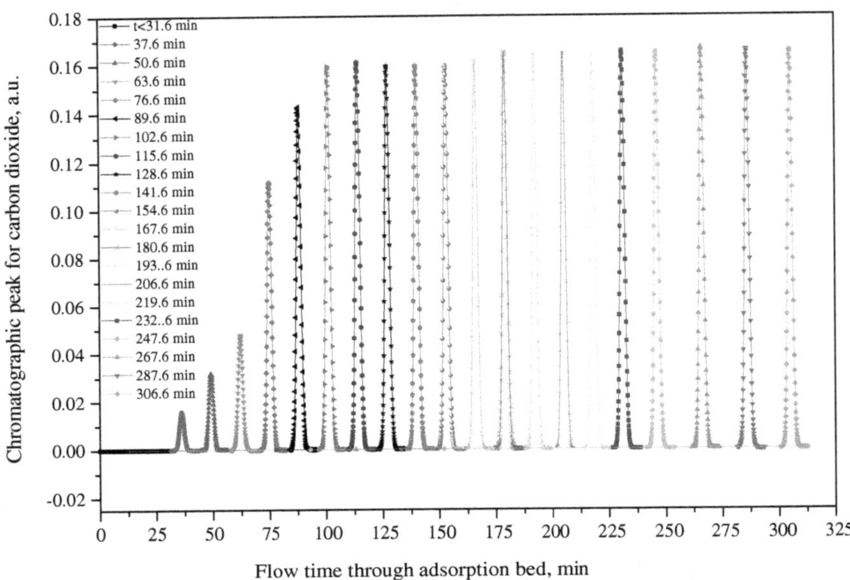

FIGURE 2.15 Chromatographic peak strength of carbon dioxide at different instances of flow time through adsorption bed.

signals that the bed is absorbing almost 100% of carbon dioxide in 31.7 min. In the subsequent time (t>31.7 min), the adsorbent bed capacity starts to decline and the carbon dioxide gas starts to escape the adsorbent bed. The sudden increase in carbon dioxide signature peak strength at the bed outlet in the time interval of 31.6–102.6 min signifies the reduction in CO_2-capturing capability of the bed. In that way, the bed continues to adsorb carbon dioxide for up to 125 min until the bed is fully saturated with carbon dioxide. When the bed becomes fully saturated with the CO_2, it stops absorbing carbon dioxide molecules from the mixture gas of N_2, O_2 and carbon dioxide. A breakthrough curve is generated based on the area difference of the captured peak at different instances of time to clearly understand the three zones A, B and C of adsorption. Figure 2.16 represents the change in the ratio of carbon dioxide outlet to inlet concentration at different instances of time.

The curve obtained from the adsorption experiments can be characterized into three zones, viz. zone A: where the uppermost layer of the adsorbents rapidly adsorb almost the majority percentage of the solute gas (carbon dioxide) from the mixture and subsequently the rest of the solutes are adsorbed while travelling down the bed, zone B: the zone where outlet concentration of carbon dioxide changes within very short time instants due to saturation of almost half of the bed while adsorbing majority portion of strong solute, and zone C: where the bed becomes 100% saturated and reaches its ultimate capability of adsorbing carbon dioxide (Figure 2.17). The curve connecting the initiation point of zone B and zone C is often designated as 'breakthrough curve'. If the gas solution

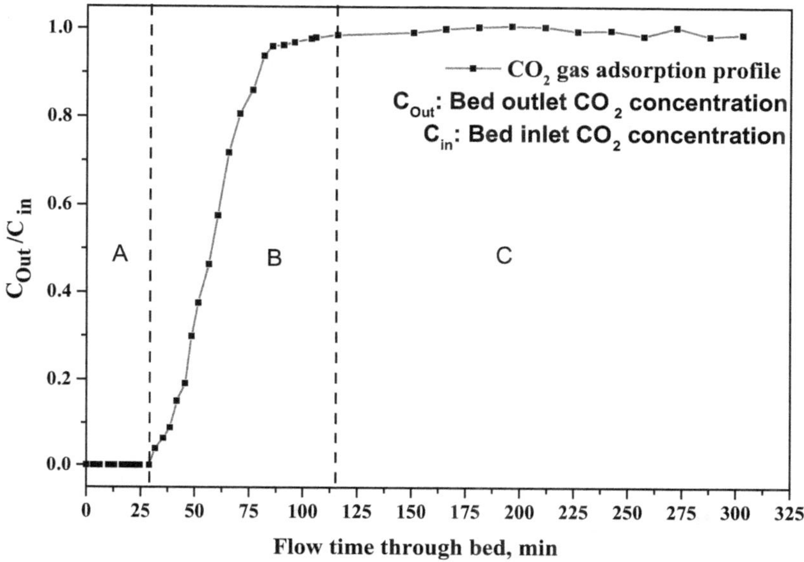

FIGURE 2.16 Breakthrough curve of carbon dioxide adsorption over molecular sieve.

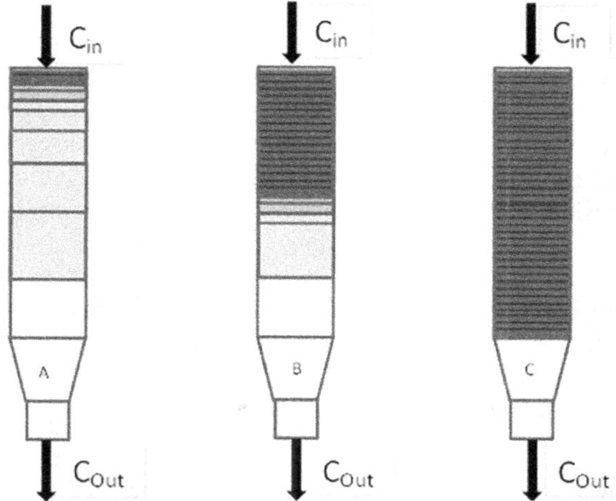

FIGURE 2.17 Level of bed saturation at different instances of time: (A) initiation of first layer of adsorbent bed saturation, (B) partial saturation of adsorption bed, (C) full saturation of absorption bed.

continues to flow through the bed after crossing the breakthrough regime, no further adsorption will take place as all adsorbent pores are filled with solute molecules. This situation can be referred to as an equilibrium condition among the adsorbed solute molecules and the solute molecules present in the unabsorbed gas mixture. The cumulative adsorption capacity of the bed is estimated based on the following equation:

$$M = \sum_{t_i=0}^{t_n} \frac{44.02 * Q_{in}}{W}\left(1 - \frac{C_{out,i}}{C_{in}}\right)\Delta t_i \text{ ----------------- (1)}$$

$M: \dfrac{g\ CO_2\ \text{adsorbed}}{g\ \text{adsorbent}}$; Q_{in} : Molar flow Rate of CO_2, $\dfrac{\text{Mol}}{\text{sec}}$;

t_i : Time at interval i; t_n : Ending time;

W : Adsorbent weight, g; Δt_i : Flow time interval, sec

$C_{out,i}$: Bed outlet concentration of CO_2 at time interval t_i;

C_{in} : Bed inlet concentration of CO_2

Experimental results show that the bed is capable of adsorbing 140–150 mg CO_2/gm of adsorbent before it attains the full saturation condition. In addition to that, the task team also recommended using the bed for 50% of the adsorption breakthrough time (zone A) beyond which the bed could not adsorb 100% carbon dioxide from the input stream. They have also recommended using a two-bed system where 50% of the time of zone A can be used for adsorption and 50% of the time can be utilized for the regeneration of the bed by reducing the system pressure to 0.1 barg. The alternative use of the adsorption and regeneration cycle is a common practice in the industry for increasing the lifetime of the bed. Based on the above considerations, the team has arrived at the following design configuration of the adsorption–desorption system.

Chemical structure and properties of molecular sieve MS-4A

MS-4A is a sodium-based nanoporous aluminosilicate structure. The approximate chemical formula for the MS-4A is $Na_{12}[(AlO_2)_{12}(SiO_2)_{12}].27H_2O$ [134]. This particular nanofunctional material is synthesized using high-pressure and high-temperature reaction of sodium, aluminium and silica precursor with supersaturated water as a solvent [135]. The reaction conducted at high pressure and high temperature results in the formation of sodium-containing aluminosilicate hydrogel with a water content of 80–90 mol percentage. Further, the crystallization of gel by drying causes the formation of nanopores in the material. While drying during the crystal formation, the number of water mole percentages reduced to ~27 mol. The reduction in water percentage results in the formation of a highly nanoporous cage-type tetrahedral structure [136]. The size of nanopores in the dried aluminosilicate structure is in the order of 4 Å (0.4 nm). The fine control on the pore size and grade of zeolite can be obtained by selective control of the Si:Al ratio. The average particle size and surface area of MS-4A are in the order of 1.5–3 nm and 450–500 m^2/g.

Mechanism of CO_2 adsorption over molecular sieve MS-4A

The mechanism by which carbon dioxide is selectively adsorbed from the mixture of nitrogen and carbon dioxide is the interaction of dipole of CO_2 and charged pores (mostly charged with sodium ion) of the zeolite. During the process of CO_2 separation from nitrogen and oxygen, the CO_2 is getting physically bonded and trapped in the nanopores (4 Å) that are larger than the size of the CO_2 molecule (3.3 Å). In the mechanism, the CO_2 is only getting bonded in the charged pores due to the charge available in metal and terminal dipole available in the O=C=O bond [137].

$(\text{Metal(Na)})^{x+}$----------------- $^{\delta-}O{=}C{=}O^{\delta+}$

CASE STUDY II: EXPERIMENTS ON CARBON DIOXIDE METHANATION REACTION OF METAL OXIDE CATALYST FOR THE REMOVAL OF CARBON DIOXIDE FROM AIR

A study of carbon dioxide methanation reaction over ruthenium/alumina-based catalyst was conducted. Based on the obtained results, it is found to be effective for carbon dioxide removal by the adsorption–desorption system discussed in Case Study I. To achieve a compact design of a carbon dioxide reduction system, multiple experiments are conducted over Ru/Al_2O_3 catalyst as per the condition given in Table 2.3. The system depicted in Figure 2.18 is used for carrying out carbon dioxide reduction reactions with hydrogen as a reducing agent.

Multiple experiments have been performed to realize a catalytic packed bed reactor system, where preheated CO_2 and H_2 gas mixture is passed over a Ru/γ-Al_2O_3 at 200°C–300°C and 2 bar pressure to produce CH_4 and H_2O as per the

TABLE 2.3
Adsorber Configuration Details

Adsorption Parameters	Values
Type	Pressure swing two-bed (A and B) adsorption–desorption system
Adsorption pressure	1 kgf/cm^2 abs
Desorption pressure	0.1 kgf/cm^2 abs
Adsorption cycle time	10 min
Desorption/regeneration cycle time	10 min
Inlet gas composition, v/v	76% Nitrogen, 4% carbon dioxide and 20% oxygen
Flow rate	600 sccm
Adsorption bed length	~600 mm
Dia	22.5 mm
Loading	120 gm
Average weight percentage of CO_2 removed per cycle	14
Adsorbent type	Molecular sieve-4A
Size	4.3 mm (L) × 4.3 mm (Φ)
Shape	Cylindrical: L/D = 1

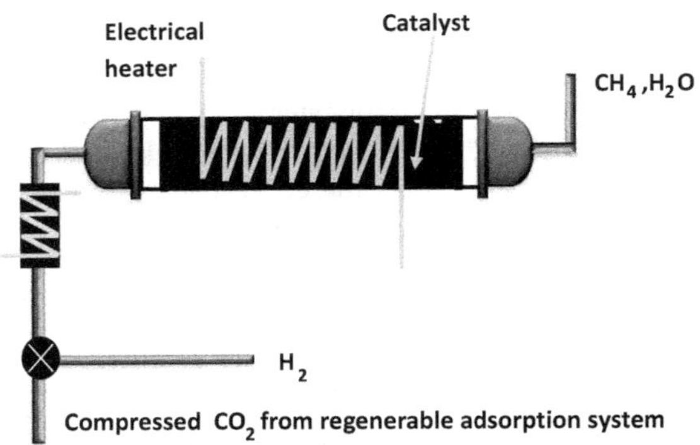

FIGURE 2.18 Configuration of catalytic carbon dioxide reduction system.

reaction given in Eq. (2.2). The objective of the experiment is to develop a suitable catalytic reactor system that can convert the CO_2 separated by the adsorption system into useful products such as methane and water which can be further used as a resource.

$$CO_2 + H_2O \rightarrow CH_4 + 2H_2O \quad \Delta HR = -165\frac{kJ}{mol} \tag{2}$$

The reaction described above is highly exothermic with the heat of the reaction in the range of 165 kJ/mol. Due to the capability of converting the industrially emitted carbon dioxide into useful products such as methane and water, the reaction has high potential for environmental remediation and industrial carbon dioxide mitigation. However, this chapter focuses only on the use of the reaction for converting adsorbed carbon dioxide from crew cabin air for long-duration human space missions. On this note, the details of the experiments performed are given in Table 2.4.

The reactant gases CO_2 and H_2 are fed in the preheater in the mole ratio of 1:4.5 with the flow rates of 100, 200, 300, 400 and 500 mL/min. The reactant feed stream (a mixture of CO_2 and H_2) is heated in the preheater and fed into a catalyst-filled reactor. The flow rates of the reactants are varied for assessing the optimized reactor design parameters for the required flow rates. It is often observed that an increase in flow rates causes enhancement in the interim mass transfer coefficients, which further helps the system to reach equilibrium faster than the reactions that are carried out in lower flow rates. The equilibrium conversion curve for the carbon dioxide reduction system is estimated based on the available thermodynamic data of the reaction. Figure 2.19 renders the equilibrium conversion of the carbon dioxide reduction reaction for different mole ratios(M). The equilibrium conversion (X_e) for the system has been computed based on the equation given in the figure 2.19 K_{eq}, i.e. equilibrium reaction rate constant, depicted in Figure 2.19 is computed based on Gibbs free energy of the reaction.

TABLE 2.4
Carbon Dioxide Reduction Reaction Parameters Used for Experiments

Reaction Parameters	Values
Reactor pressure	2 kgf/cm^2 abs
Inlet gas composition, v/v	CO_2:H_2 – 1:4.5 mole ratio
Flow rate	100–500 sccm
Reactor bed length	~340 mm
Reactor length	~740 mm
Reactor diameter	22.5 mm
Loading	35 gm
Catalyst	Ru_x-γAl_2O_3
Size	3.2 mm (L)×3.2 mm (Φ)
Shape	Cylindrical: L/D = 1

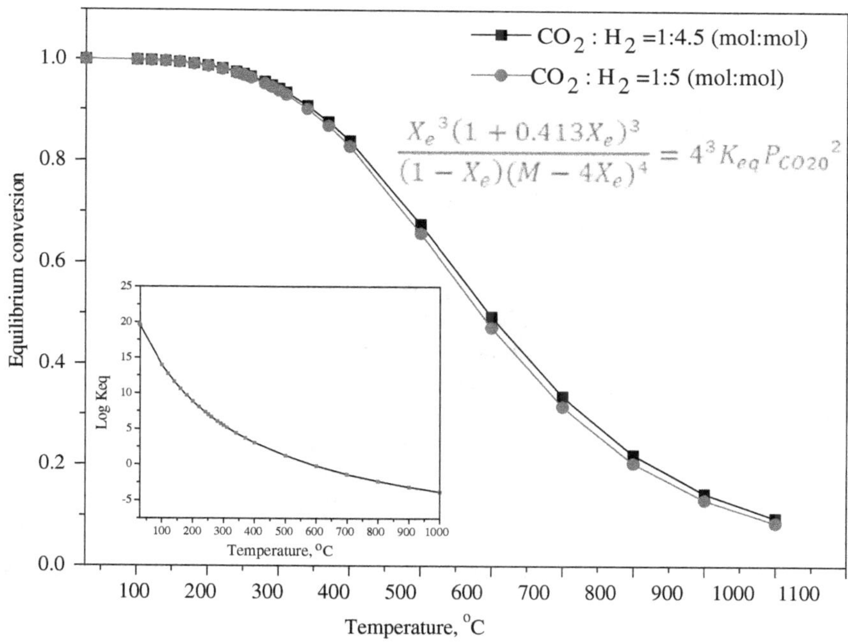

FIGURE 2.19 Equilibrium conversion of carbon dioxide reduction system.

Subsequently, five reactions are carried out in the reactor assembly depicted in Figure 6. The reactor outlet and inlet concentrations of carbon dioxide and hydrogen at each temperature between RT and 300°C are captured using gas chromatograph peaks. Further, the area of each of the curves is converted into

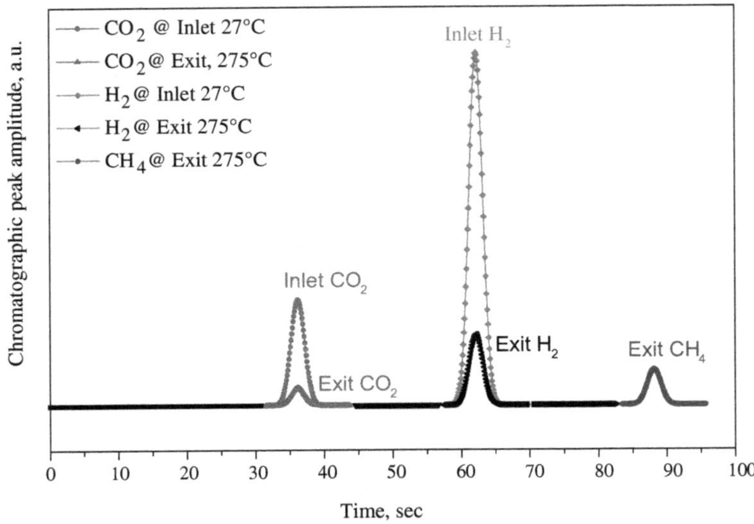

FIGURE 2.20 Chromatographic peak of reactant gas sample at the entry and exit of the reactor.

concentration using a calibration curve generated before the start of experiments. The reduction in signature peak strength with an increase in reaction temperature accelerated the system to attain the equilibrium condition shown in Figure 2.19. A typical chromate graphic peak at the entrance and outlet of the reactor is shown in Figure 2.20.

The conversion of reaction is computed based on the change in the area of chromatographic signature peak between inlet and exit peaks at different instances of temperature. Significant reduction in the chromatographic area of carbon dioxide and the appearance of a strong methane peak in the product are the indication of attainment of the maximum level of conversion at the exit reactor set temperature. The saturation of methane and CO_2 peak signature also is an indication of completion of reaction for a given set of parameters. Figure 2.21 renders the conversion temperature plot obtained for the different flow rates of the reactant gases.

The conversion temperature plot depicted above signifies that the maximum changes in the activity of the catalyst occur in the temperature window of 125°C–250°C. The temperature at which the curve changes its slope maximum is often designated as the 'light-off' temperature for the reaction system. In this particular reaction, the light-off temperature is found to be in the range of 150°C–160°C. However, the response curve captured at different flow rates does not show a significant change in the slope of the curve. The reason for the less sensitivity of the conversion temperature profile in flow rate can be contemplated to the higher mole ratio of the hydrogen in the mixture compared to the stoichiometric requirement. A higher molar ratio of reactant always helps to increase the rate of the reaction which compensates for the effect of flow rates in the reaction

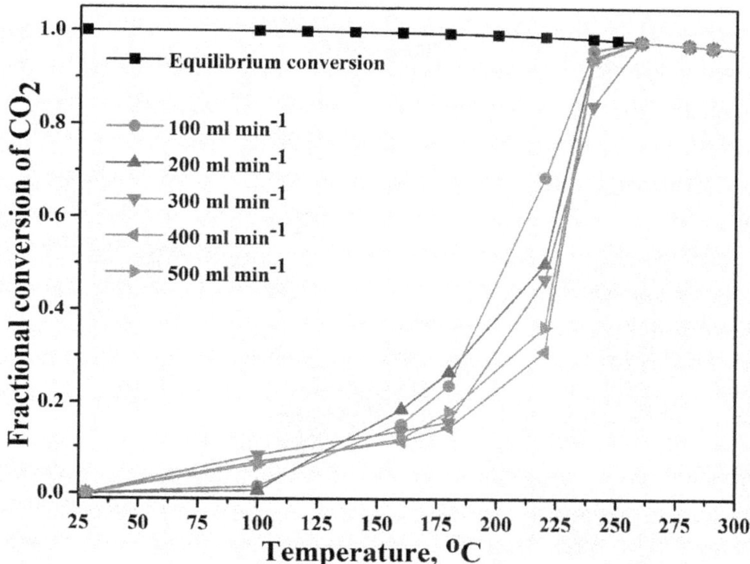

FIGURE 2.21 Conversion temperature plot at different flow rates over $Ru_x/\gamma\text{-}Al_2O_3$ at 1:4.5 mole ratio.

rate constant. However, if the reaction were to be carried out at a stoichiometric ratio, the flow would have impacted the conversion much more pronouncedly.

The activation energy and reaction rate constant for the reaction have been computed by the team based on the reactor design calculation. As the process of computation is very rigorous and is not coming under the subject matter of discussion, we are only showing the results of activation energy and reaction rate constants obtained from the computations. The apparent activation energy and reaction rate constants are computed based on the global reaction rate parameters that are often used for the representation of the reaction system.

The calculation of kinetic parameters is performed by simulating the global reaction rate expression using the experimental data of conversion vs temperature captured at each temperature and flow rate. The global rate expression [138] has been considered for the computation of reaction rate parameters based on the combination of the equations given below.

$$-r_{CO_2} = K_f{}^n \left\{ \left[P_{CO_2}\right]^n \left[P_{H_2}\right]^{4n} - \left[\frac{\left[P_{H_2O}\right]^{2n} \left[P_{CH_4}\right]^n}{\left(K_{eq}(T)\right)^n} \right] \right\} \text{---------------} (3)$$

$$K_{eq} = \frac{\left\{\left[P_{H_2O}\right]^2 \left[P_{CH_4}\right]\right\}}{\left\{\left[P_{CO_2}\right]\left[P_{H_2}\right]^4\right\}} = (\text{equlibrium constant}) \text{---------------} (4)$$

$$K_{eq}{}^n = \frac{K_f{}^n}{K_r{}^n}$$

$$K_f{}^n(T) = K_0{}^n \exp\left\{-\frac{E_a}{RT}\right\}$$

E_a: activation energy; P_{H_2O}: partial pressure of steam; P_{CH_4}: partial pressure of methane; P_{CO_2}: partial pressure of carbon dioxide; P_{H_2}: partial pressure of hydrogen; K_f: forward reaction rate constant; K_r: reverse reaction rate constant; K_{eq}: equilibrium rate constant; K_0: reaction rate constant; n: apparent reaction order; and E: activation energy of the reaction.

The activation energy and reaction rate constants are estimated based on the experimental data. Table 2.5 describes the kinetic reaction parameters for the carbon dioxide reaction conducted over the $Ru_x/\gamma\text{-}Al_2O_3$ catalyst.

Further, the computations were performed based on the kinetic parameters obtained from experimental data for sizing the rector for 4 kg/day carbon dioxide processing. The simulation is carried out, and it is found that 384 g of catalyst is required for processing 4 kg of carbon dioxide per day with 95% conversion efficiency at 260°C. The reactor pressure is considered as 3 kgf/cm^2. Table 2.6 provides a design summary of the reactor for converting 4 kg/day carbon dioxide into methane and water.

TABLE 2.5
Kinetic Parameters of Carbon Dioxide Reduction Reaction

$K_o{}^n$, bar$^{-0.45}$-g s mol^{-1}	Activation Energy (kJ/mol)	n
$(7.46 \pm 0.317\ E\text{-}4) \times 10^4$	28.167 ± 0.601	0.145 ± 0.08

TABLE 2.6
Design Summary of Carbon Dioxide Reduction Reactor

Reaction Parameters	Values
Reactor pressure	3 kgf/cm^2 abs
CO_2 processing capability	4 kg/day
Inlet gas composition, v/v	CO_2:H_2 – 1:4.5 mole ratio
Flow rate	600 sccm
Reactor bed length	~300 mm
Rector length	~600 mm
Reactor dia	50 mm
Loading	384 gm
Pressure drop in reactor	450 Pa
Catalyst	$Ru_x\text{-}\gamma Al_2O_3$
Size	3.2 mm (L) × 3.2 mm (Φ)
Shape	Cylindrical: L/D = 1

Chemical structure and properties of Ru_x-γ-Al_2O_3

The catalyst proposed to be used for performing CO_2 reduction in this case study is prepared by coating nanostructured γ-alumina with 15% ruthenium based on the impregnation technique. Nanoporous alumina is often prepared from suitable aluminium precursors such as aluminium chloride or aluminium nitrate by either sol–gel, co-precipitation, or methods such as ultrasonication. The alumina which precipitates out from this process is of highly porous nanostructure (pore diameter~1.0–1.5 nm) and high surface area (~200 m²/g).The synthesized aluminium powder is extruded as pellets by using suitable promoters and binders such as silica gel and polyethylene glycol. Finally, the prepared pellets are coated with ruthenium by using ruthenium chloride as a precursor. The final catalyst becomes highly nanoporous after drying and the separation of chloride ions from the material. The specific surface area and pore dimensions of the final prepared catalyst are, respectively, ~130 m²/g and ~2.0–2.5 nm. The structure of gamma-alumina is a close-packed cubic spinel lattice structure with a lattice side of 7.95. Å. The particle size of the catalyst also is in the range of 4–6 nm, which extensively helps in increasing the rate of the reaction.

Mechanism of CO_2 reduction over Ru_x-γ-Al_2O_3

As the reaction presented in this case study that is carried out over ruthenium/alumina-based catalyst is highly driven by its kinetics, it is necessary to understand in detail the kinetic mechanism of the reaction. The mechanism of gas–solid reactions is generally characterized using FTIR-DRIFTS (diffuse reflectance infrared Fourier transform spectroscopy) study. However, an in-detail discussion of the reaction mechanisms is beyond the scope of this chapter and we will be only discussing the probable reaction kinetic mechanism and associated steps which kinetically propel the reaction towards equilibrium. The major intermediates formed during the surface reaction that are tracked by the IR spectrum are carbonate (ν_{OCO}), bicarbonate (ν_{OCO}), formate (ν_{HCOO}), dissociated hydrogen adsorbed on Ru metal (Ru-H species – ν-Ru-H) and linearly adsorbed CO. Therefore, a nine-step probable reaction mechanism can be proposed for the reaction based on the evidenced IR signature peaks.

$$4H_2 + 4* \underset{k_{-1}}{\overset{k_1}{\rightleftarrows}} 4H_2^* \text{(Hydrogen adsorption)}$$

$$CO_2 + M(OH)_2 \underset{k_{-1}}{\overset{k_2}{\rightleftarrows}} M(OH)HCO_3 \left(\text{Bicarbonate formation}\right)$$

$$M(OH)HCO_3 \xrightarrow{k_3} MCO_3 + H_2O\left(\text{Carbonate formation}\right)$$

$$MCO_3 + H_2^* \xrightarrow{k_4} M(OH)OOCH + *\left(\text{Formate formation}\right)$$

$$M(OH)OOCH + * \xrightarrow{k_5} M(OH)_2 + CO^* \left(\text{linear CO formation}\right)$$

$$CO^* + H_2^* \xrightarrow{k_6} HCHO^* + *$$

$$HCHO^* + H_2 \xrightarrow{k_7} CH_2^* + H_2O + *$$

$$CH_2^* + H_2^* \underset{k_{-8}}{\overset{k_8}{\longleftrightarrow}} CH_4^* + *$$

$$CH_4^* \underset{k_{-9}}{\overset{k_9}{\rightleftarrows}} CH_4 + *$$

$$CO_2 + 4H_2 \underset{Ru-\gamma Al_2O_3}{\overset{270-300^\circ C, 1-4\,atm}{\rightleftarrows}} CH_4 + 2H_2O$$

* represents the active sites (mostly ruthenium) and M-Al_2O_3, and k_1–k_9 are the reaction rate constants of each step. However, the kinetic reaction intermediates and steps represented above need strict experimental validation. Furthermore, upon experimental validation, the reaction mechanism proposed in this chapter can be used for determining the kinetic rate expression for the design of the reactor for the proposed purpose. In the presented case study, the global rate expression (Eq. 2.3) is used to determine the sizing of the reactor.

SUMMARY OF THE CASE STUDIES

In the past few decades, environmental carbon dioxide remediation has become a major challenge for scientists and technologists. In the presented case studies, we have seen how actively the knowledge of adsorption and chemical catalysis can be used for controlling the CO_2 percentage inside the crew cabin during long-duration space missions. On the same note, the proposed technology in this case study also can be implemented for the control of industrial and automobile carbon dioxide emission control by capturing it through the adsorption process and subsequently converting the CO_2 into useful products such as methane and water. The automobile and industrial exhausts that mainly consist of CO_2, H_2O, CO (in trace amount), nitrogen and oxygen can be treated using selective MS to remove CO and CO_2 followed by the separation of water by condensing. The CO and CO_2 separated by the adsorption process can be passed through a catalytic oxidation chamber to convert CO into CO_2 over Pt/Pd/Rh/ceria/chromite, Cu/Co_3O_4 [139–140] catalyst. Further, the CO_2 can be fed to a Ru_x-γ-Al_2O_3-based catalytic reduction reactor to finally convert it into methane and water. Thus, an optimized combination of adsorption, oxidation and reduction processes can be exploited to design a green and closed-loop environmental remediation technique for treating industrial and automobile emissions.

Both the materials, i.e. MS-4A adsorbents and Ru_x-γ-Al_2O_3, are highly porous multifunctional nanostructured materials that are used for the capture and *in situ* resource utilization of CO_2 exhaled by the crew for environmental remediation of crew cabin atmosphere in spacecraft. Thus, it is evident that multifunctional nanomaterials form the basis of upgrading technology and its utility for myriad applications as seen in this chapter. The above case studies presented the real-world application of these materials for societal welfare and technological advancement in the area of space travel.

SPECIAL ACKNOWLEDGEMENTS

- Shri Suraj S Head, PED/PCM/VSSC: For providing technical and managerial supports for carrying out the studies presented in this chapter.
- Dr. Lakshmi V M, Deputy General Manager, PRMF/SPRE/VSSC: For critical review and comments on the book chapter.
- R Muraleekrishnan, Group Director, CSG/PCM/VSSC: For carrying out critical review of the study presented in this chapter.
- Dr. Manu SK, Group Director, PSCG/PCM/VSSC: For providing technical guidance and managerial supports for writing this chapter.
- Dr. Ilangovan S A, Deputy Director, PCM/VSSC: For providing valuable technical guidance and inputs during internal technical reviews on the topics.
- Shri. Somnath S, Director, VSSC/ISRO.

REFERENCES

[1] H. Wang, X. Liang, J. Wang, S. Jiao and D. Xue, *Nanoscale*, 2020, 12, 14–42.
[2] A. Faridi Esfanjani and S. M. Jafari, *Colloids and Surfaces B: Biointerfaces*, 2016, 146, 532–543.
[3] J. N. Israelachvili, Interfacial forces, Second Edition, *Academic Press*, 1992.
[4] W. Jiang, D. Peng, W.-R. Cui, R.-P. Liang and J.-D. Qiu, *ACS Omega*, 2020, 5, 32002–32010.
[5] E. G. Akkarat Manasilp, *Applied Catalysis B: Environmental*, 2002, 37, 17–25.
[6] S. A. Singh, S. Mukherjee and G. Madras, *Molecular Catalysis*, 2019, 466, 167–180.
[7] G. Konstantas and A. M. Stamatelos, *Proceedings of the Institution of Mechanical Engineers, Part D: Journal of Automobile Engineering*, 2007, 221, 355–373.
[8] J. A. Yancey, *Journal of Chromatographic Science*, 1994, 32, 349–356.
[9] C.H. Amberg, L. H. Little, *Canadian Journal of Chemistry*, 1962, 40.
[10] P. O. Bedolla, G. Feldbauer, M. Wolloch, S. J. Eder, N. Dörr, P. Mohn, J. Redinger and A. Vernes, *The Journal of Physical Chemistry C*, 2014, 118, 17608–17615.
[11] S. Ahmed, M. G. Rasul, W. N. Martens, R. Brown and M. A. Hashib, *Desalination*, 2010, 261, 3–18.
[12] G. Madras, R. Vinu, *Journal of the Indian Institute of Science*, 2010, 90, 2.
[13] H. Ben Mansour, D. Corroler, D. Barillier, K. Ghedira, L. Chekir and R. Mosrati, *Food and Chemical Toxicology*, 2007, 45, 1670–1677.
[14] M. Berradi, R. Hsissou, M. Khudhair, M. Assouag, O. Cherkaoui, A. El Bachiri and A. El Harfi, *Heliyon*, 2019, 5, e02711.
[15] R. Caritá and M. A. Marin-Morales, *Chemosphere*, 2008, 72, 722–725.
[16] D. Vaya and P. K. Surolia, *Environmental Technology & Innovation*, 2020, 20, 101128.
[17] N.K.R. Eswar, P C Ramamurty, G. Madras, *New Journal of Chemistry* 40(4), 3464–3475.
[18] B. Ohtani, *Journal of Photochemistry and Photobiology C: Photochemistry Reviews*, 2010, 11, 157–178.
[19] C. Sievers, Y. Noda, L. Qi, E. M. Albuquerque, R. M. Rioux and S. L. Scott, *ACS Catalysis*, 2016, 6, 8286–8307.
[20] D. Chen, Y. Cheng, N. Zhou, P. Chen, Y. Wang, K. Li, S. Huo, P. Cheng, P. Peng, R. Zhang, L. Wang, H. Liu, Y. Liu and R. Ruan, *Journal of Cleaner Production*, 2020, 268, 121725.

[21] J. Plotnikow, *Textbook of Photochemistry*, Verlag von Willhelm Knapp, Berlin, 1910, 72.
[22] A. M. B. S. E. Braslavsky, A. E. Cassano, A. V. Emeline, M. I. Litter, L. Palmisano, V. N. Parmon, N. Serpone, *Pure and Applied Chemistry*, 2011, 83, 931–1041.
[23] M. Boroski, A. C. Rodrigues, J. C. Garcia, L. C. Sampaio, J. Nozaki and N. Hioka, *Journal of Hazardous Materials*, 2009, 162, 448–454.
[24] M. S. Kari Pirkanniemi, *Chemosphere* 2002, 48, 1047–1060.
[25] M. N. Chong, B. Jin, C. W. K. Chow and C. Saint, *Water Research*, 2010, 44, 2997–3027.
[26] M. S. T. Hoffmann M.R., Choi W., Bahnemann D.W, *Chemical Reviews* 1995, 95, 69–96.
[27] J.-M. Herrmann, *Catalysis Today*, 1999, 53(1), 115–129.
[28] G. L. Amy L. Linsebigler, and J. T. Yates, Jr, *Chemical Reviews*, 1995, 95.
[29] M. Pelaez, N. T. Nolan, S. C. Pillai, M. K. Seery, P. Falaras, A. G. Kontos, P. S. M. Dunlop, J. W. J. Hamilton, J. A. Byrne, K. O'Shea, M. H. Entezari and D. D. Dionysiou, *Applied Catalysis B: Environmental*, 2012, 125, 331–349.
[30] J. Georgieva, E. Valova, S. Armyanov, N. Philippidis, I. Poulios and S. Sotiropoulos, *Journal of Hazardous Materials*, 2012, 211–212, 30–46.
[31] G. G. Bessegato, T. T. Guaraldo and M. V. B. Zanoni, *Modern Electrochemical Methods in Nano, Surface and Corrosion Science*, 2014, 271–319.
[32] I. Paramasivam, Y.-C. Nah, C. Das, N. K. Shrestha and P. Schmuki, *Chemistry - A European Journal*, 2010, 16, 8993–8997.
[33] A. Dabrowski, *Advances in Colloid and Interface Science*, 2001, 93, 135–224.
[34] J.-y. K. Fumitaka Mafune, Y. Takeda, T. Kon, and H. Sawabe, *Journal of Physical Chemistry B* 2000, 104, 8333–8337.
[35] M. A. Lopez-Quintela, *Current Opinion in Colloid and Interface Science* 2003, 8, 137–144.
[36] J. J. Brown, Z. Ke, T. Ma and A. J. Page, *ChemNanoMat*, 2020, 6, 708–719.
[37] J. Yang, X. Zhu, Q. Yu, G. Zhou, Q. Li, C. Wang, Y. Hua, Y. She, H. Xu and H. Li, *Journal of Colloid and Interface Science*, 2020, 580, 814–821.
[38] Y. Qi, L. Song, S. Ouyang, X. Liang, S. Ning, Q. Zhang and J. Ye, *Advanced Materials*, 2019, 32, 1903915.
[39] S. Chen, D. Huang, P. Xu, X. Gong, W. Xue, L. Lei, R. Deng, J. Li and Z. Li, *ACS Catalysis*, 2019, 10, 1024–1059.
[40] J. Low, J. Yu, M. Jaroniec, S. Wageh and A. A. Al-Ghamdi, *Advanced Materials*, 2017, 29, 1601694.
[41] J. Schneider, M. Matsuoka, M. Takeuchi, J. Zhang, Y. Horiuchi, M. Anpo and D. W. Bahnemann, *Chemical Reviews*, 2014, 114, 9919–9986.
[42] C. M. Teh and A. R. Mohamed, *Journal of Alloys and Compounds*, 2011, 509, 1648–1660.
[43] J. Ryu and W. Choi, *Environmental Science & Technology*, 2008, 42, 294–300.
[44] R. Vinu, S. U. Akki and G. Madras, *Journal of Hazardous Materials*, 2010, 176, 765–773.
[45] C. G. Daughton and T. A. Ternes, *Environmental Health Perspectives*, 1999, 107, 907.
[46] C. G. Daughton, ACS Publications, 2001. 10.1021/bk-2001-0791.ch001.
[47] D. Barceló, *Journal*, 2003.
[48] K. Yu, B. Li and T. Zhang, *Analytica Chimica Acta*, 2012, 738, 59–68.
[49] P. J. Ferguson, M. J. Bernot, J. C. Doll and T. E. Lauer, *Science of the Total Environment*, 2013, 458, 187–196.

[50] K. Kümmerer, *Pharmaceuticals in the Environment: Sources, Fate, Effects and Risks*, Springer Science & Business Media, 2008.
[51] N. Veldhoen, R. C. Skirrow, H. Osachoff, H. Wigmore, D. J. Clapson, M. P. Gunderson, G. Van Aggelen and C. C. Helbing, *Aquatic Toxicology*, 2006, 80, 217–227.
[52] N. Ajoudanian and A. Nezamzadeh-Ejhieh, *Materials Science in Semiconductor Processing*, 2015, 36, 162–169.
[53] C. M. C. L. Martínez, M. I. Fernández, J. A. Santaballa and J. Faria, *Applied Catalysis B: Environmental*, 2011, 102, 563–571.
[54] R. Liang, A. Hu, W. Li and Y. N. Zhou, *Journal of Nanoparticles Research*, 2013, 15, 1990.
[55] K. Sankoda, H. Matsuo, M. Ito, K. Nomiyama, K. Arizono and R. Shinohara, *Bulletin of Environmental Contamination and Toxicology*, 2011, 86, 470.
[56] C. Martínez, M. Fernández, J. Santaballa and J. Faria, *Applied Catalysis B: Environmental*, 2011, 102, 563–571.
[57] S. Dong, Y. Li, J. Sun, C. Yu, Y. Li and J. Sun, *Material Chemistry of Physics*, 2014, 145, 357–365.
[58] A. Chatzitakis, C. Berberidou, I. Paspaltsis, G. Kyriakou, T. Sklaviadis and I. Poulios, *Water Research*, 2008, 42, 386–394.
[59] A. Hu, X. Zhang, K. D. Oakes, P. Peng, Y. N. Zhou and M. R. Servos, *Journal of Hazardous Materials*, 2011, 189, 278–285.
[60] R. M. Fernández-Domene, R. Sánchez-Tovar, B. Lucas-granados, M. J. Muñoz-Portero and J. García-Antón, *Chemical Engineering Journal*, 2018, 350, 1114–1124.
[61] Z. Zheng, X. Li, L. Li and Y. Tang, *International Journal of Hydrogen Energy*, 2019, 44, 20826–20838.
[62] B. O. Orimolade, B. A. Koiki, G. M. Peleyeju and O. A. Arotiba, *Electrochimica Acta*, 2019, 307, 285–292.
[63] Y.-F. Su, G.-B. Wang, D. T. F. Kuo, M.-L. Chang and Y.-H. Shih, *Applied Catalysis B: Environmental*, 2016, 186, 184–192.
[64] M. I. Litter, *Advances in Chemical Engineering*, 2009, 36, 37–67.
[65] W. Choi, J. Yeo, J. Ryu, T. Tachikawa and T. Majima, *Environmental Science & Technology*, 2010, 44, 9099–9104.
[66] S.-H. Yoon and J. H. Lee, *Environmental Science & Technology*, 2005, 39, 9695–9701.
[67] H. Lee and W. Choi, *Environmental Science & Technology*, 2002, 36, 3872–3878.
[68] T. Xu, P. V. Kamat and K. E. O'Shea, *The Journal of Physical Chemistry A*, 2005, 109, 9070–9075.
[69] H. Fei, W. Leng, X. Li, X. Cheng, Y. Xu, J. Zhang and C. Cao, *Environmental Science & Technology*, 2011, 45, 4532–4539.
[70] D. Monllor-Satoca, R. Gómez and W. Choi, *Environmental Science & Technology*, 2012, 46, 5519–5527.
[71] K. Mukherjee, R. Saha, A. Ghosh and B. Saha, *Research on Chemical Intermediates*, 2013, 39, 2267–2286.
[72] Y. C. Zhang, J. Li, M. Zhang and D. D. Dionysiou, *Environmental Science & Technology*, 2011, 45, 9324–9331.
[73] A. Rauf, M. S. A. Sher Shah, G. H. Choi, U. B. Humayoun, D. H. Yoon, J. W. Bae, J. Park, W.-J. Kim and P. J. Yoo, *ACS Sustainable Chemistry & Engineering*, 2015, 3, 2847–2855.
[74] C. Lv, G. Chen, J. Sun and Y. Zhou, *Inorganic Chemistry*, 2016, 55, 4782–4789.
[75] X. Chen and D.-H. Kuo, *ACS Sustainable Chemistry & Engineering*, 2017, 5, 4133–4143.

[76] Y. C. Zhang, J. Li, M. Zhang and D. D. Dionysiou, *Environmental Science & Technology*, 2011, 45, 9324–9331.
[77] Q.-L. Yang, S.-Z. Kang, H. Chen, W. Bu and J. Mu, *Desalination*, 2011, 266, 149–153.
[78] J. Yang, J. Dai and J. Li, *Environmental Science and Pollution Research*, 2013, 20, 2435–2447.
[79] M.-R. Samarghandi, J.-K. Yang, O. Giahi and M. Shirzad-Siboni, *Environmental Technology*, 2015, 36, 1132–1140.
[80] S. Chakrabarti, B. Chaudhuri, S. Bhattacharjee, A. K. Ray and B. K. Dutta, *Chemical Engineering Journal*, 2009, 153, 86–93.
[81] K. Kabra, R. Chaudhary and R. Sawhney, *Journal of Hazardous Materials*, 2008, 155, 424–432.
[82] J. Saien and A. Azizi, *Process Safety and Environmental Protection*, 2015, 95, 114–125.
[83] T. S. George, *Minamata: Pollution and the Struggle for Democracy in Postwar Japan*, Harvard University Press, Cambridge, MA 2001.
[84] A. Kudo, Y. Fujikawa, S. Miyahara, J. Zheng, H. Takigami, M. Sugahara and T. Muramatsu, *Water Science and Technology*, 1998, 38, 187–193.
[85] K. Tennakone and U. Ketipearachchi, *Applied Catalysis B: Environmental*, 1995, 5, 343–349.
[86] L. Skubal and N. Meshkov, *Journal of Photochemistry and Photobiology A: Chemistry*, 2002, 148, 211–214.
[87] M. R. Prairie, L. R. Evans, B. M. Stange and S. L. Martinez, *Environmental Science and Technology*, 1993, 27, 1776–1782.
[88] M. López-Muñoz, J. Aguado, A. Arencibia and R. Pascual, *Applied Catalysis B: Environmental*, 2011, 104, 220–228.
[89] M. Emmanuel, A. G. Leyva, E. A. Gautier and M. I. Litter, *Chemosphere*, 2007, 69, 682–688.
[90] C. Miranda, J. Yáñez, D. Contreras, R. Garcia, W. F. Jardim and H. D. Mansilla, *Applied Catalysis B: Environmental*, 2009, 90, 115–119.
[91] C. Singh and R. Chaudhary, *Journal of Renewable and Sustainable Energy*, 2013, 5, 053102.
[92] D. Sud, G. Mahajan and M. Kaur, *Bioresource Technology*, 2008, 99, 6017–6027.
[93] V. N. H. Nguyen, R. Amal and D. Beydoun, *Chemical Engineering Science*, 2003, 58, 4429–4439.
[94] Xin Li, Jiaguo Yu, Mietek Jaroniec, Chem. Soc. Rev. 2016, 45, 2603.
[95] M. W. Yash Boyjoo, V. K. Pareek, Jian Liu and M. Ja, *Chemical Society Review*, 2016, 45.
[96] G. Prieto, H. Tüysüz, N. Duyckaerts, J. Knossalla, G.-H. Wang and F. Schüth, *Chemical Reviews*, 2016, 116, 14056–14119.
[97] J. Hu, M. Chen, X. Fang and L. Wu, *Chemical Society Reviews*, 2011, 40, 5472.
[98] X. Lai, J. Qi, J. Wang, H. Tang, H. Ren, Y. Yang, L. Zhang, Q. Jin, R. Yu, G. Ma, Z. Su, H. Zhaod and A. D. Wang, *Chemical Society Reviews* 2015, 44, 6749–6773.
[99] G. Liu, J. C. Yu, G. Q. Lu and H.-M. Cheng, *Chemical Communications*, 2011, 47, 6763.
[100] J. Shen, Y. Zhu, X. Yang and C. Li, *Journal of Materials Chemistry*, 2012, 22, 13341.
[101] G. Liu, H. G. Yang, J. Pan, Y. Q. Yang, G. Q. Lu and H.-M. Cheng, *Chemical Reviews*, 2014, 114, 9559–9612.
[102] B. Thokchom, P. Qiu, M. Cui, B. Park, A. B. Pandit and J. Khim, *Ultrasonics Sonochemistry*, 2017, 34, 262–272.

[103] P. Raizada, J. Kumari, P. Shandilya, R. Dhiman, V. Pratap Singh and P. Singh, *Process Safety and Environmental Protection*, 2017, 106, 104–116.
[104] S. Zhang, J. Li, M. Zeng, G. Zhao, J. Xu, W. Hu and X. Wang, *ACS Applied Materials & Interfaces*, 2013, 5, 12735–12743.
[105] L. Wang, T. Kumeria, A. Santos, P. Forward, M. F. Lambert and D. Losic, *ACS Applied Materials & Interfaces*, 2016, 8, 20110–20119.
[106] L. Liu, W. Yang, Q. Li, S. Gao and J. K. Shang, *ACS Applied Materials & Interfaces*, 2014, 6, 5629–5639.
[107] Q. Li, Y. W. Li, P. Wu, R. Xie and J. K. Shang, *Advanced Materials*, 2008, 20, 3717–3723.
[108] Y.-D. Chiou and Y.-J. Hsu, *Applied Catalysis B: Environmental*, 2011, 105, 211–219.
[109] J. Xu, T. J. K. Brenner, Z. Chen, D. Neher, M. Antonietti and M. Shalom, *ACS Applied Materials & Interfaces*, 2014, 6, 16481–16486.
[110] M. L. Liu, B. B. Chen, C. M. Li and C. Z. Huang, *Green Chemistry*, 2019, 21, 449–471.
[111] V. Blanchard, Z. Asbai, K. Cottet, G. Boissonnat, M. Port and Z. Amara, *Organic Process Research & Development*, 2020, 24, 822–826.
[112] S. Mustapha, M. M. Ndamitso, A. S. Abdulkareem, J. O. Tijani, D. T. Shuaib, A. O. Ajala and A. K. Mohammed, *Applied Water Science*, 2020, 10, 1–36.
[113] K. R. Reddy, M. S. Jyothi, A. V. Raghu, V. Sadhu, S. Naveen and T. M. Aminabhavi, 2020, 30, 139–169.
[114] https://www.who.int/airpollution/ambient/pollutants/en/.
[115] Archana Charanpahari, Sachin G Ghugal, S S Umare, R Sasikala, *New Journal of Chemistry*, 2015, 39, 3629–3638.
[116] Z.-L. Cheng and S. Han, *Chinese Chemical Letters*, 2016, 27, 467–470.
[117] B. Ge, G. Ren, X. Miao, X. Li, T. Zhang, X. Pu, C. Jin, L. Zhao and W. Li, *Journal of the Taiwan Institute of Chemical Engineers*, 2019, 102, 233–241.
[118] O. A. Attallah, M. Rabee, *RSC Advances*, 2020, 10, 40697.
[119] K. D. Dibyananda Majhi, R. Bariki, S. Padhan, A. Mishra, and P. D. Rohan Dhiman, Bismita Nayak and B. G. Mishra, *Journal of Materials Chemistry A*, 2020.
[120] A. Zhang, *Journal of Materials Chemistry A*, 2020.
[121] M. H. S. Najmeh Ahmadpour and A. S. Homaeigohar, *RSC Advance*, 2020, 10, 29808.
[122] Y. Laipan, J. Yu, R. Zhu, J. Zhu, and H. He, D. O'Hare and L. Sun, *Materials Horizons* 2019.
[123] C. S. Xiaoxue Zhan, J. Zhou and Z. Sun, *Nanoscale Horizons*, 2019.
[124] A. Thakur, P. Kumar, D. Kaur, N. Devunuri, R. K. Sinha, and P. Devi, *RSC Advance*, 2020, 10, 8941–8948.
[125] R. K. Sharma, P. Yadav, M. Yadav, R. Gupta, P. Rana, A. Srivastava, R. Zbořil, R. S. Varma, M. Antonietti and M. B. Gawande, *Materials Horizons*, 2020, 7, 411–454.
[126] N. Rahman, P. Varshney, *RSC Advance*, 2020, 10, 20322.
[127] K. Shaheen, Z. Shah, A. Asad, T. Arshad, S. B. Khan and H. Suo, *ACS Omega*, 2020, 5, 15992–16002.
[128] G. Sharma and B. Kandasubramanian, *Journal of Chemical & Engineering Data*, 2020, 65, 396–418.
[129] R. Liu, Y. Hou, S. Jiang and B. Nie, *Langmuir*, 2020, 36, 5236–5242.
[130] L. Noureen, Z. Xie, Y. Gao, M. Li, M. Hussain, K. Wang, L. Zhang and J. Zhu, *ACS Applied Materials & Interfaces*, 2020, 12, 6343–6350.
[131] Y. Wang, S. Zhan, S. Di and X. Zhao, *ACS Applied Materials & Interfaces*, 2018, 10, 26396–26404.

[132] D. Wood, S. Shaw, T. Cawte, E. Shanen and B. Van Heyst, *Chemical Engineering Journal*, 2020, 391, 123490.

[133] J. G. Lu, *Current Opinion in Psychology*, 2020, 32, 52–65.

[134] Molecular Sieve Adsorbents, www.catalysts.basf.com/adsorbents.

[135] Ertan, A.; *'Dissertation- CO_2, N_2 and Ar Adsorption on Zeolites'*, İzmir Institute of Technology, İzmir, Turkey, 2004.

[136] Ertl, G.; Knozinger, H.; Weitkamp, J.; *'Preparation of Solid Catalysts'*, WILEY-VCH Verlag GmbH, D-69469 Weinheim (Federal Republic of Germany), 1999.

[137] Siriwardane, R. V.; Shen, M.-S.; Fisher, E. P.; "Adsorption of CO2 on zeolites at moderate temperatures," *Energy & Fuels* 19, 1153–1159, 2005.

[138] Lunde, P. J.; Kester, F. L.; "Rates of methane formation from carbon dioxide and hydrogen over a ruthenium catalyst," *Journal of Catalysis*, 30, 423–429, 1973.

[139] Nagareddy, G.; Vaibhav, P.; Vedantsingh, P.; Titiksha, B.; Asif, P.; *Design and Analysis of Catalytic Converter of Automobile Engine*, IRJET, volume 06, Issue 05, 2019.

[140] Milton, B. E.; *Control Technologies in Spark-Ignition Engines, Handbook of Air Pollution From Internal Combustion Engines, Pollutant Formation and Control*, Chapter 8, pp. 189–258, 1998.

3 Carbon Nanotubes in Wastewater Remediation

Mohamed H.H. Mahmoud
College of Science, Taif University,
Central Metallurgical Research and
Development Institute

Mohammed Alsawat and Tariq Altalhi
College of Science, Taif University

CONTENTS

DOI: 10.1201/9781003129042-3

3.1 INTRODUCTION

Deficiency of clean and fresh water is a universal challenge worldwide accompanied by the growing population, industrialization and urbanization (Elimelech et al. 2011; Werber et al. 2016; Wang et al. 2019). A lot of countries are facing a scarcity of clean and safe water (Khan et al. 2019). Efficient technologies are thus essential for decontamination of wastewater to meet the shortage of water resources. Environmental protection and legislations related to wastewater and effluent discharges became a main concern. Water contamination with heavy metals and toxic organic contaminants is a common environmental issue. Detection and removal of such pollutants from discharge water has been a compulsory measure in several industries such as metallurgical, chemicals and petrochemicals, textile and printing; otherwise, it pertains to an illegal behaviour (Kyzas et al. 2018). Extensive researches are of primary importance by the scientific community to find out safe, effective and economic processes to eliminate toxic and non-readily biodegradable pollutants. Adsorption, membrane filtration, electrochemical treatment (Mudasir et al. 2016), bioremediation, solvent extraction and ion exchange, oxidation, reduction (Shah et al. 2017) and leaching (Zhang et al. 2017b) techniques have been employed for heavy metal remediation. Adsorption is the most widely accepted technique for the treatment of diverse types of water pollutants due to its acceptable features (Moradi and Zare 2013).

Nanoscience is an interdisciplinary study of structures that are in the ultra-small length scale of 1–100 nm and the interesting and unique properties demonstrated by these materials. Ten adjacent side-by-side hydrogen atoms measure 7 nm long, the width of a DNA strand measures 2.5 nm, and the measure of a red blood cell is a 7000 nm wide. Applications of nanomaterials comprise a wide interdisciplinary research branches, and expansion has grown explosively worldwide due to their reactive, optical, magnetic, mechanical and electric properties (Aigbe and Osibote 2020). Currently, nanomaterials persist as an efficient, applied and environmentally friendly choice compared to existing treatment materials from the viewpoint of environmental remediation and resource reservation (Xu et al. 2012). Since the invention of nm-sized, sp^2-bonded carbon materials such as graphene (Gr), fullerenes and carbon nanotubes (CNTs), extensive motivated studies have been conducted for diverse areas of applications (Eatemadi et al. 2014).

CNTs are novel carbon nanomaterials that have been motivating massive interest in diverse scientific fields due to their excellent chemical and physical characteristics. CNTs are evolving materials for environmental remediation (De Volder et al. 2013; Qu et al. 2018) in addition to the unlimited potential applications of CNTs in electrical devices, electronics, sensors and thermal devices. The useful environmental applications of CNTs arise from their unique features of tubular hollow assembly, huge surface area, great ratio of length to radius and hydrophobic, easily modifiable surface. Recent advances in environmental applications of CNTs and their modified derivatives related to wastewater treatment have been discussed in several reviews (Goh et al. 2017; Sarkar et al. 2018; Ma et al. 2017; Mubarak 2014a; Fiyadh et al. 2019; Bassyouni et al. 2020; Verma et al. 2020).

This book chapter is projected to highlight in brief the features and fabrication techniques of CNTs, including some of our previously published works, and then to focus mainly on environmental applications of CNTs for wastewater remediation. It collects recent information on efficient adsorption of some heavy metals and organic pollutants on CNTs. Throughout the application sections, valuable fabrication techniques are summarized and separated in the appendices. This collective fabrication and application information of CNTs will give the reader a broad and updated knowledge related to this evolutionary, advanced area of research where design of future strategic studies can be promoted.

3.2 FEATURES OF CNTs

The paper published by Iijima in 1991 created extraordinary attention to the nanostructures of carbon and has driven extensive research in nanotechnology, although CNTs had been detected prior to his discovery (Iijima 1991). Sumio Iijima, the Japanese scientist, Japan Science and Technology Agency, NEC Corporation (Photo 1), was awarded the Benjamin Franklin Medal in Physics (2002) for his invention and explanation of the helical character and atomic structure of CNTs in their multi-walled and single-walled forms, which have had a massive impact on the quickly rising field of nanoscale science and technology.

CNTs are tubular ultra-small molecules and are the hardest substance known to humanity. They are thousand times more conductive than copper metal. Carbon atoms are joined in a hexagonal pattern to form a honeycomb sheet structure of one atom thickness. These sheets are rolled, where edges join to form a

PHOTO 3.1 Sumio Iijima, the founder of CNTs.

carbon tube in the nanoscale. They comprise one or more self-folded graphene sheets forming a cylindrical structure with a radius of less than 100 nm and a length of more than 20 µm (Zhu et al. 2002). These tubes might contain only one sheet (single-walled, SWCNTs) or comprise two or more tubes of different diameters – one inside another (multi-walled, MWCNTs). CNTs have few nanometres diameter and several millimetres length. Figure 3.1 illustrates the shape and dimensions of SWCNTs and MWCNTs. The SWCNTs have large diameter (0.4–3 nm) and more than 10 nm length and thus are considered as 1D molecules (Jahanshahi et al. 2013).

Experimental TEM images of MWCNTs show that they have a spacing of 3.4 Å (Ebbesen and Ajayan, 1992). The unusual structure of CNTs is the reason

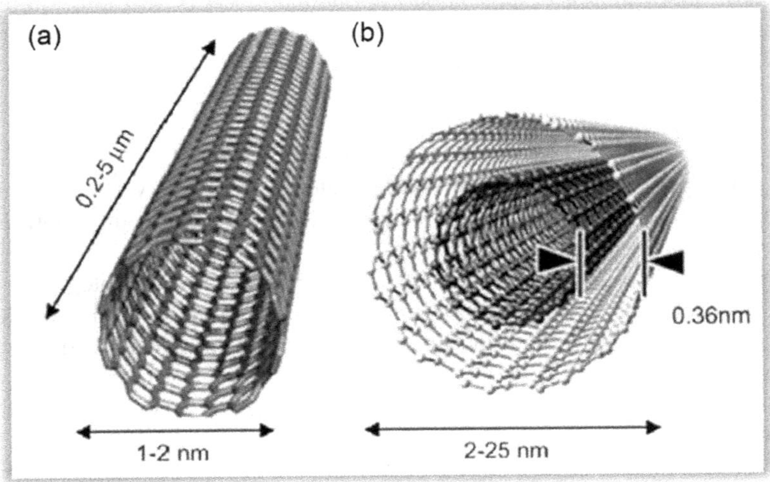

FIGURE 3.1 Shapes and dimensions of (a) single-walled and, (b) multi-walled CNT structures.

for amazing chemical and physical characteristics. CNTs are globally considered as one of the strongest materials based on the sp^2-bonded carbon atoms (Thostenson et al. 2001). They have strength and Young's modulus 10–100 times larger than steel.

CNTs have extraordinary thermal conductivity: 8–350 K for SWCNTs (Hone et al. 1999) and 4–300 K for MWCNTs (Yi et al. 1999), that is much higher than other materials (Collins and Avouris 2000) due to the proven hexagonal ring structure arrangement along the cylindrical surface.

The active locations in CNTs are positioned around the defective sites such as pentagons arranged against the body of the tube consisting of hexagons. These active sites may give CNTs their excessive capability to react with other species. Inserting functional groups in the surface of CNTs is a vital step to modify their characteristics. Functional groups can be chemically attached to the surfaces of CNTs or just cover the walls of CNTs non-covalently.

3.3 SYNTHESIS AND FUNCTIONALIZATION OF CNTs

Different structures, morphologies and characteristics of CNTs are determined by the method of preparation. Three main approaches can be applied for the synthesis of CNTs, namely chemical vapour deposition (CVD), arc discharge and laser deposition (Prasek et al. 2011; Anzar et al. 2020). CVD further opened up new routes for controlled synthesis and device integration. Compared with other techniques, the low-temperature CVD is considered feasible, economical and suitable for mass production. Carbon sources, including methane, acetylene, benzene, xylene and carbon monoxide ethylene, are utilized to prepare CNTs. MWCNTs are more easily produced from most hydrocarbons by low-temperature CVD, while SWCNTs are only generated by high-temperature CVD (900°C–1200°C) from methane and carbon monoxide (Awotunde et al. 2019).

3.3.1 Laser Ablation Technique

In the laser ablation technique (Figure 3.2), a graphite target is heated till vaporization by laser radiation in a flow of inert gas at 1200°C (Zhang et al. 1999; Herrera-Ramirez et al. 2019). Carbon species are produced, moved by a flowing gas and then accumulated on a collector that is water-cooled. Uniform SWCNTs are produced from graphite target doped with metals such as Ni or Co.

3.3.2 Arc Discharge Technique

MWCNTs were first produced by Iijima in 1991 by the arc discharge technique. Under an inert medium, arc is generated between two carbon electrodes at high temperature (>3000°C) to evaporate carbon atoms into plasma (Figure 3.3). Soft deposits of CNTs and other carbon deposits are formed on the cathode. Electrodes are loaded with an appropriate catalyst, such as Ni–Co or Co–Y, to

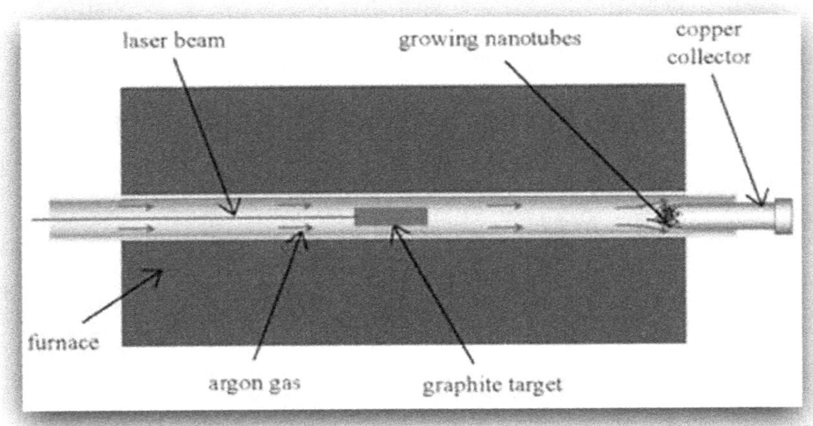

FIGURE 3.2 Schematic illustration of laser ablation apparatus. (See Thostenson et al. 2001.)

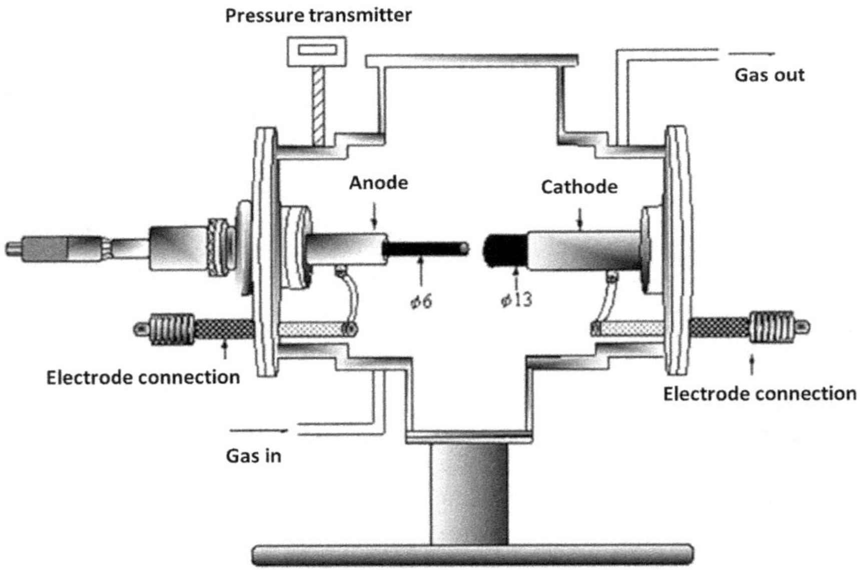

FIGURE 3.3 Schematic illustration of arc discharge apparatus. (From Szabó et al. 2010.)

produce SWCNTs. However, for the fabrication of MWCNTs, the presence of catalyst is not required. The CNTs prepared by this technique are of exceptional quality, with slight occurrence of defects, and with varied dimensions of diameter and length (Journet et al. 1997; Shi et al. 2000; Herrera-Ramirez et al. 2019).

3.3.3 Chemical Vapour Deposition (CVD) Technique

Filaments and fibres of carbon have been fabricated by the CVD technique since 1960s. A gas containing carbon is allowed to flow and dissociate on a substrate in the presence of a metal catalyst powder at 600°C or higher (Figure 3.4). This fairly low applied temperature, compared with other techniques, reduces the production costs. In addition, novel structures can be formed through the coating of catalysts on the substrate. However, a large number of defects were found on CNTs made by this method (Daenen et al. 2003; Herrera-Ramirez et al. 2019).

3.3.4 Functionalization of CNTs

Chemical modification of CNTs can control their physicochemical properties to become adapted for various applications. This can be achieved by carrying out functionalization processes through which the surface of CNTs is turned to be enriched with specific functional groups attached to the carbon atoms. Several methods have successfully been implemented for the functionalization of CNTs. Oxidation, the most common functionalization route, is performed through refluxing CNTs with a mineral acid or with a strong oxidizing agent such as $KMnO_4$ (Salam 2013). Carboxylic functional groups would be fixed on CNTs in the oxidation process. This oxidized form of CNTs can be used for further functionalization (Hirsch and Vostrowsky 2005). The best active functional groups formed on CNTs are hydroxyl (–OH) or carboxyl (–COOH) groups (Ma et al. 2010). Polar groups attached to the surface tolerate CNTs to be more efficiently dispersed in organic solvents and then increase their hydrophilic nature. Several other treating methods for the functionalization of CNTs, including arylation and alkylation, biomolecules, esterification, thiolation and polymer grafting, have been used (Hirsch and Vostrowsky 2005; Ma et al. 2006).

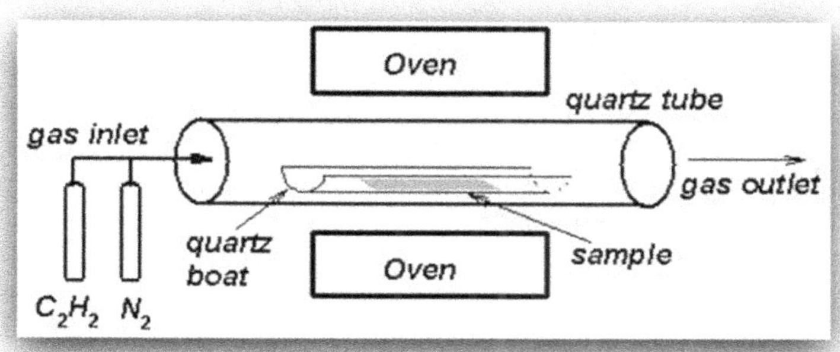

FIGURE 3.4 Schematic representation of carbon vapour deposition apparatus. (See Thostenson et al. 2001.)

3.3.5 Fabrication of CNT-Reinforced Metal Matrix Composites (CNT-MMCs)

Lightweight and high-strength functional materials can be adopted by a composite of appropriate matrix and reinforcing materials. The metal matrix composites (MMCs) were developed in 1950s to fulfil the needs of diverse potential industrial applications, mainly at elevated temperatures. The features of composites are dependent on the type, amount and geometry of matrix and reinforcing materials and their dispersed phase. The reinforcements comprise ceramic, carbon or fibres. Reinforcements such as Al_2O_3, SiC and TiC are common in reinforced aluminium matrix composite and other composites (Zhang et al. 2017; Shorowordi et al. 2003). Reinforcements with CNTs are highly attractive compared with the ceramic materials due to their high damping capacity, high thermal conductivity and self-lubricating character (Baughman 2002). Because of the nanosize of CNTs, the fabrication became relatively easy compared with reinforcements with microsized ceramic. Several reports have been published in recent years on adapting the mechanical properties of MMCs by incorporating nano-reinforcements such as graphene (Kim et al. 2013) nanofibre (Subramaniam et al. 2013) and CNTs (Hjortstam et al. 2004). CNTs are the most widely acceptable reinforcement for the large-scale production of MMCs due to their superior properties (Zhao et al. 2013).

The CNTs look twisted and entangled in a scanning electron microscope (SEM) image. Uniform spreading of CNTs in the matrix of metal and alloy is the major challenge for successful processing. At elevated temperatures, applied stresses or interaction with the molten metal of matrix material can damage the CNT structure. The selection of fabrication technique should consider these factors. The fabrication processes of CNT-MMCs are described herein.

3.3.6 Microwave-Assisted Fabrication of CNTs

Pure (98%) CNTs were produced by single-stage gas-phase tubular microwave chemical vapour deposition (TM–CVD) utilizing hydrogen (H_2) and acetylene (C_2H_2) as precursors with ferrocene catalyst under the optimum conditions: radiation time of 35 min, reaction power of 900 W and gas ratio of C_2H_2/H_2 of 0.6 (Mubarak et al. 2014c). TEM investigation revealed uniform dispersion of MWCNTs with 16–23 nm diameters. Figure 3.5 shows the diagram of CNT production by a horizontal tube microwave reactor composed of a quartz tube of OD of 55 mm, ID of 50 mm and length of 615 mm. A quartz boat was located at the middle of reaction chamber, and the ferrocene catalyst was located at the entrance. Argon gas was first pumped into the system to take away any oxygen gas. The C_2H_2 and H_2 flow gases were mixed before entering the microwave chamber; after that, the gas mixture was allowed to enter the chamber. The desired time period was set, and finally, the produced CNTs were obtained in the quartz boats.

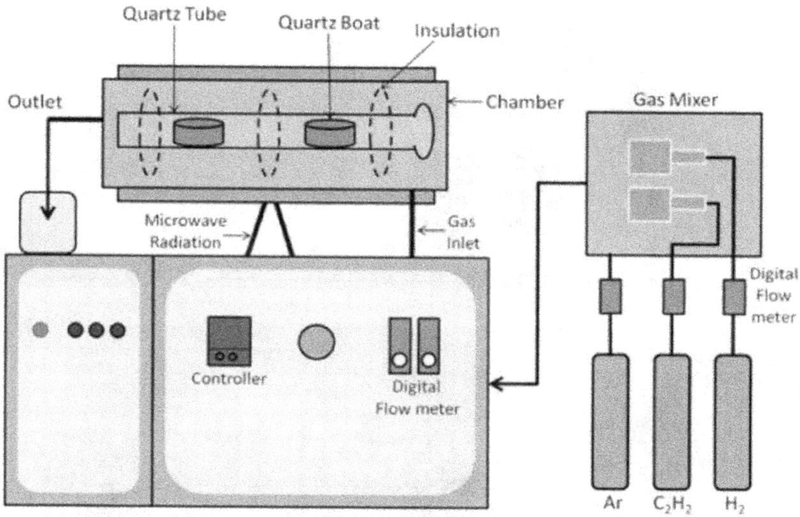

FIGURE 3.5 Schematic diagram of tubular microwave chemical vapour deposition (TM–CVD) for the production of CNTs. (See Mubarak et al. 2014c.)

3.3.7 Fabrication of CNT Membranes

Developments in nanotube-based science will remain growing to trust on the further improved highly controlled CNTs for various applications. Considerable attention has been paid to membrane science as attractive solutions to practical separation complications and challenges, and cost-effective and energy-effective solutions applied to simple, continuous and scalable systems (Wang et al. 2018). Membranes perform as a barrier that allows selective transport of certain species. Adding nanomaterials inside porous ceramic or polymeric membrane matrices enhances material properties. By this way, a novel membrane class called mixed matrix membranes (MMMs) has widely been explored (Vinoba et al. 2017). The great electrical conductivity of CNTs enables large-scale fabrication processes of CNT-based membranes (Rashed et al. 2021). Based on the method of fabrication, main classes of CNT membranes are categorized into mixed matrix (MM), vertically aligned (VA) and buckypaper (BP) membranes. Evolution in CNT-constructed membrane design has led to excessive improvement in the membrane selectivity, and liquid and gas permeance (Rashed et al. 2021).

3.3.7.1 Synthesis of Carbon Nanotubes/Porous Alumina (PA) Composite Membrane

The preparation of CNT membranes using single-walled CNT arrays is well established by polymer film implanting of CNTs. However, there are several difficulties associated with this approach (Chen et al. 2006; Holt et al. 2004). Aligning great amounts of CNTs and formulating a strong assembly of the membrane in the

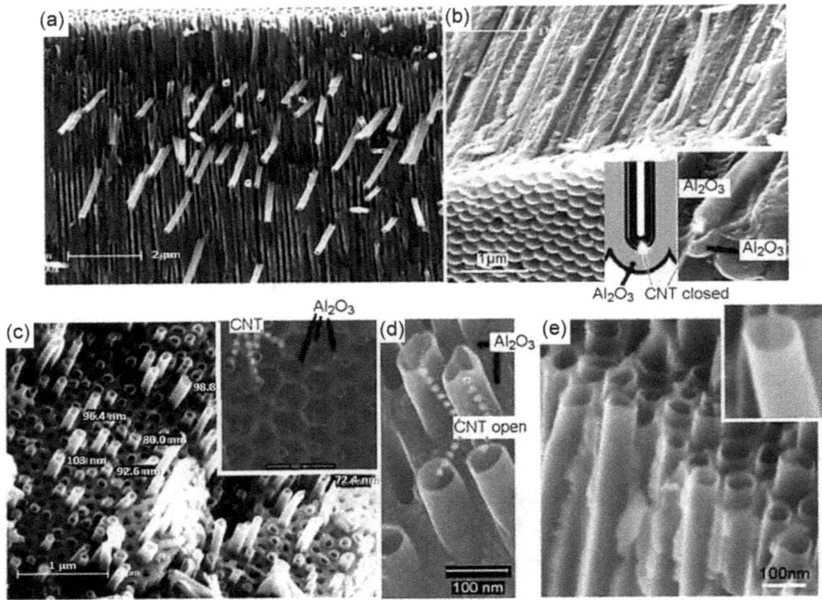

FIGURE 3.6 SEM images of CNTs prepared inside pores of PA membranes. (a) Cross-sectional image of CNTs/PA membrane showing CNTs inside pores; (b) cross-sectional image from the bottom, when PA bottom was closed; (c) CNT structure inside the pores achieved from fractured membranes; and (d, e) high-resolution images of identical shape CNTs inside PA pores. (From Altalhi et al. 2010.)

nanometre-scale dimension is a great challenge. Manufacturing such membranes is expensive, difficult to scale-up and time-consuming. Growing CNTs in pores of porous alumina (PA) membranes is a recent approach directed towards making the fabrication process of CNT membranes more mechanically stronger, less expensive and more feasible.

The growth of MWCNTs inside pores of PA template membranes was performed utilizing a CVD system (Altalhi et al. 2010). Ultrafine carbon precursor particles were inserted into the furnace tube by utilizing a producer using argon gas as a carrier. Catalyst-based ferrocene/toluene and, toluene/ethanol, ethanol, toluene, phenothiazine and catalyst-free pyridine were utilized as precursors of carbon. The optimum temperature was 850°C with varied reaction time from 20 min to 2 h for growing the CNTs. Controllable nanotube dimensions of 30–150 nm diameter and 5–100 μm thickness were confirmed (Figure 3.6).

3.3.7.2 Synthesis of Carbon Nanotubes–Nanoporous Anodic Alumina Membranes (CNTs–NAAMs)

Two approaches have been applied for the fabrication of vertically aligned CNT (VA-CNT) membranes: by utilizing either a polymer base or silicon nitride. The main drawback is the use of costly tools and facilities, small-output preparation methods and high time consumption, i.e. restricted to laboratory scale (Hinds et al. 2004; Seah et al. 2011). Another disadvantage of these synthesis techniques is the challenging control of the geometrical shape of CNTs assembly and the capability to adapt the transport. Template direct synthesis facilitates capability to avoid this restriction by building CNTs in openings of templates that have exactly organized shape and geometry.

Nanoporous anodic alumina membranes (NAAMs) containing a nanoporous hexagonal cell structure have widely been used to fabricate highly ordered CNT nanostructures (Masuda et al. 1995; Md et al. 2013). Kyotani et al. confirmed that nanoporous alumina membranes can be used as a host substrate for the controlled growth of CNTs using the catalyst-free CVD technique (Kyotani et al. 1995, 1996). The fabricated composite membranes (CNTs–NAAMs) exhibit hexagonally arranged stacks of CNTs with controlled length and inner diameter. This approach allows not only precise control of the geometry of the CNTs, but also careful modification of the surface chemistry of the inner CNT wall, leaving the outer surface unchanged. Recently, our group (Altalhi et al. 2013) has established the fabrication of CNTs–NAAMs by catalyst-free CVD using an innovative carbon source from non-degradable plastic food bags and presented their transport performance (discussed in Section 3.7.3 of this chapter).

Alsawat et al. (2015) presented the preparation of composite membranes from CNTs with controlled dimensions and investigated their effects on chemical selectivity and transport properties. These membranes were prepared by building vertically aligned multi-walled carbon nanotubes (VA-MWCNTs) in nanoporous anodic alumina membranes (NAAMs) using a catalyst-free CVD technique. The process diagram of CNTs–NAAMs and their modifications is shown in Figure 3.7.

3.3.7.3 Synthesis of Carbon Nanotube Membranes from Non-Degradable Plastic Bags

An innovative CVD fabrication process was achieved to synthesize well-organized CNTs utilizing nanoporous anodic alumina membrane (NAAM) templates and carbon source of common plastic bags without using solvents or catalysts (Altalhi et al. 2013). The fabrication scheme is presented in Figure 3.8, and the experimental procedures are summarized in Appendix 1. This approach prevents the structural defects of CNTs, reduces the formation of poisonous compounds and introduces prospective recycling procedures for commonly utilized plastic bags. The produced CNTs–NAAMs had a precisely controlled shape with open ends and flat surfaces. These structures are considered very promising for the fabrication of efficient separation/filtration membranes. Successive separation of a two-dye mixture was performed in order to prove this concept. The proposed fabrication process of CNT membranes provides several advantages, such as reprocessing of challenging wasted plastic bags and avoiding the use of expensive and poisonous chemicals, besides conservation of natural ecosystems.

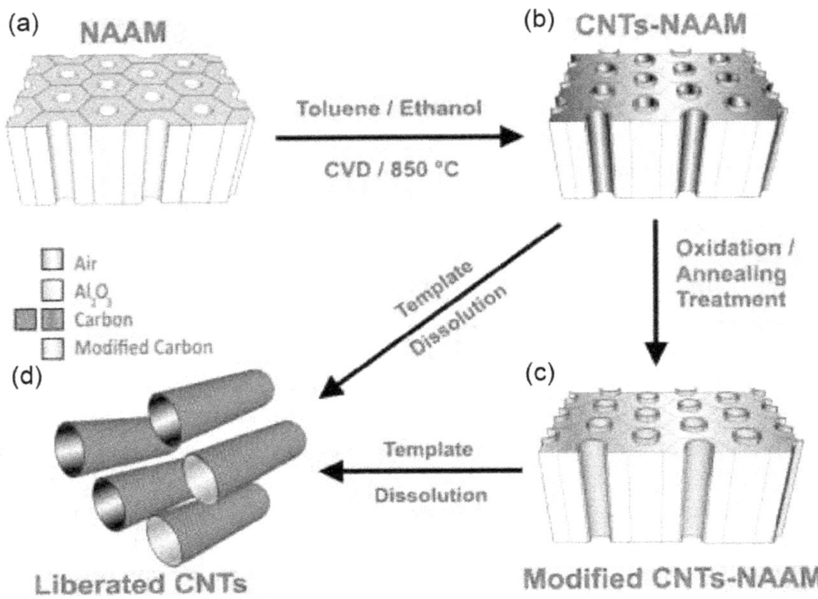

FIGURE 3.7 Fabrication process diagram of CNTs–NAAMs and their modifications. (a) NAAMs prepared by electrochemical anodization of Al foil; (b) prepared CNTs–NAAMs with CNTs inserted in NAAMs after the CVD; (c) CNTs–NAAMs after oxidation or annealing; and (d) liberated CNTs after dissolution of NAAMs by wet chemical etching. (See Alsawat et al. 2015.)

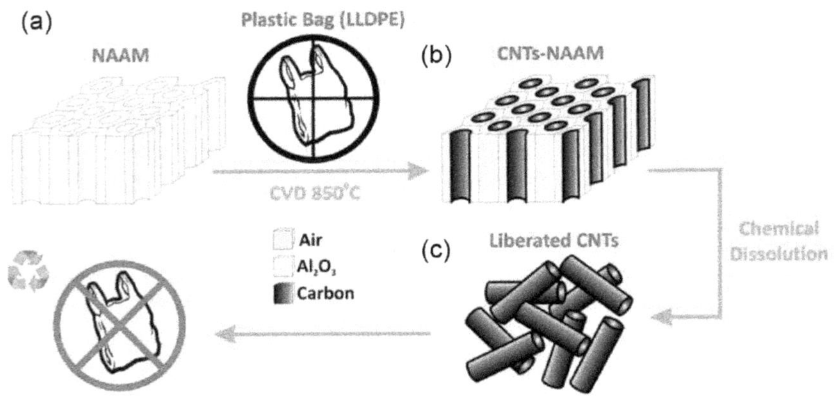

FIGURE 3.8 Schematic diagram for the fabrication of CNTs–NAAMs by catalyst-free CVD process using non-degradable plastic bags as carbon source. (a) As-produced NAAMs prepared by electrochemical anodization of Al chips. (b) Prepared CNTs–NAAMs with CNTs embedded in the alumina matrix after the CVD process. (c) Liberated CNTs by dissolution of alumina by chemical etching.

3.4 CNTs FOR REMOVAL OF HEAVY METALS FROM WASTEWATER

Contamination of water sources with heavy metals poses a significant hazard due to their accumulation in the environment for a longer time compared to other pollutants (Huang et al. 2017). Exposure to heavy metals causes frequent diseases in humans, such as neurological disorders, kidney failure and cancers (Khlifi and Hamza-Chaffai 2010). Several techniques have been employed for the remediation of heavy metals (as demonstrated in Figure 3.9), such as bioremediation, oxidation, reduction, solvent extraction, ion exchange (Shah et al. 2017), leaching (Zhang et al. 2017b), electrochemical treatment, adsorption and membrane filtration (Mudasir et al. 2016). These techniques have several advantages and disadvantages. Adsorption is the most widely accepted technique for treating aqueous heavy metals due to its flexibility, simplicity, low cost, varied pH working values and profitable recognition (Moradi and Zare 2013).

In recent years, the industrial utilization of nanomaterials for green treatment of the environment has extensively been explored (Gupta et al. 2016). Rolled sheets of graphene into hollow tubes form, what we call CNTs, with single- or multi-walled structures (Naghizadeh et al. 2012). Primarily, CNTs are of excessive attention to the researchers and have been applied in the treatment of wastewater because of their unique characteristics such as large pore volumes, high specific surface areas, mechanical flexibility and the ability to attach diverse forms of functional groups (Li et al. 2018; Zhou et al. 2018; Farghali et al. 2013). In addition to these valuable properties, the low density, high porosity

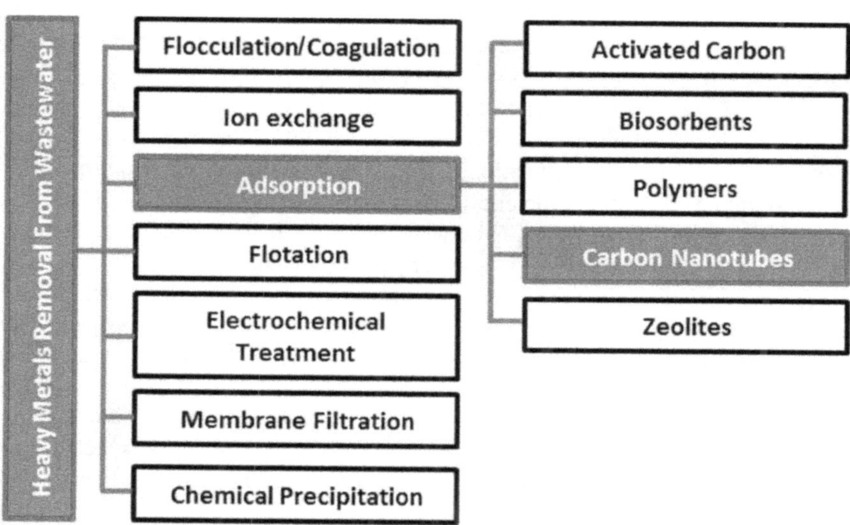

FIGURE 3.9 Employed techniques for the removal of heavy metals from wastewater. (See Verma et al. 2020.)

and hollow assembly make them an appropriate choice for interactions with water contaminants and application in wastewater treatment. In this context, numerous reviews have focused on removing heavy metals from polluted wastewater (Mubarak et al. 2014; Fiyadh et al. 2019; Bassyouni et al. 2020; Verma et al. 2020).

Diverse kinds of adsorbents have been employed, such as alumina (Wang et al. 2018), metal oxide, biomaterials, zeolite bagasse (Nakamoto et al. 2017), activated carbon, sewage sludge ash (Ihsanullah et al. 2016a), graphene oxide, kaolinite (Alagappan et al. 2017), zero-valent iron (Gong et al. 2018a), recycled alum sludge, silica, crab shell, magnetic carbon doped with fluorine and nitrogen, chitosan (Huang et al. 2018), magnetic carbon from activated sludge (Gong et al. 2018b), resins and polymers. Compared with these materials, CNTs' performance as adsorbents was found to be much more powerful.

Surface functionalization of CNTs produces modified surface chemistry, which resulted in better achievements. The adsorption mechanisms of pollutants on CNTs appear to be according to sorption-precipitation, chemical interaction with the surface functional groups and electrostatic attraction (Mubarak 2014b).

The main shortcoming of CNTs is their slight dispersion in the aqueous media, which can be overcome by functionalization using diverse types of functional groups such as SH, NH_2 and OH (Vesali-Naseh et al. 2021; Farghali et al. 2017). These modified surfaces will be rich in functional groups and enhance the hydrophilicity of the modified CNTs through the attached organic and inorganic moieties (Naseh et al. 2010). Therefore, these modified CNTs are more adapted for applications in wastewater remediation and other environmental applications.

The point of zero charge (isoelectric point, IEP) of CNTs is considered as one of the vital parameters during applications in separation science (Herrera-Herrara et al. 2012). The surface of CNT is negatively charged when the pH is higher than the IEP where electrostatic attractions can be recognized as an adsorption mechanism for cationic species. On the contrary, when the pH falls below IEP, protons compete with existed cations for the negative sites on CNTs, providing a decrease in the adsorption efficiency. The pH can be controlled to a specific value for the adsorption of metals to CNTs, which can later be eluted using more acidic solutions, allowing for efficient separation of such metals.

Elimination of major metal contaminants from synthetic or real wastewater aqueous solutions by CNTs and their multi-walled and/or surface-modified structures is surveyed. The as-received or synthesized CNTs are usually characterized by several techniques such as Brunauer, Emmett and Teller (BET) for specific surface area, scanning electron microscope (SEM) and transmission electron microscope (TEM) for size and morphology, X-ray photoelectron spectroscopy (XPS) and Fourier transform infrared spectroscopy (FTIR) for the identification of surface functional groups, vibrating-sample magnetometry (VSM) for magnetic properties, and thermogravimetric analysis (TGA). Some detailed synthesis and preparation techniques are collected in the appendices.

3.4.1 Lead

Industrial wastes and corrosion of plumbing piping are major reasons for the pollution of water with lead. Drinking water polluted with lead causes serious health problems due to accumulation in the brain, kidneys, bones and muscles (Ngueta et al. 2014; Sahraei et al. 2017; Denizli et al. 2000). Contact and exposure to lead can lead to disorders of the brain and nervous system (Bulgariu and Bulgariu 2018). Therefore, the removal of lead from water sources has been a major concern of widespread researches. This section discusses the separation of lead from water by adsorption by CNT-based sorbents.

Sorption of Pb^{2+} from water onto MWCNTs was found to be three to four times greater than other sorbents (Li et al. 2003; Chawla et al. 2015). Chawla et al. (2015) collected valuable data for adsorption of lead ions by pristine and functionalized CNTs and MWCNTs. The collected data summarized preparation methods, surface area (m^2/g), isotherm, Q_{max} mg/g, and the effect of pH, temperature and other metal ions. Surface modification of the adsorption materials included oxidation, loading with metal ions or metal oxides, and coupling with organic compounds to increase the adsorption ability. Functional groups such as carboxyl, lactones and phenols introduce negative charges on the surface of CNTs. The metal adsorption capacity on CNTs is thus improved through enhancing the exchange of protons in the functional groups with metal ions.

The adsorption reaction of Pb^{2+} with acidified MWCNTs was proposed by Wang et al. (2007) as follows:

1. Pb(II) reacts with the functional groups on the surface of MWCNTs to form a complex:
   $MWCNT\text{-}COOH + Pb^{2+} \rightarrow MWCNT\text{-}COOPb^{+} + H^{+}$
   $(MWCNT\text{-}COOH)_2 + Pb^{2+} \rightarrow (MWCNT\text{-}COO)_2Pb + 2H^{+}$
   $MWCNT\text{-}OH + Pb^{2+} \rightarrow MWCNT\text{-}OPb^{+} + H^{+}$
   $(MWCNT\text{-}OH)_2 + Pb^{2+} \rightarrow (MWCNT\text{-}O)_2Pb + 2H^{+}$
2. Pb(II) reacts with carbon dioxide and water to form inorganic deposition adsorbed on the surface of MWCNTs:
   $Pb^{2+} + CO_2 + H_2O \rightarrow PbCO_3 + 2H^{+}$
   $Pb^{2+} + H_2O \rightarrow Pb(OH)_2 + 2H+,$
   $Pb(OH)_2 \rightarrow PbO + H_2O.$

A schematic diagram of the possible interactions of Pb^{2+} ions with oxygen- and nitrogen-donor groups is presented in Figure 3.10 (Vesali-Naseh et al. 2021).

Different Pb^{2+} sorption mechanisms proposed include electrostatic attraction, van der Waals and ion–π interactions, surface precipitation, ion exchange and complexation reaction. The adsorption is a multi-step process that consists of liquid film diffusion, intra-particle diffusion and chemisorption/physisorption on the interior sites of the adsorbent (Yang et al. 2013). The uptake of Pb^{2+} on the CNT surface is mostly fitted with the Langmuir model representing a homogenous and

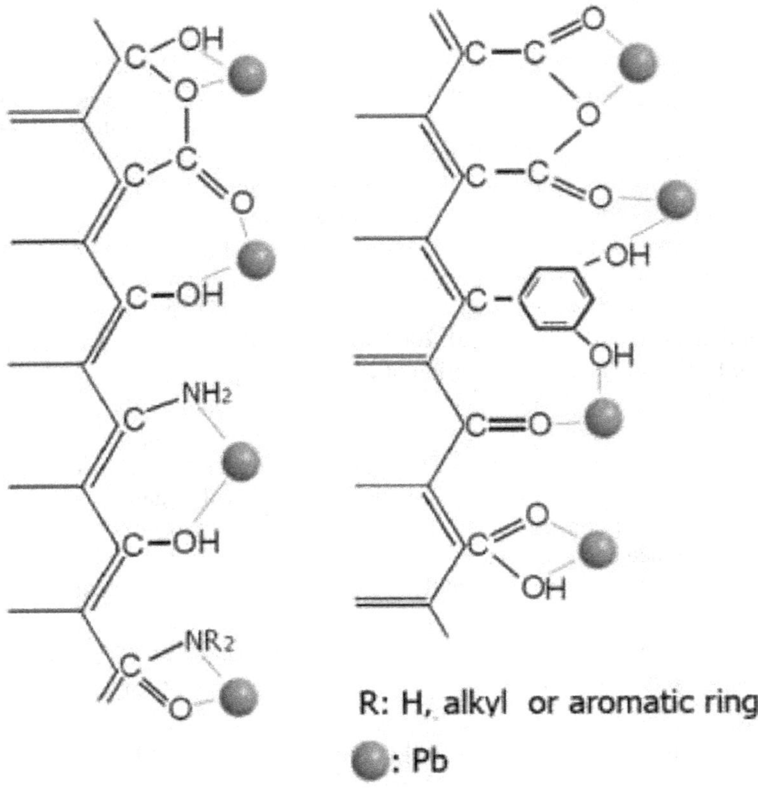

FIGURE 3.10 Proposed interactions of Pb^{2+} with surface functional groups. (See Vesali-Naseh et al. 2021.)

monolayer process (Li et al. 2002; Vesali-Naseh et al. 2021). The spontaneous and endothermic adsorption process can be considered with positive entropy. Pseudo-second-order model is the best-fit model, and the metal adsorption is mainly controlled by the second-order chemisorption.

The mechanism of Pb^{2+} adsorption on the acidified MWCNTs was classified into contribution of the specific surface area (24.7%) and contribution of the functional groups (75.3%) (Figure 3.11) (Wang et al. 2007). The adsorption with the contribution of the specific surface area is divided into the Pb in the deposition form of PbO, $Pb(OH)_2$ and $PbCO_3$ (3.4%) and in the salt form of $Pb(NO_3)_2$ (21.3%). Because the oxygenous functional groups formed on acidified MWCNTs can ionize to make the acidified MWCNTs negative, the adsorption of Pb^{2+} to acidified MCWNTs can happen through electrostatic adsorption. The adsorption with the contribution of functional groups can also be further divided into electrostatic adsorption and the other in which the Pb^{2+} reacts with the functional groups to form different types of complexes.

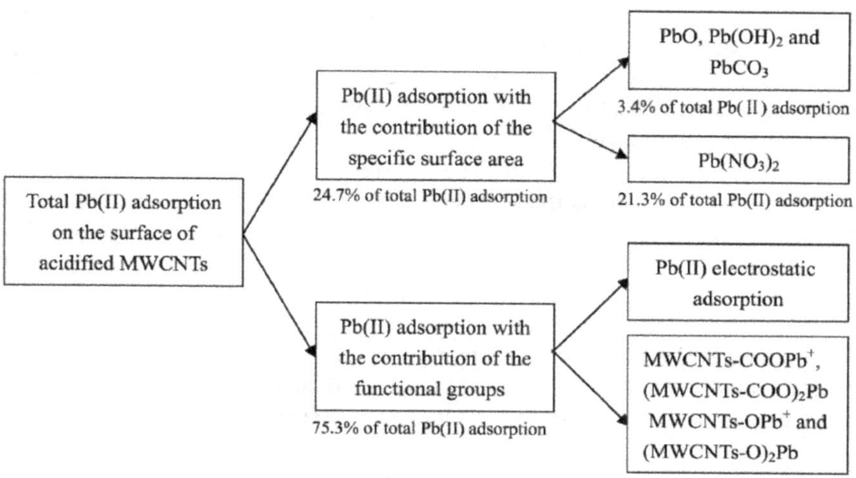

FIGURE 3.11 Classification of Pb(II) adsorbed on the surface of acidified MWCNTs. (Reproduced with permission, Wang et al. 2007.)

Carbon nanotube/metal oxide (CNT/MO) composites combine the properties of CNTs and metal oxide nanoparticles and show unique properties as a result of interaction between them (Chu et al. 2010). The latter may be synthesized using pressureless sintering, in situ CVD, ultrasonication or sol–gel method with spark plasma sintering. The adsorption capacity of CNT-based sorbents under a given set of conditions (pH 5–7) follows the following order: pristine MWCNTs < oxidized MWCNTs < polyethylene glycol (PEG)-MWCNTs < MWCNTs-Al_2O_3 < MWCNTs-tris(2-aminoethyl)amine (TAA) < CNTs/MnO_2 < MWCNTs-ethylenediamine-grafted (EDA-II) < acidified MWCNTs < MWCNTs-EDA-I. This order indicates that oxidation and modification on the surface of CNTs make them more efficient adsorbents for lead ions from water in the pH range of 5–7. Adsorption of lead by CNT was found to mainly depend on the types of nanotubes and their method of fabrication. Several cases of lead adsorption on CNTs and their derivatives are discussed hereunder.

CNTs fabricated by catalytic pyrolysis of hydrogen–propylene mixture showed that the adsorption is influenced significantly by the value of pH and the status of CNT surface. The Pb^{2+} adsorption on the CNTs depends on many factors such as the surface functional groups, the specific surface area and the solution components. The most important factor is the surface functional groups, which can be generated with oxidized acids. Oxidation of CNTs by nitric acid improves the cation exchange capacity and reduces the anion exchange capacity. Acidic functional groups attached to the CNT surfaces can dissociate at different pH values. The lead adsorption increased with increasing pH values in the range of 3–7 and reached a maximum of 49.95 mg/g at pH 7. A higher pH leads to metal precipitation, and adsorption is preferred under acidic conditions (Li et al. 2002). A remarkable increase in adsorption capacity was observed with nitric acid-treated CNTs. The modified surface of CNTs can make it more negatively charged, hydrophilic and form various functional groups which promote adsorption. In another study, a removal percentage of lead of nearly 100% is achieved using modified CNTs at pH 7, while the percentage removal at pH 4 and 5 was 17% and 25%, respectively (Atieh et al. 2010a).

TM–CVD technique was utilized for the synthesis of MWCNTs, using acetylene and hydrogen as the precursor gases and ferrocene as a catalyst. The MWCNTs were tested for the removal of Pb^{2+} from an aqueous solution. The highest removal (99.9%) of Pb(II) is at pH 5, MWCNT dosage 0.1 g, agitation speed 160 r/min and time of 22.5 min with the initial concentration of 10 mg/L (Mubarak et al. 2016).

Modifying MWCNTs by oxidation using plasma enhanced the adsorption capacity of lead (Yu et al. 2011). This resulted in introducing more oxygen-containing functional groups (e.g. COOH and OH) on the surface of the nanotubes, increasing the specific surface area and the number of surface defects, removing some amorphous carbon and MWCNTs caps and improving the dispersion of CNTs in water. Xu et al. also confirmed these results by applying XPS spectra for freeze-dried samples of Pb^{2+} adsorbed on oxidized MWCNTs at pH 6.05 and 8.86 (Figure 3.12) (Xu et al. 2008).

There was no difference in C1s spectra of the two samples, which appeared at 284.7 eV. This was explained to be due to the fact that the species of carbon is not influenced by pH values. Figure 3.12b shows the high-resolution XPS O1s spectra of the sample around 532.5 eV. It is apparent that there is a difference between the oxygen peaks at pH 6.05 and pH 8.86. This difference is due to the presence of oxygen-containing functional groups on the surface of oxidized MWCNTs (e.g. carboxyl oxygen [–O–C=O(H), 533.6 eV] and carbonyl oxygen [=C=O, 530.7 eV]), which enhance the adsorption of lead ions. The oxygen peak is shifted by 0.55 eV between pH 8.86 and 6.05 as $Pb(OH)_2$ pellets begin to form and cover the adsorbent surface at pH 8.86.

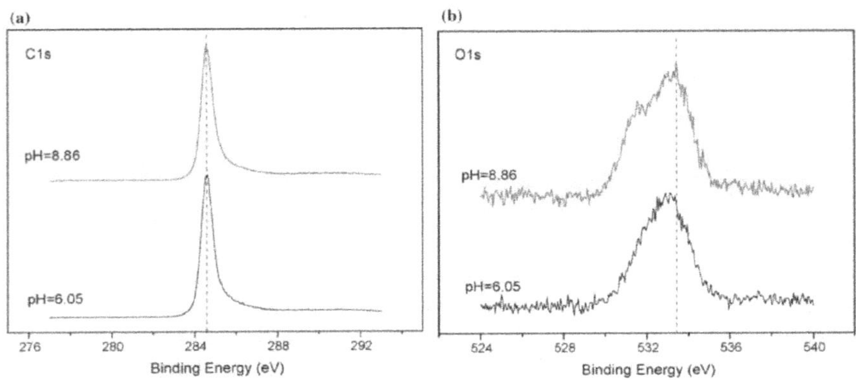

FIGURE 3.12 XPS spectra for freeze-dried samples of Pb(II) adsorbed on MWCNTs. (a) C 1s; (b) O 1s. (Reproduced with permission, Xu et al. 2008.)

The MWCNTs/MnO_2 nanocomposite was used for the adsorption of lead ions from model and wastewater solutions (Salam 2013). The adsorption reached its maximum within 5 min and was enhanced greatly by raising the solution temperature and by increasing the pH to 9.0. The adsorption was found to be chemical, spontaneous and endothermic in nature and was mainly controlled by a liquid film diffusion mechanism. The removal of lead ions from a spiked wastewater sample with MWCNTs/MnO_2 nanocomposite reached 97.3% at the optimum conditions.

It was reported, in another work, that the adsorption capacity of MnO_2-coated CNTs was much higher than that of CNTs (Wang et al. 2007). The adsorption capacity of lead increased with increasing MnO_2 load due to the presence of more binding sites. The optimum reported load for best performance is 30 wt% MnO_2.

MWCNTs of three different diameters were produced by the catalytic CVD method, using ethanol as the carbon feedstock and thiophene as the growth promoter at 1000°C–1150°C by adjusting the concentration of ferrocene catalyst (Figure 3.13) (Yu et al. 2013). The samples consisted of a fibre-like substance, with the fibres tangled with each other to form a network with diameters <10, 15–20 and >20 nm with specific surface areas of 471, 140 and 123 m^2/g, and oxygen contents (%O) of 3.2, 4.7 and 5.9, respectively. The adsorption of lead increased remarkably with decreasing outer diameter due to larger specific surface area and increase in oxygen content.

Magnetic hydroxyapatite-immobilized oxidized multi-walled carbon nanotubes (mHAP-oMWCNTs) were fabricated (see Appendix 2) and applied for lead removal from water (Wang et al. 2017). The adsorption might be attributed to the ion exchange property of the hydroxyapatite and the oxygenic functional groups on oMWCNT surface. Besides, their excellent magnetic property significantly simplified their separation from the treated water (Figure 3.14). The maximum adsorption capacity was 698.4 mg/g for Pb(II).

Biomass materials, which are low-cost by-products from agricultural or industrial processes, are renewable, environmentally friendly resources and show efficient adsorption capacity for the separation of heavy metals from aqueous solutions (Tan et al. 2012). Sugarcane bagasse, a green biosorbent, is inexpensive and a widely available agricultural waste. It is composed of 37% cellulose, 28%

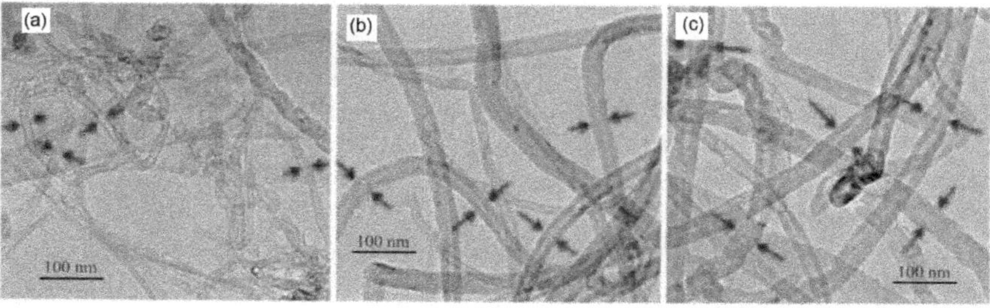

FIGURE 3.13 TEM images of multi-walled carbon nanotubes with different outer diameters: <10 nm (a), 15–20 nm (b) and 20 nm (c). (See Yu et al. 2013.)

FIGURE 3.14 Magnetic hydroxyapatite-immobilized oxidized multi-walled carbon nanotubes and their magnetic separation. (Reproduced with permission, Wang et al. 2017.)

hemicellulose and 21% lignin, and it has hydroxyl, carboxyl and phenol functional groups (Walford 2008). To enhance the efficiency of adsorption of heavy metals from aqueous solutions, a composite containing sugarcane bagasse and MWCNTs was designed. Hamza et al. (2013) examined the Pb(II) adsorption from an aqueous solution on a sugarcane bagasse/MWCNT composite. The Pb(II) adsorption capacity was found to be 56.6 mg/g at 28°C and pH 4.5 for the bagasse/MWCNT composite compared with 23.8 mg/g for bagasse. The adsorption data were best fit with the Langmuir isotherm and Elovich kinetic model. A 0.1 mol/dm^3 HCl was sufficient for efficient desorption of the lead from the loaded sorbents.

Oxidized MWCNTs have been shown to be fast and efficient for the adsorption of Pb^{2+} from an aqueous solution (Xu et al. 2008). Oxygen-containing functional groups can be attached to the surface of the CNTs through oxidation, increasing the potential for complexation with metal ions and also making the CNTs more hydrophilic and more dispersed in aqueous solutions.

3.4.2 Arsenic

The inorganic form of arsenic is extremely toxic and has been documented as the most poisonous heavy metal. Arsenic is commonly found in soils and waters and has become a huge hazard to public health (Luo et al. 2017). About 10 million people have been affected by severe or prolonged arsenic poisoning (Lee et al. 2021). Continuous exposure to arsenic causes cancer and skin pigmentation, and lesions. Arsenic has also been associated with cardiovascular diseases and diabetes (WHO). Arsenic can be found in a range of different forms and levels of toxicity. Water resources can be contaminated by arsenic either through human activity or naturally. The maximum allowable level of arsenic in drinking water is 10 mg/L, according to the World Health Organization (WHO) (Tawabini et al. 2011).

Exposure to arsenic has been linked with several dangerous and fatal diseases such as cancers of the urinary tract, bladder and skin (Sharma and Sohn 2009). Consequently, several studies have recently been devoted to finding new solutions for the removal of As from wastewater and drinking water, including oxidation/precipitation, coagulation/electro-coagulation/co-precipitation, membrane, foam flotation, solvent extraction, ion exchange, bioremediation and adsorption technologies (Lee et al. 2009; Yan et al. 2010; Mou et al. 2011; Khin et al. 2012; Mohan et al. 2007; Addo Ntim et al. 2011). Most of these techniques are well established and have their merits and inherent limitations such as the generation of toxic waste, high-tech operation and maintenance, low arsenic removal efficiencies and/or high cost (Mohan et al. 2007).

Both organic and inorganic forms of arsenic exist in the environment, while the latter is of higher toxicity and frequently found in groundwater (Styblo et al. 2000; Mandal and Suzuki 2002). Inorganic arsenic occurs in two valences: arsenite (As(III)) and arsenate (As(V)). Natural groundwaters contain predominantly As(III) since reducing conditions prevail. In natural surface waters, As(V) is the dominant species (Naghizadeh et al. 2012). In the typical range of groundwater pH (6.0–8.5), As(III) mainly exists as the neutral molecule (H_3AsO_3), while As(V) exists as anions ($H_2AsO_4^-$ and $HAsO_4^{2-}$) (Luo et al. 2017).

SWCNTs and MWCNTs were synthesized using the known CVD method with methane as the carbon source, hydrogen as the carrier gas and Co–Mo/MgO as the catalyst. As shown in Figure 3.15, the peaks at 25° and 43° are related to

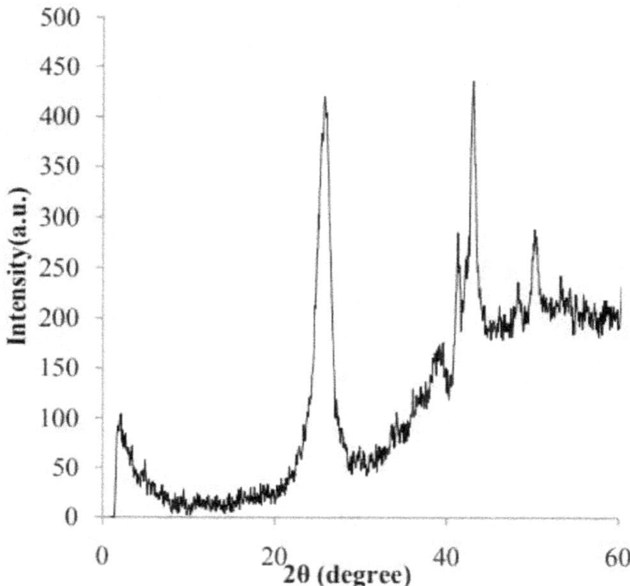

FIGURE 3.15 X-ray diffraction spectra of CNTs. (Reproduced with permission, Naghizadeh et al. 2012.)

the graphene structure of CNTs. The adsorption capacity of As by SWCNTs and MWCNTs was 148 and 95 mg/g, respectively, and the adsorption data were fitted with Freundlich and Langmuir isotherms (Naghizadeh et al. 2012).

Several iron oxide-based adsorbents (such as amorphous hydrous ferric oxide (FeO-OH), goethite (α-FeO-OH), hematite (α-Fe_2O_3) and magnetite (Fe_3O_4)) are very effective in the removal of arsenic from water due to the high affinity of iron towards arsenic (Dixit and Hering 2003; Giles et al. 2011). The potential for side-wall functionalization of MWCNTs and their surface modification make them attractive as support phases for water treatment.

SEM images showed that the MWCNTs had diameter and length in the range of 20–40 nm and 10–30 μm, respectively, with no detectable change in morphology after acid treatment (Figure 3.16a) or iron oxide coating (Figure 3.16b). The EDS data confirmed the presence of iron oxide on the surface of the CNTs (Figure 3.3c and d).

The functionalization of MWCNT and its linkage with iron oxide was confirmed by FTIR spectra. The carboxylic stretching frequency in f-MCWNT appeared at 1715 cm^{-1} (C=O), 1218 cm^{-1} (C–O) and 3424 cm^{-1} (O–H), which is clearly absent in the MWCNT and Fe-MWCNT spectra. The disappearance of the C–O and O–H peaks in the Fe-MWCNT spectra is evidence that the iron

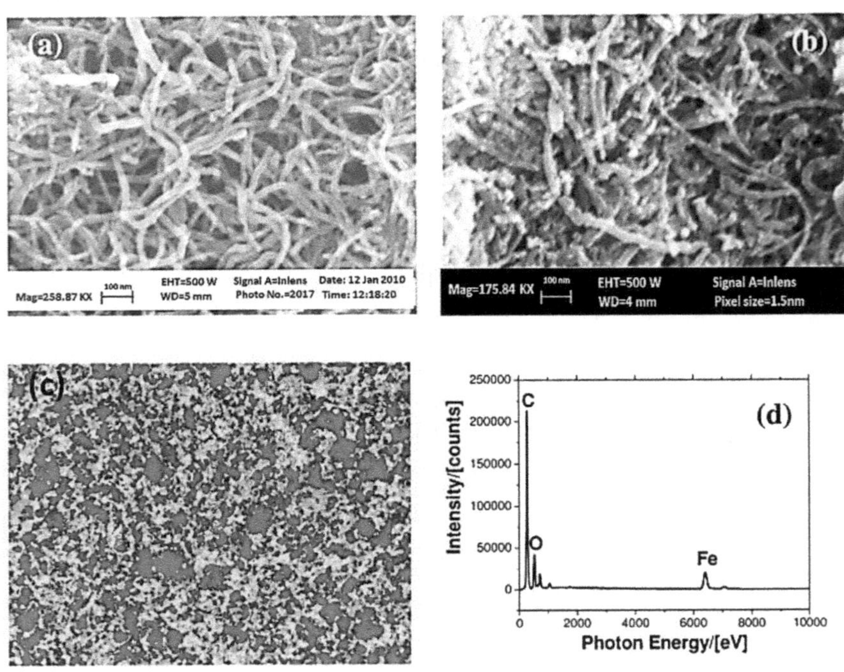

FIGURE 3.16 SEM image of (a) acid-functionalized MWCNTs, (b) the Fe-MWCNT composite, (c) EDS map (red – carbon, green – oxygen and yellow – iron), (d) EDS spectra (carbon (C), oxygen (O) and iron (Fe)). (See Addo Ntim and Mitra 2011.)

oxide particles are anchored to the MWCNTs surface by an ester-like bond. The peak around 1576 cm^{-1} was assigned to the C=C stretching of the carbon skeleton, and the two bands at 636 and 565 cm^{-1} confirmed the formation of the Fe–O bonds.

Fe-MWCNT was fabricated (see Appendix 3) and applied as a sorbent for the removal of trace level arsenic from water (Addo Ntim and Mitra 2011). The Fe-MWCNT was effective in arsenic removal to below the drinking water standard level of 10 μg/L. The adsorption capacity of the composite was 1723 and 189 μg/g for As(III) and As(V), respectively. The adsorption of As(V) was faster than that of As(III). The pseudo-second-order rate equation was found to effectively describe the kinetics of arsenic adsorption. The adsorption isotherms for As(III) and As(V) fitted both the Langmuir and Freundlich models. Adsorption of As(V) was more favourable at lower pH (>7), and that of As(III) was more favourable at higher pH (<7). The difference in sorption characteristics may be due to their ionic forms: As(V) existed in the anionic form ($H_2AsO_4^-$), whereas As(III) existed in the molecular form (H_3AsO_3). Negatively charged arsenic species may get adsorbed on protonated iron oxide sites on the adsorbent surface by covalent ligand exchange, resulting in their removal from water.

In another work, MWCNTs were functionalized and coated with zirconia to form hybrid MWCNTs-ZrO_2 and applied for the removal of arsenic from water (Addo Ntim and Mitra 2012). The adsorption capacities of the composite were 2000 and 5000 μg/g for As(III) and As(V), respectively, which were significantly higher than those reported previously for iron oxide-coated MWCNTs. Moreover, the adsorption of As(V) on MWCNT-ZrO_2 was faster than that of As(III) and the adsorption capacity was not a function of pH.

Nanosized Fe_3O_4 was suggested as one of the promising candidates for its highly specific surface area and selective removal of As. The magnetic response feature of Fe_3O_4 makes it possible for adsorbent separation by simply applying an external magnetic field after adsorption (Yavuz et al. 2006). A 2D Fe_3O_4 ultrathin nanosheet was synthesized applying the solvothermal method (Sun et al. 2014) on a surfactant solution of iron salt and applied for arsenic removal from water (Luo et al. 2017). The optimal pH value was 8.27, with valued adsorption capacity and removal efficiency being 18.55 mg/g and 93.69%, respectively. However, the weak self-supporting nature of the nanosized Fe_3O_4 could be unsuitable for the real purification in the continuously flowing water. The attachment of the prepared Fe_3O_4 nanoparticles (NPs) to host structures enables the full recovery from the flowing water with high performance. Agglomeration-free geometry of Fe_3O_4 NPs anchoring on a host structure would impact the enhancement of adsorption capability. The Fe_3O_4 NPs attached to N-doped CNTs (NCNTs) were fabricated (Figure 3.17) and applied for arsenic adsorption. The removal of As in the form of As_2O_3 or As_2O_5 by the NPs-NCNTs is proven to be highly selective and ~6 times higher than that by pristine magnetite NPs, in addition to providing a great potential for the removal of contaminated As even from the continuously flowing water.

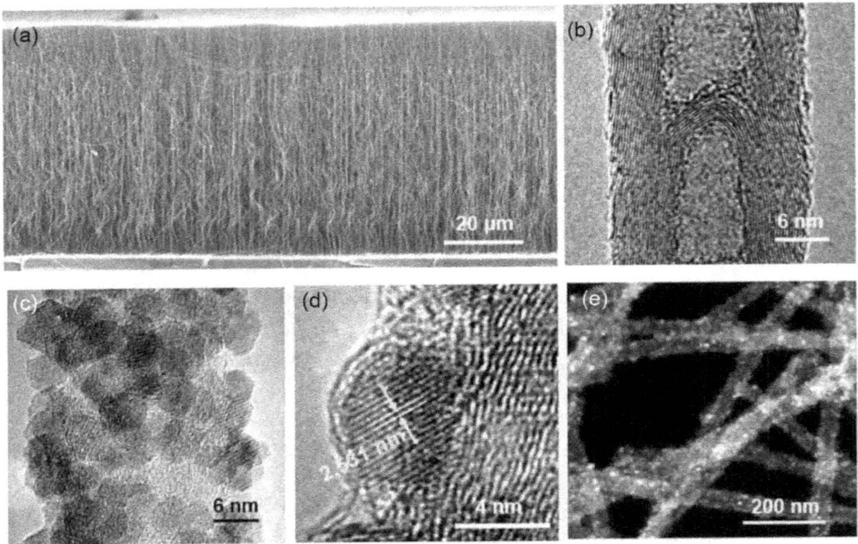

FIGURE 3.17 (a) SEM micrograph of the as-grown NCNTs. (b) HRTEM micrograph of NCNTs. (c, d) High-resolution transition electron microscopy (HRTEM). (e) High-angle annular dark field (HAADF) images of the Fe_3O_4 NP-NCNT hybrid. (See Lee et al. 2021.)

3.4.3 Mercury

Mercury (Hg) is one of the most toxic heavy metals found in nature. It can occur either in vapour form or in liquid form. The renal organs, gastrointestinal (GI) tract and neurologic systems are the most affected by exposure to mercury. Mercury (Hg) can be found in three forms: inorganic salts, organic salts and a metallic element. These elements exist in the soil, fresh water and seawater. Mercury can also be found in some industrial waste processes and products, such as the production of wiring devices, various switches, fossil fuels, dental work, lighting, and control and measuring devices (Gupta et al. 2014). The maximum allowable mercury concentration in water laid down by the WHO is 1 mg/L, due to mercury's serious effects even at deficient concentrations (Mohan et al. 2001).

The removal of mercury from water was performed by operating a number of technologies such as ion exchange (Chiarle et al. 2000), chemical precipitation (Blue et al. 2010), coagulation (Henneberry et al. 2011), solvent extraction (Reddy and Francis 2001) and adsorption (Chen et al. 2019). Adsorption is a common technique due to its simplicity, high efficiency and flexibility of materials design (Shetty et al. 2019; Deng et al. 2015; Bower et al. 2008; Jainae et al. 2015).

MWCNTs were found to be efficient in adsorption of as high as 1.0 mg/L of Hg^{2+} from aqueous solutions (Tawabini et al. 2010; Yaghmaeian et al. 2015). The uptake of Hg^{2+} increased to 100% with an increase in pH from 4 to 8. The adsorption followed a pseudo-second-order reaction and was well described by the Langmuir isotherm model with a maximum adsorption capacity of 13.16 mg/g.

Amino- and thiol-functionalized MWCNTs were fabricated (see Appendix 4) and applied for the separation of mercury ions from synthetic and real chloralkali wastewater (Hadavifar et al. 2014). The synthesized adsorption material was characterized by FTIR, XPS, TGA and SEM analyses. Covalent linking of the functional groups on the MWCNTs has been confirmed. The maximum adsorption capacity at the batch system was 84.66 mg/g, at pH 6 and adsorbent dose of 400 mg/L, by fitting with Langmuir isotherm model. The physisorption mechanism was confirmed by the negative values of ΔG that increased with increasing temperature. However, the reaction mechanism of the -SH functional group and mercury ions species is proposed by the following equations (Vroman et al. 2000):

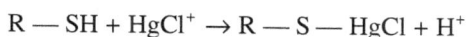

$$2R - SH + Hg^{2+} \rightarrow R - S - Hg - S - R + 2H^{+}$$

$$R - SH + HgCl^{+} \rightarrow R - S - HgCl + H^{+}$$

These reactions indicate that the adsorption is favourable with increasing pH of the solution. The overall adsorption can be described better by the pseudo-second-order than by the pseudo-first-order kinetic model. A loss in the adsorption capacity of 7.2% was detected after five cycles of adsorption–desorption tests. In the continuous mode, the adsorption capacity reached 105.65 mg/g. The MWCNTs-SH was a good adsorbent for the removal of Hg ions (88.7%) from real chloralkali wastewater.

The XPS was used to confirm the thiolation of the MWCNTs because the S–S or C=S bonds were hardly detectable in the FTIR spectra. The XPS data show the peaks corresponding to C, O and N 1s, and S 2p, confirming the successful synthesis of the MWCNTs-SH and MWCNTs-ethylenediamine (EDA) (Figure 3.18).

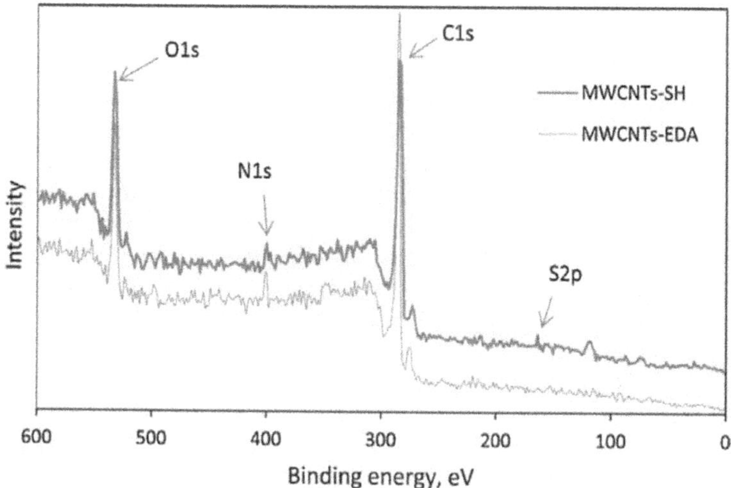

FIGURE 3.18 XPS spectra of wide scan of the MWCNTs-EDA and MWCNTs-SH. (Reproduced with permission, Hadavifar et al. (2014).)

Sulphur-containing materials exhibit soft Lewis acid-based interaction and hence excellent Hg removal efficiency (Huang and Shuai 2019). Although MWCNTs are effective adsorbents of Hg(II) (Fiyadh et al. 2019), their difficult separation limited their practical applications in water remediation. The removal of metal pollutants with magnetic adsorbents is of great interest due to their facilitated solid–liquid separation from aqueous matrix using an external magnetic field (Mehta et al. 2015). Magnetic multi-walled carbon nanotubes (M-MWCNTs prepared by a co-precipitation method) coated with sulphur (S-M-MWCNTs) were synthesized via coating a thin S layer on M-MWCNTs via a simple heating process (Appendix 5). The fabricated superparamagnetic adsorbent was applied for the removal of Hg(II) from aqueous solutions (Fayazi 2020). The adsorbent material was separated magnetically. The adsorption of Hg(II) increased with increasing pH and remained constant in the pH range of 4.5–8.0. The adsorption kinetics followed the pseudo-second-order model, and the equilibrium was attained within 3 h. The maximum adsorption capacity was 62.11 mg/g, and the isotherm data obeyed the Langmuir model. The adsorption of Hg(II) is controlled by the interaction between Hg(II) as a soft acid and the elemental sulphur coated on MWCNTs as a soft base. The coexistence of several metals (such as Cu(II), Cd(II), Co(II), Pb(II), Mn(II), Zn(II) and Cr(III)) had no notable effects on the removal of Hg(II). The S-M-MWCNT composite could be reused after successive Hg(II) removal without any loss of adsorption capacity.

Magnetic properties of the M-MWCNT and S-M-MWCNT composites were monitored with VSM. It was found that the synthesized M-MWCNTs have a high value of saturation magnetization (Ms) of 13.7 emu/g, and this value decreased to 8.2 emu/g in the S-M-MWCNTs. The drop in the Ms value in S-M-MWCNTs is due to the coating of sulphur layers on some magnetic sites of the M-MWCNTs.

3.4.4 Chromium

Chromium is one of the highly toxic elements present in groundwater and surface water due to its widespread use in diverse industrial applications such as leather tanning, metallurgy, refractory, dyes and pigments, and chrome electroplating (Aigbe and Osibote 2020). Chromium(III) occurs naturally in the environment and does not cause any health danger at common concentrations found in water sources, while chromium(VI) is a known carcinogen (Vincent et al. 2000; Dimos et al. 2012). The fatal inhalations of Cr(VI) lead to respiratory disorders, liver damage, internal haemorrhage, skin ulceration, dermatitis and chromosome abnormalities (Di et al. 2004). The accepted concentration level of chromium in drinking water is less than 0.05 mg/L (Anastopoulos et al. 2017).

Numerous research works have been conducted to remove chromium from wastewaters showing some drawbacks such as low efficiency and toxicity of the utilized materials (Anagnostopoulos et al. 2010; Han et al. 2000). The efficient removal of Cr(VI) by different cost-effective adsorbents such as treated clays, silica zeolites and lignocellulosic materials and their chemically modified forms with chemical ligating groups has been reported (Rosales-Landeros et al. 2013;

Miretzky and Cirelli 2010). Diverse types of nanomaterials have shown efficient adsorption of Cr(VI) from wastewater owing to their unique characteristics such as higher capacity, higher specificity, higher affinity and great reactivity, making them superior to adsorption with traditional materials. The sorption capacity of various nanomaterials for Cr(VI) applying the Langmuir model ranged from 3.197 to 666.67 mg/g for carbon nanomaterials, 8.67–1052.63 mg/g for nanomaterials modified with metals and metals oxides and 59.17–854.7 mg/g for polymers composite. The sorption of Cr(VI) onto various nanomaterials or functionalized nanomaterials was found to depend on the pH, quantity of sorbent, contact time and temperature (Aigbe and Osibote 2020). Nanosorbents such as magnetic nano-iron oxide have been studied for the successful removal of chromium from water (Ilankoon 2014; Gupta et al. 2015; Aigbe and Osibote 2020).

More specifically, maximum adsorption of Cr(III) and Cr(VI) was found to occur at pH ranges 5–8 and 1–4, respectively. Langmuir isotherm model and pseudo-second-order model were found to best fit the experimental adsorption data.

The redox potential (Eh) and pH determine the chemical and physical states of Cr species in the aqueous layer surrounding CNTs during the adsorption process. The solution Eh determines the distribution of Cr between Cr^{3+} and Cr^{6+}, while the pH defines the distribution of Cr(III) as Cr^{3+} and $Cr(OH)_3$ (Figures 3.19 and 3.20) (Aigbe and Osibote 2020).

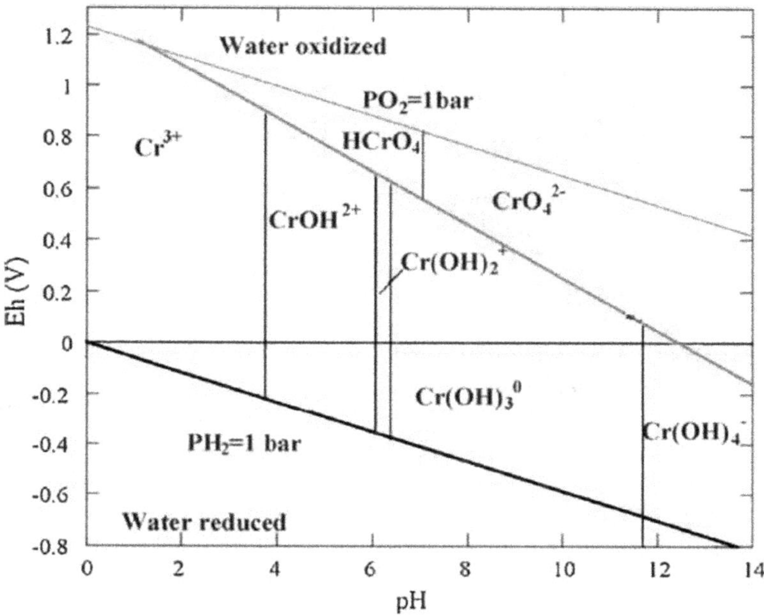

FIGURE 3.19 Eh–pH diagram of chromium. (Reproduced with permission, Aigbe and Osibote 2020.)

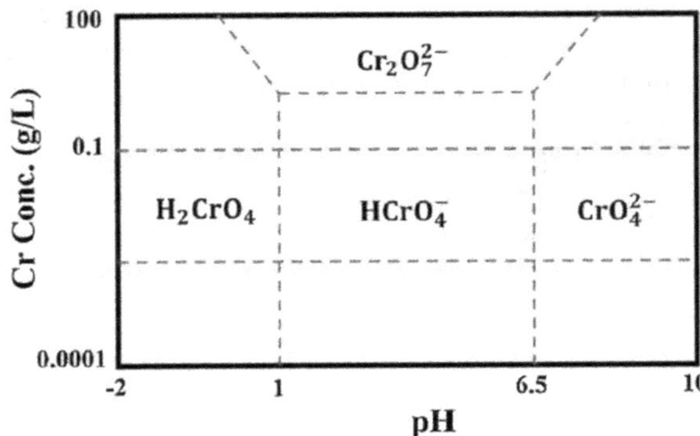

FIGURE 3.20 Speciation of chromium at different values of pH. (Reproduced with permission, Aigbe and Osibote 2020.)

In an acid medium, Cr(VI) ions sorption is favourable when the sorbent surface contains elevated numbers of hydronium ions (H^+), leading to the electrostatic attraction between charged negative chromium oxyanions and protonated sorbent.

Therefore, the sorption capacity of chromium on CNTs is greatly affected by the pH value due to the diverse types of chromium species and also to the state of protonation and deprotonation of the surface. Nearly all studies showed that the perfect pH for Cr(VI) sorption to various nanomaterials is within pH 2–3, where the $HCrO_4^-$ is the predominant species. At acidic pH 1, more protons encourage the reduction of Cr(VI) to Cr(III). When Cr(VI) pH is between 2.0 and 6.5, anionic chromium species ($HCrO_4^-$) is predominant. In acidic solutions, the sorbent will be protonated and adsorption of Cr oxoanions takes place via electrostatic attraction. In alkaline solutions, chromate anion (CrO_4^{2-}) predominates, the sorbent is deprotonated, and thus, the electrostatic adsorption declined (Aigbe and Osibote 2020).

Anastopoulos et al. (2017) concluded in their review that the monolayer adsorption capacities of Cr(III) and Cr(VI) ions on CNTs were in the range of 0.39–238.09 and 1.26–370.3 mg/g, and the values of ΔG^0 and ΔH^0 were in the range of 0.237–48.62 and 0.16–58.43 kJ/mol, respectively.

A comparative study of the adsorption of Cr(III) with raw CNTs (R-CNTs) and CNTs modified with -COOH (M-CNTs) from aqueous media showed that M-CNTs had faster and higher adsorption abilities (Qm=0.3718 and 0.5 mg/g for R-CNTs and M-CNTs, respectively) (Atieh et al. 2010b). The adsorption increased with pH increasing from 4 to 7. An increase in the agitation rate enhanced the removal capacity due to better diffusion.

The separation of Cr(III) by MnO_2/MWCNTs was tested (Mohammadkhani et al. 2015). The adsorption was found to be spontaneous and exothermic, with maximum adsorption reached at pH 5 and 0.005 g adsorbent dose at 35°C. The fabrication procedure is shown in Appendix 6.

FIGURE 3.21 FESEM images of CNTs-AC. (See Atieh 2011.)

Microwave-assisted fabrication of CNTs was performed, and the prepared material was studied for the removal of Cr(III) (Mubarak et al. 2016). The highest removal of Cr(III) was observed at pH 8, a dosage of 0.09 g, 60 min and 150 rpm.

The removal of Cr(VI) from wastewater utilizing CNTs and CNTs supported by activated carbon (CNTs-AC) showed acceptable adsorption capacities reaching 9.0 mg/g with CNTs-AC (Nassereldeen et al. 2009; Muataz et al. 2009; Atieh 2011). Figure 3.21 shows the field emission SEM (FESEM) images of CNTs-AC.

3.5 CNTs FOR REMOVAL OF ORGANIC POLLUTANTS FROM WASTEWATER

3.5.1 Removal of Monoaromatic Pollutants

Groundwater can be polluted by monoorganic molecules such as benzene (B) and toluene (T) due to their leaks from numerous industrial productions of chemicals and improper discharge of solvents from organic synthesis into underground storage tanks (Shim et al. 2002; Wibowo et al. 2007). The existence of these pollutants in water seriously harms human health (Aivalioti et al. 2010). The limit standard of benzene in drinking water is set to 5 μg/L by the US EPA (Mathur et al. 2007).

CNTs are considered more efficient materials for the elimination of several kinds of organic materials compared with other common adsorbents (Lu et al. 2007). The adsorption performance of pristine and NaOCl-oxidized MWCNTs towards the removal of B and T is superior in comparison with many other reported types of silica and carbon adsorbents (Lu et al. 2008; Su et al. 2010). The existence of a diverse range of functional groups causes large capacity of organic pollutants adsorption on CNTs. The mechanism of B and T adsorption on CNTs is ascribed to be due to the π–π donor–acceptor electron coupling between surface carboxylic groups of CNTs and aromatic rings of benzene and toluene (Lu et al. 2008). The equilibrium amount of B and T detached by SWCNTs was found to be much larger than that by MWCNTs and hybrid carbon nanotubes with silica (HCNTs) under similar experimental conditions (Pourzamani et al. 2012). The adsorption of toluene on CNTs was found to be greater than that of benzene. This behaviour of these compounds is considered to be related to the growing solubility in water and their low molecular weights.

3.5.2 Removal of Pharmaceutical and Personal Care Products by CNTs

The manufacture of pharmaceutical and personal care products (PPCPs) and endocrine-disrupting compounds (EDCs) can cause pollution of surface water at ng/L to μg/L concentrations levels (Kurwadkar et al. 2019; Murgolo et al. 2015). The modern strategies of wastewater treatment can't eliminate these pollutants totally from wastewater effluents.

Based on their exceptional characteristics of huge surface area, controllable porosity, high mechanical strength and photocatalytic activity, CNTs are established as the most efficient materials for the removal of PPCPs and EDCs from water (Kurwadkar et al. 2019; Jung et al. 2015). About 95% of these contaminants can be removed utilizing CNTs/membrane filtration systems. A variety of contaminants including PPCPs and EDCs to CO_2, and H_2O can be oxidized by CNT as a photocatalyst due to generation of active oxygen moieties.

The elimination of antibiotics from water was done utilizing several adsorbing agents and CNTs. The capacity of adsorption of sulfamethoxazole follows the trend: biochars (BCs) > MWCNTs > graphite = clay minerals, and that for tetracycline is: SWCNTs > graphite > MWCNTs = activated carbon (AC) > bentonite = humic substances = clay minerals (Ahmed et al. 2015). Antibiotics such as β-lactams/amoxicillin, lincomycin/lincosamides and sulfamethoxazole can be removed successfully (≥90%) applying CNTs and their derivatives (Tian et al. 2013; Kim et al. 2014).

Functional groups (such as a (–OH), (–C=O) or (–COOH)) can be formed on the surface of CNTs by acidification with H_2SO_4 and HNO_3 (Jung et al. 2015; Hu et al. 2015; Celik et al. 2011), making them more hydrophilic. As shown in Figure 3.22, the MWCNTs before modification are assembled and curved in a definite direction, and after modification, they are shorter, have thinner walls and have their ends opened (Hu et al. 2015). Moreover, their surface area increased from 71–123 m^2/g of the unmodified to 137–190 m^2/g of the modified (Hu et al. 2015). These modifications in surface area, and hydrophilicity of the MWCNTs, resulted in more efficient diclofenac removal (Hu et al. 2015).

The elimination of pharmaceutical compounds has been studied using several adsorbents, including CNTs, activated carbon, ion exchangers, biochars and bentonite (Kurwadkar et al. 2019). The adsorption mechanism on CNTs has been explained based on a π–π electron interaction between the benzenoid rings of

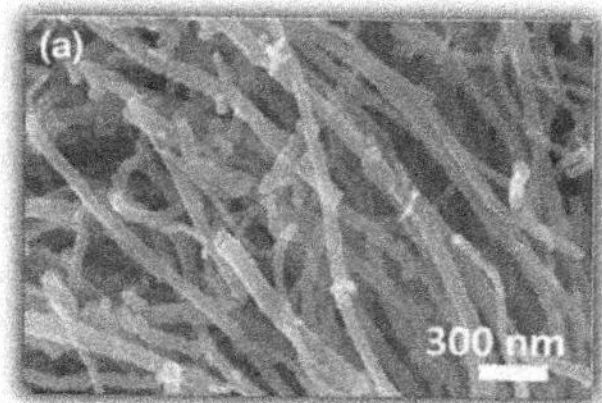

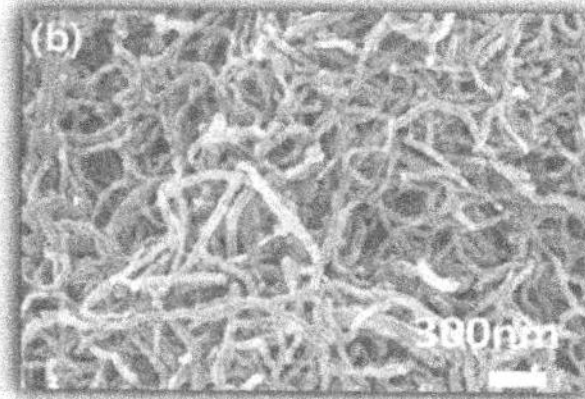

FIGURE 3.22 SEM images of multi-walled carbon nanotubes: (a) before and (b) after nitric acid modification. (Modified from Hu et al. 2015.)

CNTs and aromatic compounds (Joseph et al. 2011; Long et al. 2001). The sp^2-hybridized carbon atoms in CNTs confirm this assumption.

3.5.3 Graphene–CNT Aerogels for Oil Removal from Water

The leakage of oil into water sources causes continuing pollution of the environment and appreciable losses of energy sources. A cost-effective oil spill separation for water purification represents a challenging approach. Porous natural materials such as wool fibre (Annunciado et al. 2005; Bayat et al. 2005) and zeolite (Bi et al. 2012) have widely been used for oil elimination of oil spill, but with poor adsorption efficiency. Hydrophobic porous polymers have been fabricated and successfully applied for the separation of oils from water, but the preparation technique is lengthy and costly (Li et al. 2011; Farag and Saeed 2008). The removal of oil from water was done utilizing a three-dimensional (3D) graphene–carbon nanotube aerogel with interconnected networks (Kabiri et al. 2014). The aerogel was synthesized via a simple interaction of GO (synthesized from graphite ore) and acid-modified CNTs in the presence of ferrous ion (see Appendix 7). CNTs are combined with the network to increase the hydrophobicity and strength, leaving the porosity almost unchanged. CNTs act as spacers between graphene sheets and generate aerogel with porous structure. The functional moieties attached to the GO surface were found to be essential during the preparation of the aerogel. The prepared aerogel showed an exceptional adsorption capacity of 28 L of oil per gram.

A 3D network graphene–CNT aerogel is presented in the SEM image (Figure 3.23a). Higher-resolution SEM images in Figure 3.23b and c show the sub- to several micrometres pore sizes of the hierarchically structured graphene sheets with pore walls, networked with CNTs. Graphene sheets alone in Figure 23e show elongated shapes of iron oxide nanoparticles. Regular spreading of nanoparticles and CNTs on the graphene sheets is clearly shown in the higher-magnification TEM image of the graphene–CNT aerogels (Figure 23f). Images in Photo 2 show the progress in oil removal from water–oil system in combination with a vacuum. The hydrophobicity of the graphene–CNT aerogels quickly repelled the water and absorbed the oil.

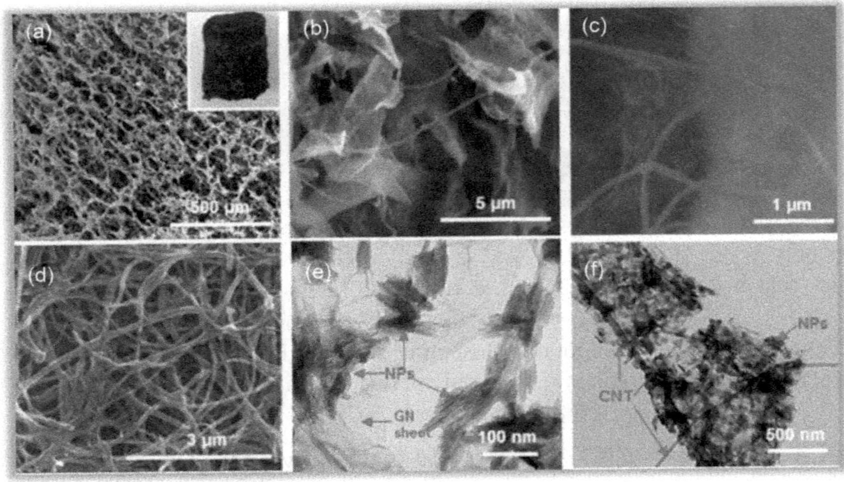

FIGURE 3.23 (a) Network structure of graphene–CNT hydrogel, (b and c) high-resolution image of graphene–CNT hydrogel, (d) CNTs prepared by the CVD process, (e) TEM images of graphene sheets with iron oxide nanoparticles (NPs) and (f) TEM images graphene–CNT aerogel. (Reproduced with permission, Kabiri et al. 2014.)

PHOTO 3.2 (a) Graphene–CNT aerogels and (b–e) progress of gasoline removal from a non-turbulent water–oil system. (Reproduced with permission 4, Kabiri et al. 2014.)

3.6 CONCLUSIONS AND FUTURE PROSPECTS

Contamination of drinking water is a result of continuous industrial development worldwide. The existing water remediation technologies require significant improvement to benefit from the evolution in the synthesis and fabrication of advanced materials. The numerous innovative and outstanding physical and chemical properties of CNTs made them novel materials for advanced technological applications in diverse fields, including water remediation. They are classified into different types, including SWCNTs, MWCNTs and functionalized CNTs, with diameter ranging from 1 nm to several hundred nm and length ranging from a few μm to 2 mm. The MWCNTs are the most commercially available type in the market.

This chapter summarized the fabrication techniques and the potential applications of the aforementioned types of CNTs, their derivatives and membranes in water treatment. Attention has been paid to the removal of heavy metals and some organic materials. The published works indicated that CNTs and their derivatives are considered promising materials for potential removal of heavy metals such as lead, arsenic, mercury and chromium, and organic and pharmaceutical pollutants. Researches are still growing in this area to cover widespread contaminants. However, studies on metal desorption and recycling of CNTs for multiple uses are mostly lacking. Moreover, some important issues, discussed below, should be addressed and enforced in future research works.

Despite the widespread applications of CNTs, there is accidental release into the environment. Exposure to CNTs can cause unfavourable conflicting effects on the human pulmonary system, irritation and oxidative stress. Assured procedures are necessary to be settled to overcome the toxicity of CNTs.

Although many research results emphasized in this chapter remain in laboratory level, some of them have been accepted and tested for commercialization. The high price of CNTs is considered the major constraint that hinders the large-scale application. The growth in demand will encourage capital investment in mass production of CNTs, which may certainly result in better economies when the material is produced at economically feasible prices. The main challenges of industrial production of MWCNTs by the commonly applied catalytic CVD technique are the high operating temperature and safety of the operation. The selection of a suitable catalyst is a critical issue from the morphological point of view. The production cost is affected by processing technology and cheap inputs.

APPENDIX 1

Preparation of Graphene Oxide (GO)

A 9:1 mixture of concentrated sulphuric acid and phosphoric acid was added to 1 g natural graphite powder and 6 g potassium permanganate, and the resulting mixture was heated at 50°C for 12 h to form a thick paste. The paste was cooled in ice with 1 mL 30% H_2O_2, filtered and washed with DW, with 32% HCl, with ethanol, centrifuged at 4200 rpm for 2 h and then vacuum-dried overnight (Kabiri et al. 2014).

Synthesis of Multi-Walled Carbon Nanotubes (MWCNTs)

MWCNTs were synthesized by the CVD technique with ferrocene/toluene as precursor carbon sources, at 850°C for 20–60 min (Altalhi et al. 2010).

Preparation of Graphene–CNT Aerogel

A 4:1 graphene–CNT aerogel was prepared as follows: 10 mg of CNTs mixed with 10 mL DW was sonicated for 2 h. Twenty millilitres of GO (2 mg/mL) was mixed with CNTs dispersion, sonicated for 30 min and mixed with 1 mmol ferrous sulphate. The pH was adjusted to 3.5, and the mixture was heated for 8 h at 90°C, washed with DW and freeze-dried (Altalhi et al. 2010).

APPENDIX 2

Synthesis of magnetic hydroxyapatite-immobilized oxidized multi-walled carbon nanotubes (mHAP-oMWCNTs)

Mesoporous Fe_3O_4 nanoparticles were prepared by dissolving 1 g $FeCl_3.6H_2O$ in 20 mL ethylene glycol (EG) followed by addition of 3 g sodium acetate (NaAc) and 10 mL 1,2 ethylenediamine (EDA); the mixture was stirred vigorously and heated in an autoclave at 200°C (Guo et al. 2009). Oxidized MWCNTs (oMWCNTs) were prepared by reaction of 0.5 g MWCNTs and 100 mL strong acid (the volume ratio of HNO_3 to H_2SO_4 was 1:3) for 3 h at 40°C (Wang et al. 2017). For the synthesis of magnetic hydroxyapatite (mHAP), 0.1 g Fe_3O_4 nanoparticles were dispersed in 40 mL ammonium bicarbonate (0.3 mol/L), 170 mL diammonium phosphate (0.3 mol/L) and 300 mL calcium nitrate (0.025 mol/L), and then the pH was kept at 11 by adding ammonium hydroxide (Cui et al. 2009). To prepare mHAP-oMWCNTs, 0.1 g oMWCNTs dispersed in 20 mL of ultrapure water was added to the mHAP system and stirred for 2 h and the solid mHAP-oMWCNTs product was collected by magnetic separation (Wang et al. 2017).

APPENDIX 3

Synthesis of Iron Oxide Multi-Walled Carbon Nanotube (Fe-MWCNT) Hybrid

The MWCNTs were first functionalized using a microwave accelerated reaction system fitted with internal temperature and pressure controls. Pre-weighed amounts of purified MWCNTs were treated with a mixture of concentrated H_2SO_4 and HNO_3 solution, and the mixture was placed under microwave radiation at 120°C for 20–40 min. This led to the formation of carboxylic groups on the surface along with some sulphonation and nitration. The iron oxide-coated MWCNT (Fe-MWCNT) hybrid was synthesized by dispersing a weighed amount of the functionalized MWCNT (f-MCWNT) in a 2:1 aqueous solution of $FeCl_3$: $FeSO_4$ under ultrasonication at room temperature. The dispersion was gently stirred at 70°C for 5 min, after which 5 M NaOH solution was added to adjust the pH to 10. The mixture was stirred vigorously at 85°C for 1 h.

APPENDIX 4

Synthesis of Carboxylated MWCNTs (MWCNTs-COOH)

Five grams of MWCNTs was treated with 10% HCl for 3 h, sonicated for 15 min, filtered and washed with distilled water (DW) till neutral pH (Figure 24a). The sample was suspended in 400 mL 10 N H_2SO_4 and HNO_3 (3:1 by volume) and refluxed at 175°C for 18 h, filtered and washed with hot DW (Figure 3.24b) (Hadavifar et al. 2014).

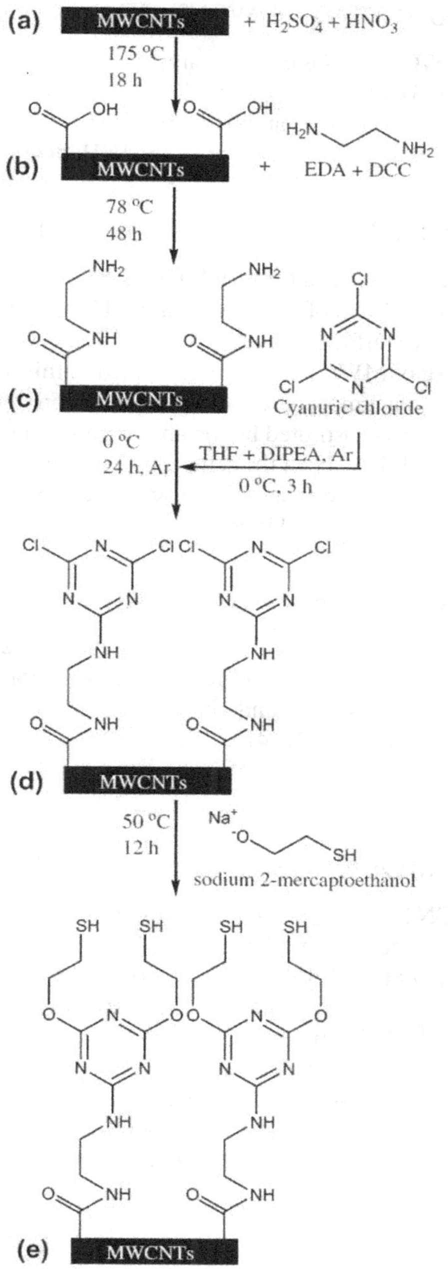

FIGURE 3.24 Functionalization scheme of CNTs. (a) pure MWCNTs, (b) chemically oxidized MWCNTs, (c) amino functionalized MWCNTs, (d) triazine functionalized MWCNTs, and (e) thiol functionalized MWCNTs. (See Hadavifar et al. 2014.)

Synthesis of amino MWCNTs (MWCNTs-EDA)

Four grams of MWCNTs-COOH was mixed with 150 mL ethylenediamine (EDA) at 25°C, then 5 g of n,n-dicyclohexylcarbodiimide (DCC) was added, and the mixture was refluxed for 48 h at 78°C. The filtered product was washed with ethanol and dried (Figure 3.24c) (Zang et al. 2009; Hadavifar et al. 2014).

Synthesis of triazine MWCNTs (MWCNTs-triazine)

In order to blend the triazine into the MWCNTs-EDA, 8.86 g of cyanuric chloride was dissolved in 480 mL of tetrahydrofuran (THF) and then 11.5 mL of n,n-diisopropylethylamine (DIPEA) was added and stirred at 0°C for 3 h under argon atmosphere. Then 3 g of MWCNTs-EDA was added to this mixture and stirred at 0°C for 48 h under argon atmosphere; during this reaction time, the chlorine atom in cyanuric chloride was substituted by the amine group in EDA and the released HCl was trapped by DIPEA. Then the sample was filtered and washed with dried THF repeatedly to remove the excess cyanuric chloride and dried at 80°C for 8 h (Figure 3.24d) (Shahbazi et al. 2011; Hadavifar et al. 2014).

Synthesis of thiol MWCNTs (MWCNTs-SH)

Two grams of MWCNTs-triazine was added to a solution of 25 mL sodium 2-mercaptoethanol and 30 mL methanol and refluxed at 50°C for 12 h with low-speed stirring. The sample was filtered and rinsed with methanol and dried at 80°C (Figure 3.24e) (Hadavifar et al. 2014).

APPENDIX 5

Synthesis of S-M-MWCNT composite

Oxidation of MWCNTs can enhance the interaction with sulphur (Ji et al. 2011). MWCNTs were refluxed in conc. HNO_3 for 5 h at 60°C, washed with DW and dried at 85°C (Fayazi et al. 2015). The magnetic MWCNT (M-MWCNT) composite was prepared by co-precipitation of a ferrous salt ($FeSO_4.7H_2O$) and a ferric salt ($FeCl_3.6H_2O$), in the presence of oxidized MWCNTs (Fayazi et al. 2016). The mixture was sonicated, alkalinized with NaOH at 70°C under N_2 and stirred for 4 h. The black powder was collected with a magnet, washed with DW and dried at 70°C under vacuum. The S-M-MWCNT composite was prepared by heating a mixture of 1.3 g elemental sulphur and 0.65 g of M-MWCNTs at 155°C for 24 h in Ar gas.

APPENDIX 6

Preparation of MnO_2/MWCNT nanocomposite using co-precipitation approach MWCNTs were dispersed in 65 wt% HNO_3 for 24 h, then rinsed with DW and dried in the air at 100°C. An amount of 0.1 g of the treated MWCNTs was vigorously stirred with 160 mL 0.1 M $KMnO_4$ solution at 40°C for 2 h. Twenty millilitres of 0.5 M $MnSO_4$ was added to the above mixture dropwise keeping intensive stirring, kept at 40°C for 24 h, filtered and washed with DW and alcohol several times, and the collected powder was dried at 100°C for 12 h (Mohammadkhani et al. 2015).

APPENDIX 7

Fabrication of nanoporous anodic alumina membranes (NAAMs)

High-purity Al chips of 1.5 cm diameter were cleaned by sonication in ethanol, washed with DW and electropolished in 1:4 (v:v) $HClO_4$–ethanol mixture at 20 V and 5°C for 3 min. The first anodization stage was carried out in 0.3 M oxalic acid at 40 V, 6°C for 20 h. The resulting alumina layer was selectively dissolved in a mixture of 0.4 M phosphoric acid and 0.2 M chromic acid at 70°C for 3 h. The second anodization stage was carried out for 17 h under similar anodization conditions as the first one. The remaining aluminium substrate was removed in a saturated cupric chloride–hydrochloric acid solution. A pore opening process was carried out by etching in 5 wt% H3PO4 at 35°C under current control conditions (Altalhi et al. 2013).

Synthesis of CNTs–NAAMs

The CNTs–NAAMs were synthesized by template and catalyst-free CVD process inside pores of NAAMs. A CVD system of a two-stage furnace with a quartz tube of diameter and length of 43 and 1000 mm, respectively, was utilized. Plastic bags are manufactured from high-density (HDPE), low-density (LDPE) or linear low-density (LLDPE) polyethylene. Plastic bags were collected from a local grocery shop and used as a carbon source for the production of CNTs. The thoroughly washed and dried plastic bags were cut into 1 cm^2 squares. Small pieces of plastic bag were placed in a ceramic crucible and introduced into the pyrolysis zone of the CVD reactor, and argon gas was then passed at 1 dm^3/min flow rate. Deposition of CNTs took place in the deposition zone of the CVD reactor where the fabricated NAAMs were placed (Altalhi et al. 2013).

REFERENCES

Abbas, A., Al-Amer, A. M., Laoui, T., Al-Marri, M. J., Nasser, M. S., Khraisheh, M., Atieh, M. A. (2016), Heavy metal removal from aqueous solution by advanced carbon nanotubes: critical review of adsorption applications. *Sep. Purif. Technol.* 157, 141–161.

Addo Ntim, S. A., Mitra, S. (2011), Removal of trace arsenic to meet drinking water standards using iron oxide coated multiwall carbon nanotubes. *J. Chem. Eng. Data* 56(5), 2077–2083.

Addo Ntim, S. A., Mitra, S. (2012), Adsorption of arsenic on multiwall carbon nanotube–zirconia nanohybrid for potential drinking water purification. *J. Colloid Interface Sci.* 375(1), 154–159.

Ahmed, M. B., Zhou, J. L., Ngo, H. H., Guo, W. (2015), Adsorptive removal of antibiotics from water and wastewater: progress and challenges. *Sci. Total Environ.* 532, 112–126.

Aigbe, U. O., Osibote, O. A. (2020), A review of hexavalent chromium removal from aqueous solutions by sorption technique using nanomaterials. *J. Environ. Chem. Eng.* 104503.

Aivalioti, M., Vamvasakis, I., Gidarakos, E. (2010), BTEX and MTBE adsorption onto raw and thermally modified diatomite. *J. Hazardous Mater.* 178(1–3), 136–143.

Alagappan, P. N., Heimann, J., Morrow, L., Andreoli, E., Barron, A. R. (2017), Easily regenerated readily deployable absorbent for heavy metal removal from contaminated water. *Sci. Rep.* 7(1), 1–7.

Albakri, M. A., Abdelnaby, M. M., Saleh, T. A., Al Hamouz, O. C. S. (2018), New series of benzene-1,3,5-triamine based cross-linked polyamines and polyamine/CNT composites for lead ion removal from aqueous solutions. *Chem. Eng. J.* 333, 76–84.

Alsawat, M., Altalhi, T., Kumeria, T., Santos, A., Losic, D. (2015), Carbon nanotube-nanoporous anodic alumina composite membranes with controllable inner diameters and surface chemistry: Influence on molecular transport and chemical selectivity. *Carbon 93*, 681–692.

Altalhi, T., Ginic-Markovic, M., Han, N., Clarke, S., Losic, D. (2010), Synthesis of carbon nanotube (CNT) composite membranes. *Membranes* 1(1), 37–47.

Altalhi, T., Kumeria, T., Santos, A., Losic, D. (2013), Synthesis of well-organised carbon nanotube membranes from non-degradable plastic bags with tuneable molecular transport: Towards nanotechnological recycling. *Carbon* 63, 423–433.

Anagnostopoulos. V. A., Symeopoulos B. D., Soupioni, M. J. (2010), Effect of growth conditions on biosorption of cadmium and copper by yeast cells. *Global NEST J.* 12, 288–295.

Anastopoulos, I., Anagnostopoulos, V. A., Bhatnagar, A., Mitropoulos, A. C., Kyzas, G. Z. (2017), A review for chromium removal by carbon nanotubes. *Chem. Ecol.* 33(6), 572–588.

Annunciado T. R., Sydenstricker T. H. D., Amico, S. C. (2005), Experimental investigation of various vegetable fibers as sorbent materials for oil spills. *Mar. Pollut. Bull.* 50(11), 1340–1346.

Anzar, N., Hasan, R., Tyagi, M., Yadav, N., Narang, J. (2020), Carbon nanotube-a review on synthesis, properties and plethora of applications in the field of biomedical science. *Sens. Int.* 1, 100003.

Atieh, M. A. (2011), Removal of chromium (VI) from polluted water using carbon nanotubes supported with activated carbon. *Procedia Environ. Sci.* 4, 281–293.

Atieh, M. A., Bakather, O. Y., Al-Tawbini, B., Bukhari, A. A., Abuilaiwi, F. A., Fettouhi, M. B. (2010a), Effect of carboxylic functional group functionalized on carbon nanotubes surface on the removal of lead from water. *Bioinorg. Chem. Appl.* 603978.

Atieh, M. A., Bakather, O. Y., Tawabini, B. S. (2010), Removal of chromium (III) from water by using modified and nonmodified carbon nanotubes. *J. Nanomater.* 2010.

Atieh, M. A., Bakather, O. Y., Tawabini, B. S., Bukhari, A. A., Khaled, M., Alharthi, M., Abuilaiwi, F. A. (2010b), Removal of chromium (III) from water by using modified and nonmodified carbon nanotubes. *J. Nanomater.* 2010, 9.

Awotunde, M. A., Adegbenjo, A. O., Obadele, B. A., Okoro, M., Shongwe, B. M., Olubambi, P. A. (2019), Influence of sintering methods on the mechanical properties of aluminium nanocomposites reinforced with carbonaceous compounds: a review. *J. Mater. Res. Technol.* 8(2), 2432–2449.

Bassyouni, M., Mansi, A. E., Elgabry, A., Ibrahim, B. A., Kassem, O. A., Alhebeshy, R. (2020), Utilization of carbon nanotubes in removal of heavy metals from wastewater: A review of the CNTs' potential and current challenges. *Appl. Phys. A* 126(1), 1–33.

Baughman, R. H. (2002), Carbon nanotubes--the route toward applications. *Science* 297, 787.

Bayat, A., Aghamiri, S. F., Moheb, A., Vakili-Nezhaad, G. R. (2005), Oil spill cleanup from sea water by sorbent materials. *Chem. Eng. Technol.* 28(12), 1525–1528.

Berber, S., Kwon, Y. K., Tománek, D. (2000), Unusually high thermal conductivity of carbon nanotubes. *Phys. Rev. Lett.* 84(20), 4613.

Bi, H., Xie, X., Yin, K., Zhou, Y., Wan, S., He, L. (2012), Spongy graphene as a highly efficient and recyclable sorbent for oils and organic solvents. *Adv. Funct. Mater.* 22(21), 4421–4425.

Blue, L. Y., Jana, P., Atwood, D. A. (2010), Aqueous mercury precipitation with the synthetic dithiolate, BDTH2, *Fuel* 89, 1326–1330.

Bower, J., Savage, K. S., Weinman, B., Barnett, M. O., Hamilton, W. P., Harper, W. F. (2008), Immobilization of mercury by pyrite (FeS2). *Environ. Pollut.* 156, 504–514.

Bulgariu, L., Bulgariu, D. J. (2018), Functionalized soy waste biomass- a novel environmental-friendly biosorbent for the removal of heavy metals from aqueous solution, *J. Cleaner Prod.* 197, 875–885.

Celik, E., Park, H., Choi, H., Choi, H. (2011), Carbon nanotube blended polyethersulfone membranes for fouling control in water treatment. *Water Res.* 45(1), 274–282.

Chawla, J., Kumar, R., Kaur, I. (2015). Carbon nanotubes and graphenes as adsorbents for adsorption of lead ions from water: a review. *J. Water Supply: Res. Technol.—AQUA* 64(6), 641–659.

Chen, H., Sholl, D. S. (2006), Prediction of selectivity and flux for CH_4/H_2 separations using single walled carbon nanotubes as membranes. *J. Membr. Sci.* 269, 152–160.

Chen, K., Zhang, Z., Xia, K., Zhou, X., Guo, Y., Huang, T. (2019), Facile synthesis of thiol-functionalized magnetic activated carbon and application for the removal of mercury (II) from aqueous solution. *ACS Omega* 4, 8568–8579.

Chiarle, S, Ratto, M., Rovatti M. (2000), Mercury removal from water by ion exchange resins adsorption. *Water Res.* 34, 2971–2978.

Chu, H., Wei, L., Cui, R., Wang, J., Li, Y. (2010), Carbon nanotubes combined with inorganic nanomaterials: preparations and applications. *Coord. Chem. Rev.* 254, 1117–1120.

Collins, P. G., Avouris, P. (2000), Nanotubes for electronics. *Sci. Am.* 283(6), 62e69.

Cui, J., Liu, Y., Hao, J. (2009), Multiwalled carbon-nanotube-embedded microcapsules and their electrochemical behavior. *J. Phys. Chem. C* 113(10), 3967–3972.

Cvjetko Bubalo, M., Vidović, S., Radojčić Redovniković, I., Jokić, S. (2015), Green solvents for green technologies. *J. Chem. Technol. Biotechnol.* 90(9), 1631–1639.

Daenen, M., de Fouw, R. D., Hamers, B., Janssen, P. G. A., Schouteden, K., Veld, M. A. J. (2003), The wondrous world of carbon nanotubes a review of current carbon nanotube technologies, *Tech. Univ. Eindhoven* 1–96.

De Volder, M. F. L., Tawfick, S. H., Baughman, R. H., Hart, A. J. (2013), Carbon nanotubes: present and future commercial applications. *Science* 339, 535–539.

Deng, S., Zhang, G., Wang, X., Zheng, T., Wang, P. (2015), Preparation and performance of polyacrylonitrile fiber functionalized with iminodiacetic acid under microwave irradiation for adsorption of Cu (II) and Hg (II). *Chem. Eng. J.* 276, 349–357.

Denizli, A., Büyüktuncel, E., Tuncel, A., Bektas, S., Genç, Ö. (2000), Batch removal of lead ions from aquatic solutions by polyethyleneglycol-methacrylate gel beads carrying cibacron blue F3GA. *Environ. Technol.* 21(6), 609–614.

Di, Z., Li, Y., Luan, Z. Liang, J. (2004), Adsorption of chromium (VI) ions from water by carbon nanotubes, *Adsorpt. Sci. Technol.* 22(6) 467–474.

Dimos, V., Haralambous, K. J., Malamis, S. (2012), A review on the recent studies for chromium species adsorption on raw and modified natural minerals. *Crit. Rev. Environ. Sci. Technol.* 42, 1977–2016.

Dixit, S., Hering, J. (2003), Comparison of arsenic(V) and arsenic(III) sorption onto iron oxide minerals: implications for arsenic mobility. *Environ. Sci. Technol.* 37(18), 4182–4189.

Eatemadi, A., Daraee, H., Karimkhanloo, H., Kouhi, M., Zarghami, N., Akbarzadeh, A., Abasi, M., Hanifehpour, Y., Joo, S. (2014), Carbon nanotubes: properties, synthesis, purification, and medical applications, *Nanoscale Res. Lett.* 9 (1) 393.

Ebbesen, T., Ajayan, P. M. (1992), Large-scale synthesis of carbon. *Nature* 358, 16.

Elimelech, M., Phillip, W. A. (2011), The future of seawater desalination: Energy, technology, and the environment. *Science* 333, 712–717.

Farag R. K., El-Saeed S. M. (2008), Synthesis and characterization of oil sorbers based on docosanyl acrylate and methacrylates copolymers. *J. Appl. Polym. Sci.* 109(6), 3704–3713.

Farghali, A. A., ElRouby, W. M. A., Khedr, M. H. (2013), Decoration of multi-walled carbon nanotubes (MWCNTs) with different ferrite nanoparticles and its use as an adsorbent. *J. Nanostruct. Chem.* 3(1), 50.

Farghali, A. A., Tawab, H. A. A., Moaty, S. A. A., Khaled, R. (2017), Functionalization of acidified multi-walled carbon nanotubes for removal of heavy metals in aqueous solutions. *J. Nanostruct. Chem.* 7(2), 101–111.

Fayazi, M. (2020), Removal of mercury (II) from wastewater using a new and effective composite: sulfur-coated magnetic carbon nanotubes. *Environ. Sci. Pollut. Res. 27*(11), 12270–12279.

Fayazi, M, Taher, M. A., Afzali, D., Mostafavi, A. (2015), Preparation of molecularly imprinted polymer coated magnetic multi-walled carbon nanotubes for selective removal of dibenzothiophene. *Mater. Sci. Semicond. Process.* 40, 501–507.

Fayazi, M., Taher, M. A., Afzali, D., Mostafavi, A., Ghanei-Motlagh, M. (2016), Synthesis and application of novel ion-imprinted polymer coated magnetic multi-walled carbon nanotubes for selective solid phase extraction of lead (II) ions. *Mater. Sci. Eng. C* 60, 365–373.

Fiyadh, S. S., AlSaadi, M. A., Jaafar, W. Z., AlOmar, M. K., Fayaed, S. S., Mohd, N. S., El-Shafie, A. (2019), Review on heavy metal adsorption processes by carbon nanotubes. *J. Cleaner Prod.* 230, 783–793.

Giles, D., Mohapatra, M., Issa, T., Anand, S., Singh, P. (2011), Iron and aluminium based adsorption strategies for removing arsenic from water. *J. Environ. Manage.* 92(12), 3011–3022.

Goh, K., Chen, Y. (2017), Controlling water transport in carbon nanotubes. *Nano Today* 14, 13–15.

Gong, K., Hu, Q., Xiao, Y., Cheng, X., Liu, H., Wang, N., et al., (2018a), Triple layered core–shell ZVI@carbon@polyaniline composite enhanced electron utilization in Cr(vi) reduction. *J. Mater. Chem. A* 6(24), 11119–11128.

Gong, K., Hu, Q., Yao, L., Li, M., Sun, D., Shao, Q., et al., (2018b), Ultrasonic pretreated sludge derived stable magnetic active carbon for Cr(VI) removal from wastewater. *ACS Sustainable Chem. Eng.* 6(6), 7283–7291.

Guo, S., Li, D., Zhang, L., Li, J., Wang, E. (2009), Monodisperse mesoporous superparamagnetic single-crystal magnetite nanoparticles for drug delivery. *Biomaterials* 30(10), 1881–1889.

Gupta, V. K., Moradi, O., Tyagi, I., Agarwal, S., Sadegh, H., Shahryari-Ghoshekandi, R., et al., (2016), Study on the removal of heavy metal ions from industry waste by carbon nanotubes: effect of the surface modification: a review. *Crit. Rev. Environ. Sci. Technol.* 46(2), 93–118.

Gupta, V. K., Tyagi, I., Sadegh, H., Ghoshekandi, R. S., Makhlouf, A. H. (2015), Nanoparticles as adsorbent; a positive approach for removal of noxious metal ions: a review. *Sci. Technol. Dev.* 34, 195–214.

Hadavifar, M., Bahramifar, N., Younesi, H., Li, Q. (2014), Adsorption of mercury ions from synthetic and real wastewater aqueous solution by functionalized multi-walled carbon nanotube with both amino and thiolated groups. *Chem. Eng. J.* 237, 217–228.

Hamza, I. A., Martincigh, B. S., Ngila, J. C., Nyamori, V. O. (2013), Adsorption studies of aqueous Pb (II) onto a sugarcane bagasse/multi-walled carbon nanotube composite. *Phys. Chem. Earth Parts A/B/C*, 66, 157–166.

Han, I., Schlautman, M. A., Batchelor, B. (2000), Removal of hexavalent chromium from groundwater by granular activated carbon. *Water Environ. Res.* 72, 29–39.

Henneberry, Y. K., Kraus T. E., Fleck J. A., Krabbenhoft D. P., Bachand P. M., Horwath, W. R. (2011), Removal of inorganic mercury and methylmercury from surface waters following coagulation of dissolved organic matter with metal-based salts. *Sci. Total Environ.* 409, 631–637.

Herrera-Herrara, A. V., Gonzalez-Curbelo, M. A., Hernandez-Borges, J. (2012), Carbon nanotubes applications in separation science: a review. *Anal. Chim. Acta* 734, 1–30.

Herrera-Ramirez, J. M., Perez-Bustamante, R., Aguilar-Elguezabal, A. (2019), An overview of the synthesis, characterization, and applications of carbon nanotubes. In *Carbon-Based Nanofillers Their Rubber Nanocomposites*. pp. 47–75.

Hinds, B. J., Chopra, N., Rantell, T., Andrews, R., Gavalas, V., Bachas, L. G. (2004), Aligned multiwalled carbon nanotube membranes. *Science* 303(5654), 62–5.

Hirsch, A., Vostrowsky, O. (2005), Functionalization of carbon nanotubes. *Funct. Mol. Nanostruct.* 193–237.

Hjortstam, O., Isberg, P., Söderholm, S., Dai, H. (2004), Can we achieve ultra-low resistivity in carbon nanotube-based metal composites?. *Appl. Phys. A* 78(8), 1175–1179.

Holt, J. K., Noy, A., Huser, T., Eaglesham, D., Bakajin, O. (2004), Fabrication of a carbon Nanotube-embedded silicon nitride membrane for studies of nanometer-scale mass transport. *Nano Lett.* 4, 2245–2250.

Hone, J., Whitney, M., Piskoti, C., Zettl, A. (1999), Thermal conductivity of single walled carbon nanotubes. *Phys. Rev. B* 59(4), R2514.

https://www.epo.org/news-events/events/europeaninventor/finalists/2015/iijima.html.

Hu, Z., Cheng, Z. (2015), Removal of diclofenac from aqueous solution with multi-walled carbon nanotubes modified by nitric acid. *Chin. J. Chem. Eng.* 23, 1551–1556.

Huang, J., Cao, Y., Peng, F., Cao, Y., Shao, Q., et al., (2018), Hexavalent chromium removal over magnetic carbon nanoadsorbents: synergistic effect of fluorine and nitrogen co-doping. *J. Mater. Chem. A* 6(27), 13062–13074.

Huang, J., Yuan, F., Zeng, G., Li, X., Gu, Y., Shi, L., et al., (2017), Influence of pH on heavy metal speciation and removal from wastewater using micellar-enhanced ultrafiltration. *Chemosphere* 173, 199–206.

Huang, L., Shuai, Q. (2019), Facile approach to prepare sulfur functionalized magnetic amide-linked organic polymers for enhanced Hg (II) removal from water. *ACS Sustainable Chem. Eng.* 7, 9957–9965.

Ihsanullah Al-Khaldi, F. A., Abu-Sharkh, B., Abulkibash, A. M., Qureshi, M. I., Laoui, T., Atieh, M. A. (2016a), Effect of acid modification on adsorption of hexavalent chromium (Cr(VI)) from aqueous solution by activated carbon and carbon nanotubes. *Desalin. Water. Treat.* 57(16), 7232–7244.

Iijima, S. (1991), Helical microtubules of graphitic carbon. *Nature* 354(6348): 56–58.

Ilankoon, N. (2014), Use of iron oxide magnetic nanosorbents for Cr(VI) removal from aqueous solutions: a review. *J. Eng. Res. Appl.* 4, 55–63.

Jahanshahi, M., Kiadehi, A. (2013), Fabrication, purification, and characterization of carbon nanotubes: arc-discharge in liquid media (ADLM). In *Syntheses and Applications of Carbon Nanotubes and Their Composites*, pp. 55–76.

Jainae, K., Sukpirom, N., Fuangswasdi, S., Unob, F. (2015), Adsorption of Hg (II) from aqueous solutions by thiol-functionalized polymer-coated magnetic particles. *J. Ind. Eng. Chem.* 23, 273–278.

Jani, A. M., Losic, D., Voelcker, N. H. (2013), Nanoporous anodic aluminium oxide: advances in surface engineering and emerging applications. *Prog. Mater. Sci.* 58(5), 636–704.

Ji, L., Rao, M., Zheng, H., Zhang, L., Li, Y., Duan, W., Guo, J., Cairns, E.J., Zhang, Y. (2011), Graphene oxide as a sulfur immobilizer in high performance lithium/sulfur cells. *J. Am. Chem. Soc.* 133, 18522–18525.

Jiang, H., Liu, B., Huang, Y., Hwang, K. C. (2004). Thermal expansion of single wall carbon nanotubes. *J. Eng. Mater. Technol.* 126(3), 265–270.

Joseph, L., Heo, J., Park, Y. G., Flora, J. R., Yoon, Y. (2011), Adsorption of bisphenol A and 17 α-ethinylestradiol on single-walled carbon nanotubes from seawater and brackish water. *Desalination* 281, 68–74.

Journet, C., Maser, W. K., Bernier, P., Loiseau, A., de La Chapelle, M. L., Lefrant, D. S., Fischer, J. E. (1997), Large-scale production of single-walled carbon nanotubes by the electric-arc technique. *Nature 388*(6644), 756–758.

Jung Son A., Her, N. Zoh, K. D., Cho, J., Yoon, Y. (2015), Removal of endocrine disrupting compounds, pharmaceuticals, and personal care products in water using carbon nanotubes: a review. *J. Ind. Eng. Chem.* 27, 1–11.

Khan, S. T., Malik, A. (2019), Engineered nanomaterials for water decontamination and purification: from lab to products. *J. Hazard. Mater.* 363, 295–308.

Khin, M. M., Nair, A. S., Babu, V. J., Murugan, R., Ramakrishna, S. (2012), A review on nanomaterials for environmental remediation. *Energy Environ. Sci.* 5, 8075–8109.

Khlifi, R., Hamza-Chaffai, A. (2010), Head and neck cancer due to heavy metal exposure via tobacco smoking and professional exposure: a review. *Toxicol. Appl. Pharmacol.* 248(2), 71–88.

Kim, H., Hwang, Y. S., Sharma, V. K. (2014), Adsorption of antibiotics and iopromide onto single-walled and multi-walled carbon nanotubes. *Chem. Eng. J.* 255, 23–27.

Kim, Y., Lee, J., Yeom, M. S., Shin, J. W., Kim, H., Cui, Y., Han, S. M. (2013), Strengthening effect of single-atomic-layer graphene in metal–graphene nanolayered composites. *Nat. Commun.* 4(1), 1–7.

Kurwadkar, S., Hoang, T. V., Malwade, K., Kanel, S. R., Harper, W. F., Struckhoff, G. (2019), Application of carbon nanotubes for removal of emerging contaminants of concern in engineered water and wastewater treatment systems. *Nanotechnol. Environ. Eng.* 4(1), 12.

Kyotani, T., Tsai, L. F., Tomita, A. (1995), Formation of ultrafine carbon tubes by using an anodic aluminum oxide film as a template. *Chem. Mater.* 7(8), 1427–1428.

Kyotani, T, Tsai, L. F., Tomita A. (1996), Preparation of ultrafine carbon tubes in nanochannels of an anodic aluminum oxide film. *Chem. Mater.* 8(8), 2109–2113.

Kyzas, G. Z., Deliyanni, E. A., Bikiaris, D. N. (2018), Mitropoulos, A.C. Graphene composites as dye adsorbents: review. *Chem. Eng. Res. Des.* 129, 75–88.

Lee, J. W., Jeong, H. M., Lee, G. H., Jung, Y. W., Jo, S. G., Kang, J. K. (2021), Agglomeration-free Fe_3O_4 anchored via nitrogen mediation of carbon nanotubes for high-performance arsenic adsorption. *J. Environ. Chem. Eng.* 9(3), 104772.

Lee, J. W., Viswan, R., Choi, Y. J., Lee, Y., Kim, S. Y., Cho, J., Jo, Y., Kang, J. K. (2009), Facile fabrication and superparamagnetism of silica-shielded magnetite nanoparticles on carbon nitride nanotubes, *Adv. Funct. Mater.* 19, 2213–2218.

Li, A., Sun, H. X., Tan, D. Z, Fan, W. J., Wen, S. H., Qing, X. J., Deng, W. Q. (2011), Superhydrophobic conjugated microporous polymers for separation and adsorption. *Energy Environ. Sci.* 4(6), 2062–2065.

Li, Y. H., Wang, S. G., Luan, Z. K., Ding, J., Xu, C. L., Wu, D. H. (2003), Adsorption of cadmium(II) from aqueous solution by surface oxidized carbon nanotubes. *Carbon* 41, 1057–1062.

Li, Y., Wang, S., Wei, J., Zhang, X., Xu, C., Luan, Z., Wei, B. (2002), Lead adsorption on carbon nanotubes. *Chem. Phys. Lett.* 357(3–4), 263–266.

Li, Y., Zhou, B., Zheng, G., Liu, X., Li, T., Yan, C., et al., (2018), Continuously prepared highly conductive and stretchable SWNT/MWNT synergistically composited electrospun thermoplastic polyurethane yarns for wearable sensing. *J. Mater. Chem.* C 6(9), 2258–2269.

Liu, W., Feng, Y., Tang, H. Yuan, H., He, S., Miao, S. (2016), Immobilization of silver nanocrystals on carbon nanotubes using ultra-thin molybdenum sulfide sacrificial layers for antibacterial photocatalysis in visible light. *Carbon* 96, 303–310. *Water Res.* 45, 274–282.

Long, R. Q., Yang, R. T. (2001), Communications to the editor. *J. Am. Chem. Soc.* 5112–5113.

Lu, C., Su, F. (2007), Adsorption of natural organic matter by carbon nanotubes. *Sep. Purif. Technol.* 58(1), 113–121.

Lu, C., Su F., Hu, S. (2008), Surface modification of carbon nanotubes for enhancing BTEX adsorption from aqueous solutions. *Appl. Surf. Sci.* 254(21), 7035–7041.

Luo, H., Cheng, F., Hu, W., Wang, J., Xiang, S., Fidalgo de Cortalezzi, M. (2017), 2D-Fe_3O_4 nanosheets for effective arsenic removal. *J. Contemp. Water Res. Educ.* 160(1), 132–143.

Ma, L., Dong, X., Chen, M., Zhu, L., Wang, C., Yang, F., Dong, Y. (2017) Fabrication and water treatment application of carbon nanotubes (cnts)-based composite membranes: a review. *Membranes* 7, 16.

Mandal, B., Suzuki, K. (2002), Arsenic round the world: a review. *Talanta* 58(1), 201–235.

Masuda, H., Fukuda, K. (1995), Ordered metal nanohole arrays made by a two-step replication of honeycomb structures of anodic alumina. *Science* 268, 1466–1468.

Mathur, A. K., Majumder, C. B., Chatterjee, S. (2007), Combined removal of BTEX in air stream by using mixture of sugar cane bagasse, compost and GAC as biofilter media. *J. Hazardous Mater.* 148(1–2), 64–74.

Mehta, D., Mazumdar, S., Singh, S. K. (2015), Magnetic adsorbents for the treatment of water/wastewater-a review. *J. Water Process. Eng.* 7, 244–265.

Miretzky, P., Cirelli, A. F. (2010), Cr(VI) and Cr(III) removal from aqueous solution by raw and modified lignocellulosic materials: a review. *J Hazard Mater.* 180, 1–19.

Mohammadkhani, S., Gholami, M. R., Aghaie, M. (2015), Thermodynamic study of Cr^{+3} ions removal by "MnO_2/MWCNT" nanocomposite. *Orient J Chem.* 31, 1429–1436.

Mohan, D., Pittman, C. U. Jr. (2007), Arsenic removal from water/waste water using adsorbents - a critical review. *J. Hazard. Mater.* 142, 1–53.

Moradi, O., Zare, K. (2013), Adsorption of ammonium ion by multi-walled carbon nanotube: Kinetics and thermodynamic studies. *Fuller. Nanotubes Carbon Nanostruct.* 21(6), 449–459.

Mou, F., Guan, J., Xiao, Z., Sun, Z. Shi, W., Fan, X. (2011), Solvent-mediated synthesis of magnetic Fe_2O_3 chestnut-like amorphous-core/γ-phase-shell hierarchical nanostructures with strong As(V) removal capability. *J. Mater. Chem.* 21, 5414–5421.

Muataz, A. A., Fettouhi, M., Al-Mammum, A., Yahya, N. (2009), Lead removal by using carbon nanotubes. *Int. J. Nanopart.* 2329–338.

Mubarak, N. M., Sahu, J. N., Abdullah, E. C., Jayakumar, N. S. (2014b), Removal of heavy metals from wastewater using carbon nanotubes. *Sep. Purif. Rev.* 43(4), 311–338.

Mubarak, N. M., Sahu, J. N., Abdullah, E. C., Jayakumar, N. S. (2016), Rapid adsorption of toxic Pb (II) ions from aqueous solution using multiwall carbon nanotubes synthesized by microwave chemical vapor deposition technique. *J. Environ. Sci.* 45, 143–155.

Mubarak, N. M., Sahu, J. N., Abdullah, E. C., Jayakumar, N. S., Ganesan, P. (2014c). Single stage production of carbon nanotubes using microwave technology. *Diamond Relat. Mater.* 48, 52–59.

Mubarak, N. M., Thobashinni, M., Abdullah, E. C., Sahu, J. N. (2016), Comparative kinetic study of removal of Pb^{2+} ions and Cr^{3+} ions from waste water using carbon nanotubes produced using microwave heating. *C—J. Carbon Res.* 2(1), 7.

Mudasir, M., Karelius, K., Aprilita, N. H., Wahyuni, E. T. (2016), Adsorption of mercury(II) on dithizone-immobilized natural zeolite. *J. Environ. Chem. Eng.* 4(2), 1839–1849.

Murgolo, S., Petronella, F., Ciannarella, R., Comparelli, R., Agostiano, A., Curri, M. L., Mascolo, G. (2015), UV and solar-based photocatalytic degradation of organic pollutants by nano-sized TiO_2 grown on carbon nanotubes. *Catal. Today* 240, 114–124.

Naghizadeh, A., Yari, A. R., Tashauoei, H. R., Mahdavi, M., Derakhshani, E., Rahimi, R., Bahmani, P., Daraei, H., Ghahremani, E. (2012), Carbon nanotubes technology for removal of arsenic from water. *Arch. Hyg. Sci.* 1(1), 6–11.

Nakamoto, K., Ohshiro, M., Kobayashi, T. (2017), Mordenite zeolite - polyethersulfone composite fibers developed for decontamination of heavy metal ions. *J. Environ. Chem. Eng.* 5(1), 513–525.

Nassereldeen, A., Muataz, A. A. Abdullah, A., Mohamed, E. S., Alam, M. D., Yahya, N. (2009), Kinetic adsorption of application of carbon nanotubes for Pb(II) removal from aqueous solution, *J. Environ. Sci.* 21, 539–544.

Ngo, C. L., Le, Q. T., Ngo, T. T., Nguyen, D. N., Vu, M. T. (2013), Surface modification and functionalization of carbon nanotube with some organic compounds. *Adv. Nat. Sci.: Nanosci. Nanotechnol.* 4(3), 035017.

Ngueta, G., Prevost, M., Deshommes, E., Abdous, B., Gauvin, D., Levallois, P. (2014), Exposure of young children to household water lead in the Montreal area (Canada): the potential influence of winter-to-summer changes in water lead levels on children's blood lead concentration. *Environ. Int.* 73, 57–65.

Piao, Y., Burns, A., Kim, J. Wiesner, U., Hyeon, T. (2008), Designed fabrication of silica-based nanostructured particle systems for nanomedicine applications. *Adv. Funct. Mater.* 18, 3745–3758.

Pourzamani, H., Bina, B., Amin, M. M., Rashidi, A. (2012), Monoaromatic pollutant removal by carbon nanotubes from aqueous solution. In *Advanced Materials Research*, 488, pp. 934–939.

Prasek, J., Drbohlavova, J., Chomoucka, J., Hubalek, J., Jasek, O., Adam, V., Kizek, R. (2011), Methods for carbon nanotubes synthesis. *J. Mater. Chem.* 21(40), 15872–15884.

Qu, X., Alvarez, P. J. J., Li, Q. (2013), Applications of nanotechnology in water and wastewater treatment. *Water Res.* 47, 3931–3946.

Rashed, A. O., Merenda, A., Kondo, T., Lima, M., Razal, J., Kong, L., Dumée, L. F. (2021), Carbon nanotube membranes-strategies and challenges towards scalable manufacturing and practical separation applications. *Sep. Purif. Technol.* 117929.

Reddy, M. L. P., Francis, T. (2001), Recent advances in the solvent extraction of mercury (II) with calixarenes and crown ethers. *Solvent Extr. Ion. Exch.* 19, 839–863.

Rosales-Landeros, C., Barrera-Díaz, C., Bilyeu, B., Guerrero, V. V., Ure, F. A. (2013), Review on Cr(VI) adsorption using inorganic materials. *Am. J. Anal. Chem.* 4, 8–16.

Sahraei, R., Pour, Z.S., Ghaemy, M. (2017), Novel magnetic bio-sorbent hydrogel beads based on modified gum tragacanth/graphene oxide: removal of heavy metals and dyes from water, *J. Clean. Prod.* 142, 2973–2984.

Salam, M. A. (2013), Coating carbon nanotubes with crystalline manganese dioxide nanoparticles and their application for lead ions removal from model and real water. *Colloids Surf. A: Physicochem. Eng. Aspects* 419, 69–79.

Sarkar, B., Mandal, S., Tsang, Y. F., Kumar, P., Kim, K.-H., Ok, Y. S. (2018), Designer carbon nanotubes for contaminant removal in water and wastewater: a critical review. *Sci. Total Environ.* 612, 561–581.

Seah, C. M., Chai, S. P., Mohamed, A. R. (2011), Synthesis of aligned carbon nanotubes. *Carbon* 49(14), 4613–4635.

Shah, K., Gupta, K., Sengupta, B. (2017), Selective separation of copper and zinc from spent chloride brass pickle liquors using solvent extraction and metal recovery by precipitation-stripping. *J. Environ. Chem. Eng.* 5(5), 5260–5269.

Shahbazi, A., Younesi, H., Badiei, A. (2011), Functionalized SBA-15 mesoporous silica by melamine-based dendrimer amines for adsorptive characteristics of Pb(II), Cu(II) and Cd(II) heavy metal ions in batch and fixed bed column, *Chem. Eng. J.* 168, 505–518.

Shetty, D., Boutros, S., Eskhan, A., De Lena, A. M., Skorjanc, T., Asfari, Z., Traboulsi, H., Mazher, J., Raya, J., Banat, F., Trabolsi, A. (2019), Thioether-crown-rich calix[4] arene porous polymer for highly efficient removal of mercury from water. *ACS Appl. Mater. Interfaces* 11, 12898–12903.

Shi, Z., Lian, Y., Liao, F. H., Zhou, X., Gu, Z., Zhang, Y., Zhang, S. L. (2000), Large scale synthesis of single-wall carbon nanotubes by arc-discharge method. *J. Phys. Chem. Solids* 61(7), 1031–1036.

Shim, H., Shin, E., Yang, S. T. (2002), A continuous fibrous-bed bioreactor for BTEX biodegradation by a co-culture of Pseudomonas putida and Pseudomonas fluorescens, *Adv. Environ. Res.* 7(1), 203–216.

Shorowordi, K. M., Laoui, T., Haseeb, A. S. M. A., Celis, J. P., Froyen, L., Kabiri, S., Tran, D. N., Altalhi, T., Losic, D. (2014), Outstanding adsorption performance of graphene–carbon nanotube aerogels for continuous oil removal. *Carbon* 80, 523–533.

Styblo, M., Del Razo, L. M., Vega, L., Germolec, D. R., LeCluyse, E. L., Hamilton, G. A., Reed, W., Wang, C., Cullen, W. R., Thomas, D. J. (2000), Comparative toxicity of trivalent and pentavalent inorganic and methylated arsenicals in rat and human cells. *Arch. Toxicol.* 74(6), 289–299.

Su, F., Lu, C., Hu, S. (2010), Adsorption of benzene, toluene, ethylbenzene and p-xylene by NaOCl oxidized carbon nanotubes. *Colloids Surf. A: Physicochem. Eng. Aspects* 353(1), 83–91.

Subramaniam, C., Yamada, T., Kobashi, K., Sekiguchi, A., Futaba, D. N., Yumura, M., Hata, K. (2013), One hundred fold increase in current carrying capacity in a carbon nanotube–copper composite. *Nat. Commun.* 4(1), 1–7.

Sun, Z., Liao, T., Dou, Y., Hwang, S. M., Park, M. S., Jiang, L., Dou, S. X. (2014), Generalized self-assembly of scalable two-dimensional transition metal oxide nanosheets. *Nat. Commun.* 5(1), 1–9.

Szabó, A., Perri, C., Csató, A., Giordano, G., Vuono, D., Nagy, J. B. (2010), Synthesis methods of carbon nanotubes and related materials. *Materials* 3(5), 3092–3140.

Tawabini, B., Al-Khaldi, S., Atieh, M., Khaled, M. (2010), Removal of mercury from water by multi-walled carbon nanotubes. *Water Sci. Technol.* 61(3), 591–598.

Thostenson, E. T., Ren, Z., Chou, T. W. (2001), Advances in the science and technology of carbon nanotubes and their composites: a review. *Composites Sci. Technol.* 61(13), 1899–1912.

Tian, Y., Gao, B., Morales, V. L., Chen, H., Wang, Y., Li, H. (2013), Removal of sulfamethoxazole and sulfapyridine by carbon nanotubes in fixed-bed columns. *Chemosphere* 90, 2597–2605.

Verma, B., Balomajumder, C. (2020), Surface modification of one-dimensional carbon nanotubes: a review for the management of heavy metals in wastewater. *Environ. Technol. Innov.* 17, 100596.

Vesali-Naseh, M., Naseh, M. R. V., Ameri, P. (2021), Adsorption of Pb (II) ions from aqueous solutions using carbon nanotubes: a systematic review. *J. Clean. Prod.* 125917.

Vincent, J. B. (2000), The biochemistry of chromium. *J. Nutr.* 130, 715–718.

Vinoba, M., Bhagiyalakshmi, M., Alqaheem, Y. Alomair, A. Perez, A. A., Rana, M. S. (2017), Recent progress of fillers in mixed matrix membranes for CO_2 separation: a review. *Sep. Purif. Technol.* 188, 431–450.

Vroman, I., Dujardin, M.C., Caze, C. (2000), Ion-exchange resins bearing thiol groups to remove mercury: Part 1: synthesis and use of polymers prepared from thioester supported resin. *React. Funct. Polym.* 43, 123–132.

Walford, S. N. (2008), Sugarcane bagasse: how easy is it to measure its constituents? *Proc. South African Sugarcane Technol. Assoc.* 81, 266–273.

Wang, H., Zhou, A., Peng, F., Yu, H., Yang, J. (2007), Mechanism study on adsorption of acidified multiwalled carbon nanotubes to Pb(II). *J. Coll. Interface Sci.* 316, 277–283.

Wang, X., Sun, J. Z., Tang, B. Z. (2018), Poly(disubstituted acetylene)s: advances in polymer preparation and materials application. *Prog. Polym. Sci.* 79, 98–120.

Wang, Y. P., Zhou, P., Luo, S. Z., Liao, X. P., Wang, B., Shao, Q., et al., (2018), Controllable synthesis of monolayer poly(acrylic acid) on the channel surface of mesoporous alumina for Pb(II) adsorption. *Langmuir* 34(26), 7859–7868.

Wang, Y., Hu, L., Zhang, G., Yan, T., Yan, L., Wei, Q., Du, B. (2017), Removal of Pb (II) and methylene blue from aqueous solution by magnetic hydroxyapatite-immobilized oxidized multi-walled carbon nanotubes. *J. Colloid Interface Sci.* 494, 380–388.

Wang, Y., Pan, C., Chu, W., Vipin, A. K., Sun, L. (2019). Environmental remediation applications of carbon nanotubes and graphene oxide: adsorption and catalysis. *Nanomaterials* 9(3), 439.

Werber, J. R., Osuji, C. O., Elimelech, M. (2016), Materials for next-generation desalination and water purification membranes. *Nat. Rev. Mater.* 1, 16018.

WHO (2018), Arsenic. https://www.who.int/news-room/fact-sheets/detail/arsenic.

Wibowo, N., Setyadhi, L., Wibowo, D., Setiawan, J., Ismadji, S. (2007), Adsorption of benzene and toluene from aqueous solutions onto activated carbon and its acid and heat treated forms: influence of surface chemistry on adsorption. *J. Hazardous Mater.* 146(1–2), 237–242.

Xu, D., Tan, X., Chen, C., Wang, X. (2008), Removal of Pb(II) from aqueous solution by oxidized multi-walled carbon nanotubes. *J. Hazard. Mater.* 154(1–3), 407–416.

Xu, P., Zeng, G., Huang, D., Feng, C., Hu, S., Zhao, M., Lai, C., Wei, Z., Huang, C., Xie, G., Liu, Z. (2012), Use of iron oxide nanomaterials in wastewater treatment: a review, *Sci. Total Environ.* 424, 1–10.

Yaghmaeian, K., Mashizi, R. K., Nasseri, S., Mahvi, A. H., Alimohammadi, M., Nazmara, S. (2015), Removal of inorganic mercury from aquatic environments by multiwalled carbon nanotubes. *J. Environ. Health Sci. Eng.* 13(1), 1–9.

Yan, W., Ramos, M. A. V., Koel, B. E., Zhang, W. (2010), Multi-tiered distributions of arsenic in iron nanoparticles: observation of dual redox functionality enabled by a core-shell structure, *Chem. Commun.* 46, 6995–6997.

Yang, W., Ding, P., Zhou, L., Yu, J., Chen, X., Jiao, F. (2013), Preparation of diamine modified mesoporous silica on multi-walled carbon nanotubes for the adsorption of heavy metals in aqueous solution. *Appl. Surf. Sci.* 282, 38–45.

Yi, W., Lu, L., Dian-Lin, Z., Pan, Z., Xie, S. (1999), Linear specific heat of carbon nanotubes. *Phys. Rev. B* 59(14), R9015.

Yu, F., Wu, Y., Ma, J., Zhang, C. (2013), Adsorption of lead on multi-walled carbon nanotubes with different outer diameters and oxygen contents: kinetics, isotherms and thermodynamics. *J. Environ. Sci.* 25(1), 195–203.

Yu, M. F., Lourie, O., Dyer, M. J., Moloni, K., Kelly, T. F., Ruoff, R. S. (2000). Strength and breaking mechanism of multiwalled carbon nanotubes under tensile load. *Science* 287(5453), 637–640.

Yu, X., Luo, T., Zhang, Y., Jia, Y., Zhu, B., Fu, X., Huang, X. (2011), Adsorption of lead(II) on O_2-plasma-oxidized multi-walled carbon nanotubes. *ACS Appl. Mater. Interfaces* 3(7), 2585–2593.

Zang, Z., Hu, Z., Li, Z., He Q., Chang, X. (2009), Synthesis, characterization and application of ethylenediamine-modified multiwalled carbon nanotubes for selective solid-phase extraction and preconcentration of metal ions. *J. Hazardous Mater.* 172, 958–963.

Zhang, H., Feng, P., Akhtar, F. (2017), Aluminium matrix tungsten aluminide and tungsten reinforced composites by solid-state diffusion mechanism. *Sci. Rep.* 7, 12391.

Zhang, Y., Iijima, S. (1999), Formation of single-wall carbon nanotubes by laser ablation of fullerenes at low temperature. *Appl. Phys. Lett.* 75(20), 3087–3089.

Zhang, Y., Jin, B., Ma, B., Feng, X. (2017b), Separation of indium from lead smelting hazardous dust via leaching and solvent extraction. *J. Environ. Chem. Eng.* 5(3), 2182–2188.

Zhao, Z. Y., Zhao, W. J., Bai, P. K., Wu, L. Y., Huo, P. C. (2019), The interfacial structure of Al/Al_4C_3 in graphene/Al composites prepared by selective laser melting: first-principles and experimental. *Mater. Lett.* 255, 126559.

Zhou, B., Li, Y., Zheng, G., Dai, K., Liu, C., Ma, Y., et al., (2018), Continuously fabricated transparent conductive polycarbonate/carbon nanotube nanocomposite films for switchable thermochromic applications. *J. Mater. Chem. C* 6(31), 8360–8371.

Zhu, H. W., Xu, C. L., Wu, D. H., Wei, B. Q., Vajtai, R., Ajayan, P. M. (2002), Direct synthesis of long single-walled carbon nanotube strands. *Science* 296(5569), 884–886.

4 Polymer-Based Hybrid Adsorbents for Water Remediation

Chandra Shekhar Kushwaha, S K Shukla, and NB Singh
University of Delhi
Sharda University

CONTENTS

4.1 INTRODUCTION

Polymer-based hybrid (PBH) materials, i.e. nanocomposites, hydrogels and membranes, exhibit several advanced surface area, mechanical and chemical properties for wide range of applications such as efficient water purification. These materials are comprised of both organic polymers and inorganic fillers with their indigenous properties such as processability of polymers and selective catalytic interactivity of inorganic fillers. Thus, the synergistic effect among the properties of polymer and non-polymer materials evolves unique novel features for different applications, viz. chemical sensors, automobiles, support catalysts, packaging, textile and adsorbents for water purification [1]. These properties are further

DOI: 10.1201/9781003129042-4

positively optimized with size and dimensional confinements, and different polymeric nanostructure-based adsorbents and membranes are developed for effective elimination of different impurities such as heavy metals, dyes, pharmaceuticals and hydrocarbons [2,3].

However, the use of adsorbent is an ancient route to the effective removal of different impurities present in water with several advantageous features such as reusability, low cost, easy processability and availability. In this context, several adsorbents such as activated carbon, metal oxides, agro-wastes and polymer nanocomposites are explored in water purification. However, the PBH adsorbents bear several advantageous features in terms of stability, processability, reusability and effective purification ability [4]. The importance of PBH adsorbents is supported by the consistence increase in the use of PBH for water purification. This exponential use of PBH in water purification has also posed the importance of compilation of recent findings on the topic for researchers and technologists. In continuation to above discussion, this chapter compiles the advances on the synthesis and applications of PBH adsorbents in water purification along with technical details. Although, the consistence in increase in publication frequency is a critical conditions towards selection of paper but an honest effort has been made to screen the paper on the area based on their indigenous contributions.

4.2 OVERVIEW OF POLYMER-BASED HYBRID ADSORBENTS

The integration of chemical composition, structural manipulation and size confinement has yielded different types of polymer-based adsorbents with high water purification ability due to huge surface area, water-retaining capacity, selective adsorption and catalytic nature. In general, the pristine polymers, polymer composites and their modified structures are used as adsorbents for the purification of water. The pristine polymers are functionalized as well as chemically modified to develop the specific properties such as selectivity to suitable adsorption of different impurities. For example, the chemical modification of cyclodextrin by polydopamine yields an adsorbent with carboxyl and catechol functionality for electrostatic interactive removal of different dyes, i.e. methylene blue, malachite green, crystal violet and cupric ions [5]. Similarly, the heterogeneous chemical structure and morphology introduce superior adsorption behaviours along with catalytic properties for the effective removal of metals, dyes and organic impurities. The presence of selective metals generates specific path for catalytic degradation-induced water purification of organic impurities. The grafting, branch manipulation and induction of magnetic particles add porosity and reversible interactivity for the effective purification of polluted water [6]. The processing of adsorbents is another area for improvements in effectiveness of polymer adsorbents in the form of beads, membranes and hydrogels. Some significant representative polymer adsorbents are depicted in Table 4.1 along with their advantageous properties and remarks.

TABLE 4.1
Representative Polymer-Based Hybrid Adsorbents

S. No.	Polymer Hybrids	Properties	Applications	Ref.
1	Polypyrrole/chitosan/graphene oxide	Exfoliated magnetic sheet	Removal of dyes	[7]
2	Lignin/silica hybrid	Functionalized high adsorption capacity	Methylene blue	[8]
3	2D-MoS_2, WS_2 and poly(lactic acid)	Improved adsorption capacity	Oil	[9]
5	Bimetallic PANI	A p–n junction	Photocatalytic degradation and bacteria inactivation	[10]
6	Graphene oxide, nano-polyaniline and zirconium silicate	Multifunctional	Dyes	[11]
7	Halloysite–Cyclodextrin	Electrostatic interactions-based adsorption	Rhodamine B	[12]

4.3 PREPARATION AND CHEMICAL MODIFICATIONS

Both direct and indirect methods are regularly explored for the preparation of PBH adsorbents with their inherited advantages and disadvantages [2]. In the direct method, separately prepared [7] constituents solution are mixed by mechanical or thermal energy (mechanical dispersion and melt compounding). However, in the indirect methods the monomers are polymerized in the presence of fillers using different polymerizing agents, i.e. chemicals, electricity, light, heat and enzymes. A brief comparison of different methods is given in Table 4.2 along with suitable references.

The dispersion of inorganic filler is a basic tool to improve the properties of polymer-based adsorbents. Shukla et al. prepared a highly dispersed MO-doped urea–formaldehyde composite. This obtained CuO-encapsulated urea–formaldehyde resin-based adsorbent was used for the effective removal of As(III) present in water

TABLE 4.2
Comparison of Different Polymerizing Methods

S. No.	Method	Advantages	Disadvantages
1	Chemical	Easy to perform with scale-up possibilities	Impurities of solvents
2	Electrical	Purity along with simultaneous doping	Costly and limitations in scaling up
3	Photopolymerization	Eco-friendly	Costly photosensitizer

sample by 69% along with reusability [13]. The use of both hard and soft templates is exercised for controlling the size and shape of adsorbents. The mostly used hard templates are polymer microspheres, porous membranes, plastic foam, ion exchange resins, carbon fibres and porous anodic aluminium oxides. These templates are used for the preparation of highly ordered nanohybrid structures with different sets of advantageous properties and limitation of their removed after use [14]. The soft templates are also used with flexible structure and inherited advantageous features for the preparation of different nanostructure adsorbents. The representatives of soft templates used are surfactants, polymers, organic molecules and biopolymers with broader prospective and intermolecular interaction [15]. The use of templates directly controls the agglomeration, dispersion of fillers along the controls and dimension of resultant hybrid adsorbents for better surface properties, but they all add cost during commercialization [16]. Furthermore, the composite structure of different polymers is present in different natural products such as rice husk, coir, fruit peels and straw. These composite materials are abundantly available throughout the world throughout the year. These naturally occurring polymer composite structures are also explored for water purification after requisite physical and chemical treatments. In example, the chemical treatment of rice husk develops porous structure with lesser bulk density and water-retaining capacity. Thus, the obtained chemically modified rice husk was found suitable for the effective removal of copper (II) and arsenic (III) by up to 90% and 93% with regeneration possibility [17].

4.4 APPLICATIONS IN WATER REMEDIATION

A comparison among the properties of different adsorbents shows that PBH adsorbents exhibit several advantageous features such as cost-effective availability and prone for a wide range of functionalization and modifications. Thus, the PBH adsorbents are used for the removal of different water pollutants due to their properties, as shown in Figure 4.1.

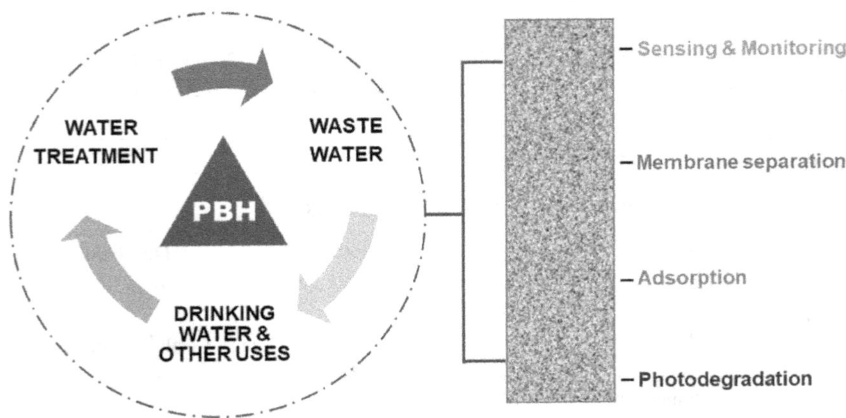

FIGURE 4.1 Use of PBH for water purification.

4.5 DYES AND ORGANIC COMPOUNDS

Dyes and frequently used organic compounds such as hydrocarbons, pharmaceuticals and agrochemicals are significant water pollutants with sever adverse effects on health and life of humans, animals and other aquatic organisms. The consumption of dyes by fish and aquatic organisms is also responsible for their biomagnification, while their atmospheric degradation with time generates secondary pollutants with serious atmospheric impacts. Therefore, the precise removal of these pollutants is very important for both society and industries. The PBH adsorbents are frequently used in this regard due to their selective interactive adsorption and oxidative degradation. For example, chitosan is a well-documented saccharide-based biopolymer with the potential to use in several applications after due modifications. A simple case study on chitosan was presented by Bilal et.al. for the removal of textile dyes along with potential details and effectiveness after chemical medications such as grafting and incorporation of metal oxides [18,19]. Grafting is another tool to modify chitosan to develop functionality and interactivity along with the retention of the significant properties of chitosan for better adsorption of dyes. Meenakshi et al. prepared a titania-impregnated chitosan-grafted hydroxyapatite ternary adsorbent for the removal of organic dyes such as methylene blue and rhodamine B with excellent photostability and reusability. The tailored surface reactivity of the hydroxyapatite due to grafting converts the adsorbed dyes into simple organic molecules, and the proposed reaction scheme is illustrated in Figure 4.2 [20].

The other PBH adsorbents for dye removal are conducting polymer composites, biopolymer-grafted hydrogels, metal polymer composites and chemically modified lignocellulosic polymers. Although significant application of PBH are reported for dye removal but complete removal is still need to achieve along with reusability of adsorbent. The hydrocarbons, agrochemicals and residual pharmaceuticals are other hazardous pollutants with weak hazards. The principle

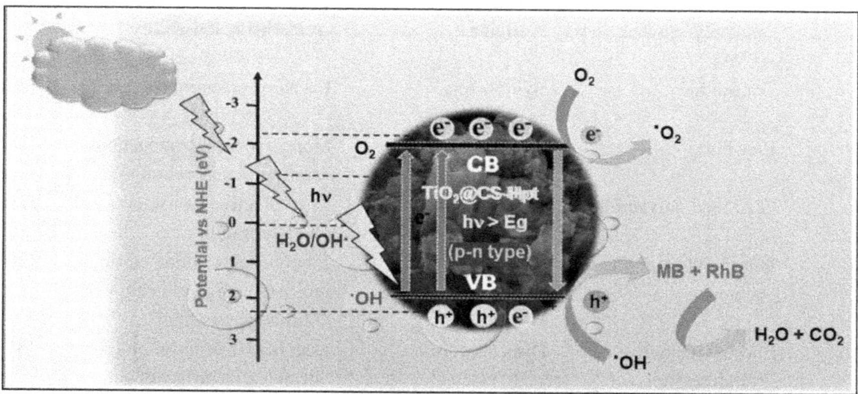

FIGURE 4.2 Diagrammatic sketch of catalytic removal of dyes over TiO_2@CS-Hpt system. (See Ref. [20].)

involved in the removal of pharmaceuticals is similar to that of organic dyes due to their similar functionality with dyes. However, the removal of hydrocarbon by π–π stacking is explored due to their nonpolar nature. In this context, the porosity in adsorbents has a significant impact on adsorption along hydrogen bonding, π–π stacking and cation–π interactions ability for the removal of different hydrocarbon like poly alicyclic hydrocarbons and other phenolic compounds. The porosity is in general incorporated by the addition of different metal oxides and different carbon nanostructures [21]. Some other significant PBH adsorbents are tabulated in Table 4.3, along with their removal properties.

TABLE 4.3
Illustration of Significant PBH Adsorbents

S. No.	Adsorbent	Pollutants	Properties	Ref.
1	Fe_3O_4/graphite oxide/ citric acid-cross-linked β-cyclodextrin	Methylene blue	Host–guest supramolecular interaction and the π–π interaction based removal	[22]
2	Graphene oxide and β-cyclodextrin	For anionic dyes, i.e. lemon yellow, sunset yellow and Ponceau 4R	Langmuir adsorption isotherm followed; excellent adsorption capacity	[23]
3	Ce(III)/ chitosan/β-cyclodextrin	Toxic dyes	Electrostatic attraction, hydrogen bonding and surface complexation interactions-based removal	[24]
4	Cellulose-grafted acrylic acid-based superadsorbing hydrogels	Methylene blue	70% Removal capacity with the Langmuir model and the pseudo-second-order.	[25]
5	Coffee-grounds/chitosan and poly(vinyl alcohol) (PVA)	Acetaminophen and caffeine	Cost-effective due to remarkable reusability	[26]
6	Tea residue	Hydralazine hydrochloride	Excellent adsorption capacity, i.e. 131.63 mg/g	[27]
7	Rice husk	Metronidazole	Monolayer and homogenous surface of adsorption	[28]
8	Fe_3O_4 and polyaniline	Polyaromatic hydrocarbons	Cost-effective preparation and high reusability	[29]
9	Poly(vinyl alcohol) and bentonite	Dyes	Improved adsorption capacities for wide range of methylene blue concentrations	[30]
10	Carboxymethyl cellulose/alginate/ poly(vinyl alcohol)/rice husk	Dyes	Degradative removal of dyes from textile effluents	[31]

4.6 HEAVY METAL REMOVAL

The charged ionic structure of adsorbents with porous nature has been shown to be effective for the removal of different heavy metal ions. The source and health impacts of different heavy metal pollutants are listed in Table 4.4.

TABLE 4.4
Limits, Sources and Health Effects of Various Heavy Metal Ion Contaminations [32]

S. No.	Heavy Metals	Common Sources	Health Impacts	WHO Limits & BIS Limits (mg/L)
1	Lead (Pb)	PVC pipes in sanitation, agriculture, lead paints, jewellery, lead batteries, lunch boxes, etc.	Penetrates through protective blood–brain barrier (BBB) and a risk factor for Alzheimer's disease and senile dementia. It decreases IQ, causes kidney damage, decreases bone growth and causes behavioural issues, ataxia, hyperirritability and stupor.	0.05 & 0.05
2	Cadmium (Cd)	Paints, pigments, batteries, plastics, synthetic rubber, photographic, photoconductors and photovoltaic cells	Renal toxicity, hypertension, weight loss, fatigue, anaemia, lymphocytosis, pulmonary fibrosis, atherosclerosis, peripheral neuropathy, lung cancer, osteoporosis and hyperuricemia.	0.005 & 0.01
3	Mercury (Hg)	Combustion of coal, municipal solid waste incineration and volcanic emissions	Impaired neurologic development, effects on digestive system, immune system, lungs, kidneys, skin and eyes, and hypertension.	0.001 & 0.001
4	Arsenic (Ar)	Wooden electricity poles, arsenic-based preservatives, pesticides, fertilizers, release of untreated effluents, oxidation of arsenopyrite (FeAsS)	Causes effects on central nervous system (CNS), peripheral nervous system (PNS) and gastrointestinal tract (GI) and causes cardiovascular and pulmonary diseases, genitourinary (GU) cancer, localized oedema and skin cancer.	0.05 & 0.05
5	Chromium (Cr)	Leather industry, tanning and chrome plating industries	Reproductive toxicity, embryotoxicity, mutagenicity, carcinogenicity, lung cancer, dermatitis, skin ulcers, perforation of septum and irritant contact dermatitis.	0.05 & 0.05
6	Copper (Cu)	Fertilizers, tanning and photovoltaic cells	Adrenocortical hyperactivity, allergies, anaemia, alopecia, arthritis, autism, cystic fibrosis, diabetes, haemorrhage and kidney disorders.	1.3 & 1.3

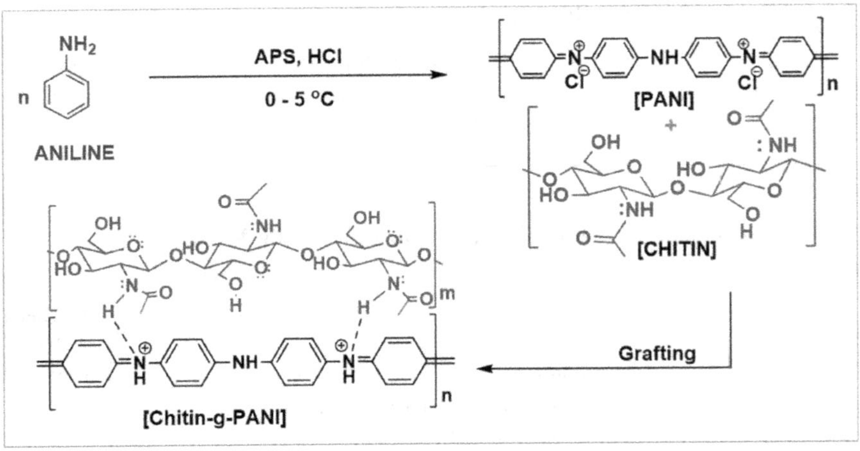

FIGURE 4.3 Synthesis scheme and interacting behaviour of chitin-g-PANI. (See Ref. [33].)

Chemical modification and impregnation of metallic ions and metal oxides liberate free coordinating group as well as generate polarity-induced ions to effectively adsorb metal ions. For example, Shukla et al. reported the liberation of carbonyl groups of chitin after grafting of polyaniline [33]. Thus, the liberated carbonyl efficiently interacted with residual cupric ions present in waste water in the concentration range of 1–1000 ppm at moderate pH. The preparative scheme and interacting behaviour of chitin-grafted polyaniline is shown in Figure 4.3 [33].

Furthermore, the agricultural biomass consists of lignocellulosic polymer compact network along with other metals and non-metal oxides. The chemical modification develops fascinating surface properties for adsorption-based removal of different heavy metals from water [34]. The development of magnetic property in adsorbents is another aspect towards advancing the heavy metal removal capacity. Xiao et al. prepared a magnetic biochar after co-precipitation of ferric and ferrous ions on rice straw. The adsorbent was found suitable for effective adsorption of cadmium ions due to ion exchange properties of the adsorbent [35]. The incorporation of magnetic properties in PBH adds another property for the effective removal of heavy metals. The general process to add magnetic properties is the incorporation of metals and their metal oxides [36–38]. Some other applications of PBH adsorbents are given in Table 4.5 along with their remediation properties.

The shortage of water in different parts of the world towards multiple uses also stimulates the scientists to invent appropriate treatments for the reuse of the consumed water. Thus, the removal of wide spectrum of metallic impurities requires advanced adsorbents. For example, Shukla et.al. proposed a proto-plant to purify sewage water with the help of a rice husk-derived adsorbent in significant proportion, with the probability to reuse the sewage water [49]. The proposed set-up is shown in Figure 4.4, along with water quality.

TABLE 4.5
Removal of Heavy Metals on PBH Adsorbents

S. No.	Adsorbent	Heavy Metal	Removal Parameters	Ref.
1	Polymer–clay composite	Cadmium	Excellent adsorption capacity, i.e. 20,400 ± 13 mg/L, and adsorption rate constant ≈$7.45 \times 10^{-3} \pm 0.0002$ L/(min mg) at 50% breakthrough	[39]
2	Chitosan, poly(methacrylic acid) and halloysite nanotube	Pb(II) and Cd(II)	The adsorption capacity of 357.4/89.4 and 341.6/85.4 for Pb(II) and Cd(II), respectively	[40]
3	Starch and ZnO	Lead	Temperature-dependent adsorption	[41]
4	Chitosan, bone powder and iron oxide	Radioactive caesium metal	Best-fitted model is Langmuir	[42]
5	Metal–organic frameworks-doped alginate beads	Chromium (VI)	Pseudo-second-order kinetics with Langmuir isotherm model	[43]
6	$SiO_2/CuFe_2O_4$/polyaniline	Cu, Fe and Mn	Electrostatic attraction-based chelation	[44]
7	Magnetic β-cyclodextrin polymer	Cr, Pb, Zn and Cu	Room-temperature regeneration	[45]
8	Bentonite clay and *M. oleifera* seed biopolymers	Cadmium, chromium and lead	Functionalized interactions	[45]
9	Nickel oxide-intercalated chitosan-grafted polyaniline	Lead	Extraction and recovery	[46]
10	Poly(amidoamine)-functionalized graphene oxide	Cd and As	Electrostatic interactions, multilayer and π–π interactions induced adsorption	[47]
11	Sodium alginate, carrageenan and bentonite	Fe^{3+}, Ni^{2+} and Cr^{3+}	Effective adsorption and reusability	[48]

4.7 MICROBES AND BACTERIA

The removal microbes needs specific types of adsorbents with the ability to oxidize or decompose the cell wall of microbes. In this regard, photoinduced catalysis has been an important tool for the removal of microbes from water [50]. The basic mechanism of photocatalysts-based microbial removal is shown in Figure 4.5, along with the basic principle of the photocatalytic degradation of microbe cells after interaction with the photocatalyst [51].

In general, the microbes engulf the photocatalysts, which in turn photodegrade the outer cell wall through redox reaction initiated by photogenerated electrons. Here, the microbicidal efficiency depends on the material parameters and properties such as band gap and exciting abilities. Therefore, several PBH photocatalysts are synthesized and used for remediation purposes [52–54]. Some of the

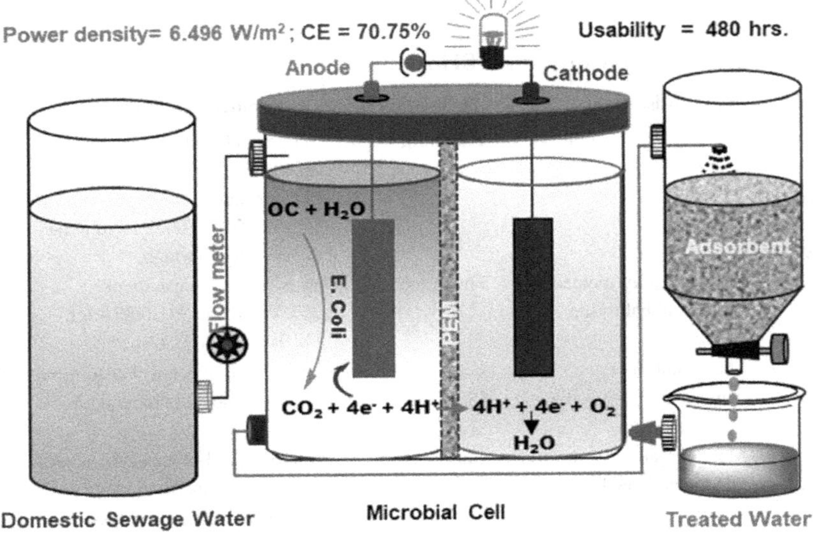

FIGURE 4.4 Water purification chamber for the treatment of water.

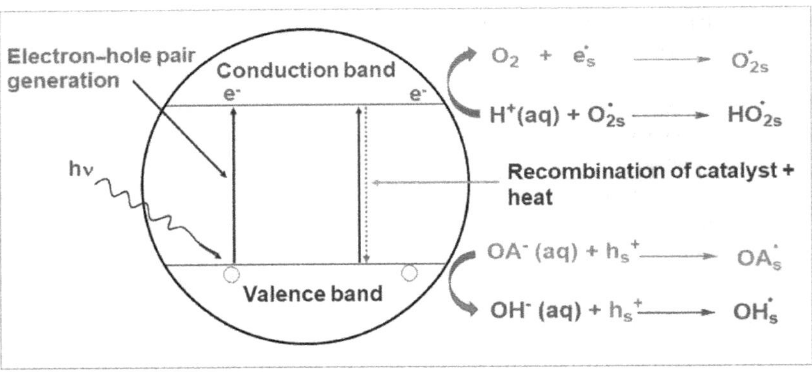

FIGURE 4.5 Schematic diagram showing photocatalysis on semiconducting surface. (See Ref. [51].)

photocatalysts used for the removal of microbes are given in Table 4.6, along with brief efficacy.

4.8 ADSORPTION BEHAVIOUR

4.8.1 Isotherm

The impregnation of different fillers in flexible polymeric matrix improves the reversibility of adsorption behaviour as well as its use in water purification. In general, the lowering in size improves the surface area for improved adsorption

TABLE 4.6
Some Typical Adsorbents for Removal of Microbes

S. No.	PBH Photocatalysts	Microbes	Remarks	Ref.
1	ZnO-ZnS@ PANI	*E. coli*	Most effective at 100 μg dose of catalyst with 600 nM light irradiation. 2 cfu/mL total bacterial count and 20±1 mm inhibition zone.	[55]
2	Fe_3O_4-PPy/ chitosan/GO	*E. coli* and *Fusarium*	Catalyst has a good inhibition activity against *E.coli* and *Fusarium* with inhibition zone (3, 1.8 cm).	[7]
3	CMC-PAA-PANI	*P. aeruginosa*, *S. aureus* and *B. subtilis*	Zones of growth inhibition (mm) by hybrid film are 21.0±0.7, 20.0±1.4, and 20.0 ±1.4, respectively.	[56]
4	CMC-chitosan	*E. coli* and *S. aureus*	Zones of growth inhibition in mm are 14.0±0.1 and 16.0±0.2, respectively.	[57]
5	PAA-Ag composite	*E. coli* and *S. aureus*	Zones of growth inhibition in mm using the hybrid polymer are 3.6±0.1 and 2.0±0.3, respectively.	[58]
6	*PANI/Au–Pd*	*Streptococcus* sp., *Staphylococcus* sp., *E. coli* and *Klebsiella* sp.	The polymer hybrid shows antibacterial activity against *E. coli* (25±0.87 mm dia) followed by *Staphylococcus* sp. (22±0.35 mm dia), *Streptococcus* sp. (21±0.33 mm dia) and *Klebsiella* sp. (21±0.31 mm dia).	[59]
7	PVP-CD hydrogels	*E. coli* and *S. aureus*	The hydrogels were found to effectively annihilate both gram-positive and gram-negative bacteria in real polluted water in less than 10 min of photoexcitation.	[60]
8	Ag/MWNTs on the surface of PAN	*E. coli*	Ag/MWNTs/PAN exhibited efficacy in killing *E. coli* bacteria (about >99%).	[61]

capacity. The nature of adsorption is represented by adsorption isotherm, which demonstrates the importance of pressure on adsorption capacity at specified temperature. The important isotherm models used in water purification are Langmuir, Freundlich, Dubinin–Radushkevich, Temkin, Flory–Huggins, Redlich–Peterson, Sips, Khan, BET and FHH [62,63].

4.9 KINETICS

The interacting nature and energy of adsorbate with adsorbent involve in the efficacy of remediation. This study used to study mass transports and adsorption-induced purification of different water pollutants. In general, the kinetics involved are different orders of reaction along with peculiar types of adsorption energy [64,65].

4.10 THERMODYNAMICS

The exothermic nature of adsorption dictates the exchange of heat during the adsorption process between pollutants and adsorbents. Here, the interesting nature synchronized the proportionality between enthalpy and entropy of reactions. Further, the extent of energy involved in adsorption process also used to calculate change in activation energy, enthalpy, entropy and Gibbs free energy, which controls the regeneration of used adsorbent [66,67].

4.11 REGENERATION

The repetitive use of adsorbents depends on the possibility of regeneration, which is the removal of adsorbed water pollutant [68,69]. The processes recommended for the regeneration of adsorbents are washing and treatment with different reagents such as acid and alkali. Currently, the irradiation of adsorbent by UV radiation is significantly reported for the removal of adsorbed organic impurities such as pesticides and pharmaceutical compounds. Wu et al. reported one-pot preparation of a polymer adsorbent with superadsorption capacity along with quenching-induced sensing and removal of organic pollutants. Further, the adsorbent exhibits excellent regeneration for its repetitive use in sensing and remediation. The sample was regenerated by simple sonication-induced repetitive washing using the desired solution for reusability [70].

4.12 CONCLUSIONS AND FUTURE PROSPECTS

The significance and properties of polymer-based hybrid adsorbents are described along with their potentials. The suitability of PBH adsorbents is described for the removal of different pollutants, i.e. heavy metals, organic molecules and microbes, along with a brief of adsorption isotherms, kinetics and thermodynamic parameters.

REFERENCES

[1] C.-S. Ha, Polymer based hybrid nanocomposites; a progress toward enhancing interfacial interaction and tailoring advanced applications, *The Chemical Record.* 18 (2018) 759–775. doi: 10.1002/tcr.201700030.

[2] N. Pandey, S.K. Shukla, N.B. Singh, Water purification by polymer nanocomposites: an overview, *Nanocomposites.* 3 (2017) 47–66. doi: 10.1080/20550324.2017.1329983.

[3] W.A. Adeosun, D.F. Katowah, A.M. Asiri, M.A. Hussein, Conducting terpolymers and its hybrid nanocomposites variable trends. From synthesis to applications. A review, *Polymer-Plastics Technology and Materials.* 60 (2021) 271–285. doi: 10.1080/25740881.2020.1811316.

[4] R. Mukhopadhyay, D. Bhaduri, B. Sarkar, R. Rusmin, D. Hou, R. Khanam, S. Sarkar, J. Kumar Biswas, M. Vithanage, A. Bhatnagar, Y.S. Ok, Clay–polymer nanocomposites: Progress and challenges for use in sustainable water treatment, *Journal of Hazardous Materials.* 383 (2020) 121125. doi: 10.1016/j.jhazmat.2019.121125.

[5] H. Chen, Y. Zhou, J. Wang, J. Lu, Y. Zhou, Polydopamine modified cyclodextrin polymer as efficient adsorbent for removing cationic dyes and Cu2+, *Journal of Hazardous Materials.* 389 (2020) 121897. doi: 10.1016/j.jhazmat.2019.121897.

[6] Q. Kong, X. Shi, W. Ma, F. Zhang, T. Yu, F. Zhao, D. Zhao, C. Wei, Strategies to improve the adsorption properties of graphene-based adsorbent towards heavy metal ions and their compound pollutants: a review, *Journal of Hazardous Materials.* 415 (2021) 125690. doi: 10.1016/j.jhazmat.2021.125690.

[7] N.A. Salahuddin, H.A. El-Daly, R.G. El Sharkawy, B.T. Nasr, Nano-hybrid based on polypyrrole/chitosan/grapheneoxide magnetite decoration for dual function in water remediation and its application to form fashionable colored product, *Advanced Powder Technology.* 31 (2020) 1587–1596. doi: 10.1016/j.apt.2020.01.030.

[8] T.M. Budnyak, S. Aminzadeh, I. V. Pylypchuk, D. Sternik, V.A. Tertykh, M.E. Lindström, O. Sevastyanova, Methylene Blue dye sorption by hybrid materials from technical lignins, *Journal of Environmental Chemical Engineering.* 6 (2018) 4997–5007. doi: 10.1016/j.jece.2018.07.041.

[9] T. Krasian, W. Punyodom, P. Worajittiphon, A hybrid of 2D materials (MoS2 and WS2) as an effective performance enhancer for poly(lactic acid) fibrous mats in oil adsorption and oil/water separation, *Chemical Engineering Journal.* 369 (2019) 563–575. doi: 10.1016/j.cej.2019.03.092.

[10] S.W. Lv, J.M. Liu, F.E. Yang, C.Y. Li, S. Wang, A novel photocatalytic platform based on the newly-constructed ternary composites with a double p-n heterojunction for contaminants degradation and bacteria inactivation, *Chemical Engineering Journal.* 409 (2021) 128269. doi: 10.1016/j.cej.2020.128269.

[11] M.E. Mahmoud, M.F. Amira, S.M. Seleim, G.M. Nabil, M.E. Abouelanwar, Multifunctionalized graphene oxide@nanopolyaniline@zirconium silicate nanocomposite for rapid microwable removal of dyes, *Journal of Nanostructure in Chemistry.* (2021) 1–18. doi: 10.1007/s40097-021-00390-0.

[12] M. Massaro, C.G. Colletti, G. Lazzara, S. Guernelli, R. Noto, S. Riela, Synthesis and characterization of halloysite-cyclodextrin nanosponges for enhanced dyes adsorption, *ACS Sustainable Chemistry and Engineering.* 5 (2017) 3346–3352. doi: 10.1021/acssuschemeng.6b03191.

[13] R.P. Rastogi, N.B. Singh, S.K. Shukla, Nano-size copper oxide encapsulated urea - formaldehyde resin film for arsenic (III) removal from aqueous solutions, *Indian Journal of Engineering and Materials Sciences.* 18 (2011) 390–392.

[14] N. Ren, Y. Tang, Template-induced assembly of hierarchically ordered zeolite materials, *Petrochemical Technology.* 34 (2005) 405–411. https://en.cnki.com.cn/Article_en/CJFDTotal-SYHG200505000.htm (accessed April 26, 2021).

[15] R.R. Poolakkandy, M.M. Menamparambath, Soft-template-assisted synthesis: A promising approach for the fabrication of transition metal oxides, *Nanoscale Advances.* 2 (2020) 5015–5045. doi: 10.1039/d0na00599a.

[16] S. Fu, Z. Sun, P. Huang, Y. Li, N. Hu, Some basic aspects of polymer nanocomposites: a critical review, *Nano Materials Science.* 1 (2019) 2–30. doi: 10.1016/j.nanoms.2019.02.006.

[17] S.K. Shukla, S. Nidhi, N. Pooja, A. Charu, M. Silvi, A.B. Rizwana, G.C. Dubey, Metal decontamination from chemically modified rice husk film, *Advanced Materials Letters.* 5 (2014) 352–355. doi: 10.5185/amlett.2014.1018.

[18] S.A. Qamar, M. Ashiq, M. Jahangeer, A. Riasat, M. Bilal, Chitosan-based hybrid materials as adsorbents for textile dyes–a review, *Case Studies in Chemical and Environmental Engineering.* 2 (2020) 100021. doi: 10.1016/j.cscee.2020.100021.

[19] D.C. da Silva Alves, B. Healy, L.A. Pinto, T.R. Cadaval, C.B. Breslin, Recent developments in chitosan-based adsorbents for the removal of pollutants from aqueous environments, *Molecules*. 26 (2021) 594. doi: 10.3390/molecules26030594.

[20] S. Vigneshwaran, P. Sirajudheen, C.P. Nabeena, S. Meenakshi, In situ fabrication of ternary TiO2 doped grafted chitosan/hydroxyapatite nanocomposite with improved catalytic performance for the removal of organic dyes: experimental and systemic studies, *Colloids and Surfaces A: Physicochemical and Engineering Aspects*. 611 (2021) 125789. doi: 10.1016/j.colsurfa.2020.125789.

[21] Z.U. Zango, N.S. Sambudi, K. Jumbri, A. Ramli, N.H.H.A. Bakar, B. Saad, M.N.H. Rozaini, H.A. Isiyaka, A.M. Osman, A. Sulieman, An overview and evaluation of highly porous adsorbent materials for polycyclic aromatic hydrocarbons and phenols removal from wastewater, *Water*. 12 (2020) 1–40. doi: 10.3390/w12102921.

[22] L.W. Jiang, F.T. Zeng, Y. Zhang, M.Y. Xu, Z.W. Xie, H.Y. Wang, Y.X. Wu, F.A. He, H.L. Jiang, Preparation of a novel Fe3O4/graphite oxide nanosheet/citric acid-crosslinked β-cyclodextrin polymer composite to remove methylene blue from water, *Advanced Powder Technology*. 32 (2021) 492–503. doi: 10.1016/j.apt.2020.12.026.

[23] Y. Wu, Z. Jia, C. Bo, X. Dai, Preparation of magnetic β-cyclodextrin ionic liquid composite material with different ionic liquid functional group substitution contents and evaluation of adsorption performance for anionic dyes, *Colloids and Surfaces A: Physicochemical and Engineering Aspects*. 614 (2021) 126147. doi: 10.1016/j.colsurfa.2021.126147.

[24] P. Sirajudheen, P. Karthikeyan, S. Vigneshwaran, M. Nikitha, C.A.A. Hassan, S. Meenakshi, Ce(III) networked chitosan/β-cyclodextrin beads for the selective removal of toxic dye molecules: adsorption performance and mechanism, *Carbohydrate Polymer Technologies and Applications*. 1 (2020) 100018. doi: 10.1016/j.carpta.2020.100018.

[25] Y. Zhou, S. Fu, H. Liu, S. Yang, H. Zhan, Removal of methylene blue dyes from wastewater using cellulose-based superadsorbent hydrogels, *Polymer Engineering & Science*. 51 (2011) 2417–2424. doi: 10.1002/pen.22020.

[26] E.F. Lessa, M.L. Nunes, A.R. Fajardo, Chitosan/waste coffee-grounds composite: an efficient and eco-friendly adsorbent for removal of pharmaceutical contaminants from water, *Carbohydrate Polymers*. 189 (2018) 257–266. doi: 10.1016/j.carbpol.2018.02.018.

[27] C.S. Patil, D.B. Gunjal, V.M. Naik, N.S. Harale, S.D. Jagadale, A.N. Kadam, P.S. Patil, G.B. Kolekar, A.H. Gore, Waste tea residue as a low cost adsorbent for removal of hydralazine hydrochloride pharmaceutical pollutant from aqueous media: an environmental remediation, *Journal of Cleaner Production*. 206 (2019) 407–418. doi: 10.1016/j.jclepro.2018.09.140.

[28] H. Azarpira, D. Balarak, Rice husk as a biosorbent for antibiotic metronidazole removal: Isotherm studies and model validation, *International Journal of ChemTech Research*. 9 (2016) 566–573. https://www.scopus.com/inward/record.uri?eid=2-s2.0-84982221268&partnerID=40&md5=15165ec23bfe7d7f06e015e3d54e6515 (accessed March 30, 2021).

[29] Q. Zhou, Y. Wang, J. Xiao, H. Fan, C. Chen, Preparation and characterization of magnetic nanomaterial and its application for removal of polycyclic aromatic hydrocarbons, *Journal of Hazardous Materials*. 371 (2019) 323–331. doi: 10.1016/j.jhazmat.2019.03.027.

[30] L.M. Sanchez, R.P. Ollier, V.A. Alvarez, Sorption behavior of polyvinyl alcohol/bentonite hydrogels for dyes removal, *Journal of Polymer Research*. 26 (2019) 1–8. doi:10.1007/s10965-019-1807-4.

[31] H.N. Bhatti, Y. Safa, S.M. Yakout, O.H. Shair, M. Iqbal, A. Nazir, Efficient removal of dyes using carboxymethyl cellulose/alginate/polyvinyl alcohol/rice husk composite: Adsorption/desorption, kinetics and recycling studies, *International Journal of Biological Macromolecules*. 150 (2020) 861–870. doi:10.1016/j.ijbiomac.2020.02.093.
[32] M.B. Gumpu, S. Sethuraman, U.M. Krishnan, J.B.B. Rayappan, A review on detection of heavy metal ions in water - An electrochemical approach, *Sensors and Actuators, B: Chemical*. 213 (2015) 515–533. doi:10.1016/j.snb.2015.02.122.
[33] V.K. Singh, C.S. Kushwaha, S.K. Shukla, Potentiometric detection of copper ion using chitin grafted polyaniline electrode, *International Journal of Biological Macromolecules*. 147 (2020) 250–257. doi:10.1016/j.ijbiomac.2019.12.209.
[34] H.A. Hegazi, Removal of heavy metals from wastewater using agricultural and industrial wastes as adsorbents, *HBRC Journal*. 9 (2013) 276–282. doi:10.1016/j.hbrcj.2013.08.004.
[35] F. Huang, S.M. Zhang, R.R. Wu, L. Zhang, P. Wang, R.B. Xiao, Magnetic biochars have lower adsorption but higher separation effectiveness for Cd2+ from aqueous solution compared to nonmagnetic biochars, *Environmental Pollution*. 275 (2021) 116485. doi: 10.1016/j.envpol.2021.116485.
[36] A. Hu, X. Yang, Q. You, Y. Liu, Q. Wang, G. Liao, D. Wang, Magnetically hyper-cross-linked polymers with well-developed mesoporous: a broad-spectrum and highly efficient adsorbent for water purification, *Journal of Materials Science*. 54 (2019) 2712–2728. doi: 10.1007/s10853-018-2967–z.
[37] G. de V. Brião, J.R. de Andrade, M.G.C. da Silva, M.G.A. Vieira, Removal of toxic metals from water using chitosan-based magnetic adsorbents: a review, *Environmental Chemistry Letters*. 18 (2020) 1145–1168. doi: 10.1007/s10311-020-01003-y.
[38] A. Mehmood, F.S.A. Khan, N.M. Mubarak, Y.H. Tan, R.R. Karri, M. Khalid, R. Walvekar, E.C. Abdullah, S. Nizamuddin, S.A. Mazari, Magnetic nanocomposites for sustainable water purification—a comprehensive review, *Environmental Science and Pollution Research*. (2021) 1–26. doi: 10.1007/s11356-021-12589-3.
[39] E.I. Unuabonah, B.I. Olu-Owolabi, E.I. Fasuyi, K.O. Adebowale, Modeling of fixed-bed column studies for the adsorption of cadmium onto novel polymer-clay composite adsorbent, *Journal of Hazardous Materials*. 179 (2010) 415–423. doi: 10.1016/j.jhazmat.2010.03.020.
[40] J. Maity, S.K. Ray, Chitosan based nano composite adsorbent—Synthesis, characterization and application for adsorption of binary mixtures of Pb(II) and Cd(II) from water, *Carbohydrate Polymers*. 182 (2018) 159–171. doi: 10.1016/j.carbpol.2017.10.086.
[41] M. Naushad, T. Ahamad, K.M. Al-Sheetan, Development of a polymeric nanocomposite as a high performance adsorbent for Pb(II) removal from water medium: Equilibrium, kinetic and antimicrobial activity, *Journal of Hazardous Materials*. 407 (2021) 124816. doi: 10.1016/j.jhazmat.2020.124816.
[42] B. Işık, A.E. Kurtoğlu, G. Gürdağ, G. Keçeli, Radioactive cesium ion removal from wastewater using polymer metal oxide composites, *Journal of Hazardous Materials*. 403 (2021) 123652. doi: 10.1016/j.jhazmat.2020.123652.
[43] S. Daradmare, M. Xia, V.N. Le, J. Kim, B.J. Park, Metal–organic frameworks/alginate composite beads as effective adsorbents for the removal of hexavalent chromium from aqueous solution, *Chemosphere*. 270 (2021) 129487. doi: 10.1016/j.chemosphere.2020.129487.
[44] M. Abu Taleb, R. Kumar, A.A. Al-Rashdi, M.K. Seliem, M.A. Barakat, Fabrication of SiO2/CuFe2O4/polyaniline composite: a highly efficient adsorbent for heavy metals removal from aquatic environment, *Arabian Journal of Chemistry*. 13 (2020) 7533–7543. doi: 10.1016/j.arabjc.2020.08.028.

[45] X. Hu, Y. Hu, G. Xu, M. Li, Y. Zhu, L. Jiang, Y. Tu, X. Zhu, X. Xie, A. Li, Green synthesis of a magnetic β-cyclodextrin polymer for rapid removal of organic micropollutants and heavy metals from dyeing wastewater, *Environmental Research.* 180 (2020) 108796. doi: 10.1016/j.envres.2019.108796.

[46] C.S. Kushwaha, S.K. Shukla, Potentiometric extractive sensing of lead ions over nickel oxide intercalated chitosan-grafted-polyaniline composite, *Dalton Transactions.* (2020). doi: 10.1039/d0dt02687e.

[47] R. V. Xikhongelo, F.M. Mtunzi, P.N. Diagboya, B.I. Olu-Owolabi, R.A. Düring, Polyamidoamine-functionalized graphene oxide-SBA-15 mesoporous composite: adsorbent for aqueous arsenite, cadmium, ciprofloxacin, ivermectin, and tetracycline, *Industrial and Engineering Chemistry Research.* (2021). doi: 10.1021/acs.iecr.0c04902.

[48] E.G. Al-Sakkari, O.M. Abdeldayem, E.E. Genina, L. Amin, N.T. Bahgat, E.R. Rene, I.M. El-Sherbiny, New alginate-based interpenetrating polymer networks for water treatment: A response surface methodology based optimization study, *International Journal of Biological Macromolecules.* 155 (2020) 772–785. doi: 10.1016/j.ijbiomac.2020.03.220.

[49] C.S. Kushwaha, S.K. Shukla, P.P. Govender, N.S. Abbas, S.K. Shukla, Sustainable water purification and energy generation over crystalline chitosan grafted polyaniline composite, *Journal of Polymers and the Environment.* (2021) 1–12. doi: 10.1007/s10924-021-02129-y.

[50] M.A. Mahmood, S. Baruah, A.K. Anal, J. Dutta, Heterogeneous photocatalysis for removal of microbes from water, *Environmental Chemistry Letters.* 10 (2012) 145–151. doi: 10.1007/s10311-011-0347-x.

[51] S. Baruah, J. Dutta, Nanotechnology applications in pollution sensing and degradation in agriculture, *Environmental Chemistry Letters.* 7 (2009) 191–204. doi: 10.1007/s10311-009-0228-8.

[52] R. Djellabi, X. Zhao, C.L. Bianchi, P. Su, J. Ali, B. Yang, Visible light responsive photoactive polymer supported on carbonaceous biomass for photocatalytic water remediation, *Journal of Cleaner Production.* 269 (2020) 122286. doi: 10.1016/j.jclepro.2020.122286.

[53] M. Liras, M. Barawi, V.A. De La Peña O'Shea, Hybrid materials based on conjugated polymers and inorganic semiconductors as photocatalysts: From environmental to energy applications, *Chemical Society Reviews.* 48 (2019) 5454–5487. doi: 10.1039/c9cs00377k.

[54] C. Duan, C. Liu, X. Meng, K. Gao, W. Lu, Y. Zhang, L. Dai, W. Zhao, C. Xiong, W. Wang, Y. Liu, Y. Ni, Facile synthesis of Ag NPs@ MIL-100(Fe)/ guar gum hybrid hydrogel as a versatile photocatalyst for wastewater remediation: photocatalytic degradation, water/oil separation and bacterial inactivation, *Carbohydrate Polymers.* 230 (2020) 115642. doi: 10.1016/j.carbpol.2019.115642.

[55] M. Anjum, M. Oves, R. Kumar, M.A. Barakat, Fabrication of ZnO-ZnS@polyaniline nanohybrid for enhanced photocatalytic degradation of 2-chlorophenol and microbial contaminants in wastewater, *International Biodeterioration and Biodegradation.* 119 (2017) 66–77. doi: 10.1016/j.ibiod.2016.10.018.

[56] N. Bagheri, M. Mansour Lakouraj, V. Hasantabar, M. Mohseni, Biodegradable macro-porous CMC-polyaniline hydrogel: synthesis, characterization and study of microbial elimination and sorption capacity of dyes from waste water, *Journal of Hazardous Materials.* 403 (2021) 123631. doi: 10.1016/j.jhazmat.2020.123631.

[57] D. Hu, H. Wang, L. Wang, Physical properties and antibacterial activity of quaternized chitosan/carboxymethyl cellulose blend films, *LWT - Food Science and Technology.* 65 (2016) 398–405. doi: 10.1016/j.lwt.2015.08.033.

[58] Y.S. Wei, K.S. Chen, L.T. Wu, In situ synthesis of high swell ratio polyacrylic acid/silver nanocomposite hydrogels and their antimicrobial properties, *Journal of Inorganic Biochemistry*. 164 (2016) 17–25. doi: 10.1016/j.jinorgbio.2016.08.007.

[59] P. Boomi, H.G. Prabu, Synthesis, characterization and antibacterial analysis of polyaniline/Au-Pd nanocomposite, *Colloids and Surfaces A: Physicochemical and Engineering Aspects*. 429 (2013) 51–59. doi: 10.1016/j.colsurfa.2013.03.053.

[60] S. Nayak, S.R. Prasad, D. Mandal, P. Das, Carbon dot cross-linked polyvinylpyrrolidone hybrid hydrogel for simultaneous dye adsorption, photodegradation and bacterial elimination from waste water, *Journal of Hazardous Materials*. 392 (2020) 122287. doi: 10.1016/j.jhazmat.2020.122287.

[61] P. Gunawan, C. Guan, X. Song, Q. Zhang, S.S.J. Leong, C. Tang, Y. Chen, M.B. Chan-Park, M.W. Chang, K. Wang, R. Xu, Hollow fiber membrane decorated with Ag/MWNTs: Toward effective water disinfection and biofouling control, *ACS Nano*. 5 (2011) 10033–10040. doi: 10.1021/nn2038725.

[62] N.B. Singh, G. Nagpal, S. Agrawal, Water purification by using adsorbents: a review, *Environmental Technology and Innovation*. 11 (2018) 187–240. doi: 10.1016/j.eti.2018.05.006.

[63] J. Chen, C. Shu, N. Wang, J. Feng, H. Ma, W. Yan, Adsorbent synthesis of polypyrrole/TiO_2 for effective fluoride removal from aqueous solution for drinking water purification: adsorbent characterization and adsorption mechanism, *Journal of Colloid and Interface Science*. 495 (2017) 44–52. doi: 10.1016/j.jcis.2017.01.084.

[64] L. Das, P. Das, A. Bhowal, C. Bhattachariee, Synthesis of hybrid hydrogel nano-polymer composite using graphene oxide, chitosan and PVA and its application in waste water treatment, *Environmental Technology and Innovation*. 18 (2020) 100664. doi: 10.1016/j.eti.2020.100664.

[65] S. Sahnoun, M. Boutahala, Adsorption removal of tartrazine by chitosan/polyaniline composite: kinetics and equilibrium studies, *International Journal of Biological Macromolecules*. 114 (2018) 1345–1353. doi: 10.1016/j.ijbiomac.2018.02.146.

[66] E. Igberase, P. Osifo, Equilibrium, kinetic, thermodynamic and desorption studies of cadmium and lead by polyaniline grafted cross-linked chitosan beads from aqueous solution, *Journal of Industrial and Engineering Chemistry*. 26 (2015) 340–347. doi: 10.1016/j.jiec.2014.12.007.

[67] H. Javadian, M. Ghaemy, M. Taghavi, Adsorption kinetics, isotherm, and thermodynamics of Hg2+ to polyaniline/hexagonal mesoporous silica nanocomposite in water/wastewater, *Journal of Materials Science*. 49 (2014) 232–242. doi: 10.1007/s10853-013-7697-7.

[68] L. Mdlalose, M. Balogun, K. Setshedi, L. Chimuka, A. Chetty, Performance evaluation of polypyrrole–montmorillonite clay composite as a re-usable adsorbent for Cr(VI) remediation, *Polymer Bulletin*. (2020) 1–13. doi: 10.1007/s00289-020-03338-6.

[69] J. Son, Y. Hong, G. Han, T.S. Nguyen, C.T. Yavuz, J.I. Han, Gold recovery using porphyrin-based polymer from electronic wastes: Gold desorption and adsorbent regeneration, *Science of the Total Environment*. 704 (2020) 135405. doi: 10.1016/j.scitotenv.2019.135405.

[70] Y. Xie, W. Huang, B. Zheng, S. Li, Q. Liu, Z. Chen, W. Mai, R. Fu, D. Wu, All-in-one porous polymer adsorbents with excellent environmental chemosensory responsivity, visual detectivity, superfast adsorption, and easy regeneration, *Advanced Materials*. 31 (2019) 1900104. doi: 10.1002/adma.201900104.

5 Magnetic Nanomaterials for Wastewater Remediation

Dimple Sharma
Lovely Professional University

Sonika Singh
Guru Amar Dass Public School

Jandeep Singh and Harminder Singh
Lovely Professional University

CONTENTS

DOI: 10.1201/9781003129042-5

ABBREVIATIONS

CR	Congo red
CTAB	Cetyltrimethylammonium bromide
CV	Crystal violet
EDTA	Ethylenediaminetetraacetate
GO	Graphene oxide
MB	Methylene blue
MG	Malachite green
MNMs	Magnetic nanomaterials
MO	Methyl orange
MWCNT	Multi-walled carbon nanotube
RhB	Rhodamine B
SDS	Sodium dodecyl sulphate
ZVI	Zero-valent iron

5.1 INTRODUCTION

Environmental pollution is a major problem worldwide. Pollutants such as heavy metals, dyes and oils from industrial, agricultural and household wastes when escaping into the environment pose a great threat. When these pollutants are released into water bodies, the quality of water is deteriorated. The treatment of wastewater has become an issue of utmost importance. However, it is a major challenge to remove pollutants from wastewater. The methods of removal of pollutants from wastewater include ion exchange, adsorption, ozonation, membrane filtration, coagulation and flocculation, nanofiltration and reverse osmosis, desalination, chemical precipitation and anaerobic–aerobic treatment [1–6]. These methods have various advantages and disadvantages such as high capital and operational cost and generation of secondary waste. A novel and economical technology using MNMs for wastewater remediation has attracted the researchers in the recent years. In this chapter, MNMs, their types, synthesis methods, applications in wastewater remediation, role as catalysts, separation, regeneration and toxicity have been discussed.

5.2 MAGNETIC NANOMATERIALS

Magnetic fields affect MNMs to a great extent. This is so because MNMs usually have a magnetic material (such as iron, cobalt and nickel), and the other component provides the chemical functionality. MNMs consist of elements of magnetic nature, i.e. iron, nickel, cobalt or their oxides such as maghemite (Fe_2O_3), magnetite (Fe_3O_4), cobalt ferrite ($CoFe_2O_4$), nickel ferrite ($NiFe_2O_4$) and zinc ferrite (Zn Fe_2O_4). MNMs have the following interesting properties:

- They have a high surface-to-volume ratio.
- They can be easily separated using magnets.
- They are specific in nature.
- Their surface chemistry is very good for the removal of pollutants.

5.3 TYPES OF MAGNETIC NANOMATERIALS

MNMs that are used for water remediation purpose include Fe, Co, Ni, their alloys or oxides [7–9], zero-valent iron (ZVI) [10] and magnetic spinel ferrites [11,12]. MNMs can be classified as oxides/ferrites and metallic with a shell.

5.3.1 Oxides/Ferrites

Iron oxide nanoparticles (maghemite or magnetite) and spinel ferrites are mostly used in the treatment of wastewater. When the size of nanoparticles becomes smaller than 128 nm, they become superparamagnetic [13].

5.3.2 Modification of MNMs

Surfactants, polymers, metals or their oxides are often used to modify the surface of the MNMs such as ferrites to increase their stability in acidic/basic solutions as well as in organic solvents and to achieve higher magnetization [14].

5.4 METHODS FOR THE SYNTHESIS OF MAGNETIC NANOMATERIALS

The morphology and structure of MNMs depend upon the method used for their synthesis. Various methods of preparation of MNMs include sol–gel, hydrothermal, co-precipitation, sonochemical, microemulsion and thermal decomposition methods. A brief description of few methods is given below.

5.4.1 Sol–Gel Method

By sol–gel method, stoichiometric, pure, monodisperse MNMs are synthesized. MNMs of size approximately 10 nm can be synthesized by this method. In this method, an initiator is required to initiate the reaction between organic and inorganic materials. Usually, citric acid is used as a reaction initiator. The preparation of aqueous solution of metal and iron salts leads to the formation of a gel in the presence of citric acid at basic pH. It is the most preferable method to synthesize MNMs [15–17]. The gel obtained is dried and calcinated in the temperature range of 450°C–800°C [18]. Adjustment of speed of stirring and conditions of reactions helps in controlling the size of the MNMs; however, an initiator is required to initiate the reaction to synthesize MNMs.

5.4.2 Hydrothermal Method

In this method, inorganic substances are dissolved in an aqueous medium at elevated temperature and vapour pressure. A phase separation mechanism then occurs. Calcination of the solution is then done by placing it in a Teflon-lined autoclave. The reaction mixture is allowed to cool after the completion of the reaction, and by using various solvents and stabilizing agents, the impurities are removed. Following are the merits and demerits of the hydrothermal method. MNMs prepared by this method can be coated with polymers to enhance their properties. The stability of the MNMs depends on the stabilizing agents, polarity of the solvent and steric factors.

5.4.3 Co-Precipitation Method

The reduction of metallic elements can be achieved by adjusting the pH of the solution at ambient temperature at elevated temperature (using NaOH), and this method is based on the same principle. $NaOH/NH_4OH$ mixture acts as a co-precipitation agent and forms an alkaline solution in which stoichiometric quantities of metal salt solutions are added. The mixture is allowed to react for 12 h. The precipitates so formed are filtered and washed several times with distilled water. Uniform-sized MNMs can be generated via homogeneous precipitation reactions by this method. It is cost-effective, is easy to implement and uses less hazardous materials. The effect of hydrothermal treatment on co-precipitation is negligible at constant pH. However, this method is not suitable for the preparation of highly pure MNMs because undesirable crystal growth takes place.

5.4.4 Microemulsion Method

A microemulsion comprises of three components:

- A polar phase (water).
- A nonpolar phase (oil).
- A surfactant.

This is optically active and thermodynamically stable. Microemulsion method comprises of two types of techniques, viz. water-in-oil (W/O) [19] and oil-in-water (O/W). Mixing two dissimilar microemulsions in the presence of a surfactant results in the formation of micro-droplets that ultimately precipitate. This method involves gentle mixing of individual counterparts and is an efficient method for loading water-soluble drugs. However, there is the requirement of hydrophilic initiator or emulsifier agents. The yield of synthesized MNMs is low. For the generation of MNMs, a large amount of solvent is required, which makes it inefficient. Finally, it is a costly method.

5.4.5 Sonochemical Method

In all the above methods discussed for the synthesis of MNMs, it is difficult to control their morphology and surface properties. Ultrasound field using sonochemical reactions can be used to synthesize MNMs. In this method, bubbles form, grow and then collapse in an aqueous solution continuously. The collapse of bubbles results in the formation of MNMs at very high temperature (>5000 K), pressure (>20 MPa) and high cooling rate (1010 K/s), in an ultrasound reactor. Thus, finer particles with narrow size distribution are synthesized [20]. It is a cost-effective method. This method is environmentally benign as compared to other synthetic methods used for the synthesis of MNMs.

5.5 APPLICATIONS OF MAGNETIC NANOMATERIALS IN WASTEWATER REMEDIATION

MNMs have applications in various fields. Figure 5.1 shows the potential applications of MNMs in various fields with special reference to environment.

MNMs are immensely used in wastewater treatment using the following techniques:

a. Adsorption.
b. Oil/water separation.
c. Catalysis.

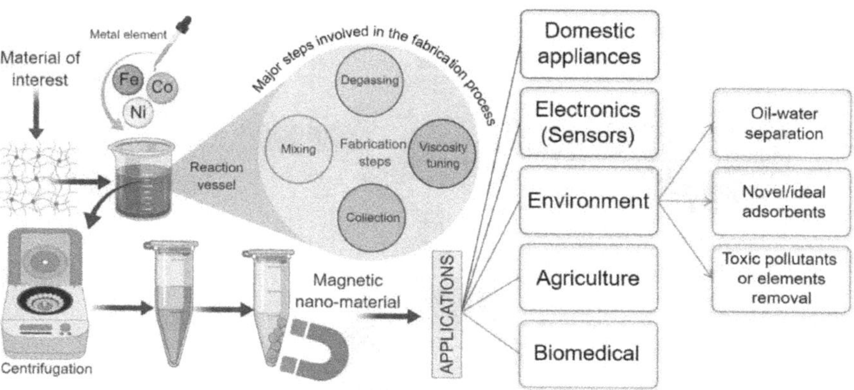

FIGURE 5.1 Applications of magnetic nanomaterials. (Adapted with permission from [21].)

5.5.1 Magnetic Nanomaterials as Adsorbents

Various adsorbents used for wastewater remediation include activated carbon, inorganic materials (silica gel and zeolites), ion exchange resins, natural materials, industrial by-products, biosorbents and hydrogels. In the past two decades, nanotubes, nanowires, nanorods, nanoplates and nanopowders have gained the immense interest of the researchers. Nanomaterials have enormous interesting properties for wastewater remediation. The benefits of using nanomaterials in wastewater purification are the following:

1. MNMs can remove the pollutants at low concentrations.
2. MNMs can take up hazardous pollutants from wastewater.
3. MNMs by altering the oxidation state of pollutants can reduce their toxicity.
4. From the surface of MNMs, speedy removal of adsorbed pollutants takes place.
5. They are cost-effective and reusable in nature.

Apart from the above merits, high surface area, size in nanoscale dimensions, high selectivity, easy removal after use by applying an external magnetic field and recyclability are the other properties that make them an excellent choice for wastewater treatment. The various advantages of MNMs in the adsorption of pollutants from wastewater are summarized in Figure 5.2.

However, the stability of these particles especially when these are exposed to air remains a concern. So suitable solution is required to increase the stability,

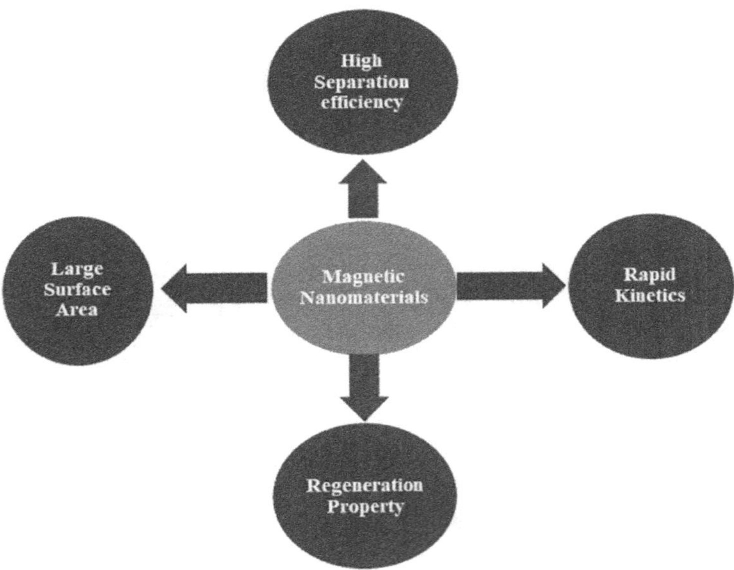

FIGURE 5.2 Advantages of magnetic nanomaterials.

and coating of magnetic nanoparticles is one of the options. This may include surfactant or polymer coating and doping with various metals and their oxides. Coating/doping is beneficial in the following ways:

- It helps in protecting the core–shell.
- It adds the functional groups that provide the specificity to the MNMs for the removal of pollutants from wastewater.

Thus, there is need for the modification of MNMs.

5.5.1.1 Doping with Metals and Their Oxides

As the concentration of doping agent increases, the crystalline structure and size of MNMs are affected. Their adsorption behaviour can be explained by ion exchange mechanism. Due to the difference in atomic sizes of metals and their oxides (doping agents) and parent metal ions, a greater number of sites are activated on the doped surface. Free electrons present on the crystal lattice surface reduce the pollutants from higher oxidation state to lower ones. Hence, they are easily adsorbed on the surface of MNMs. Table 5.1 illustrates the adsorption capacities of different MNMs doped with metals or their oxides.

5.5.1.2 Surfactant Coating

The general mechanism of surfactant-coated nanoferrites for the removal of pollutants [29] is represented in Figure 5.3. A single or double layer is formed on the surface of MNMs due to coating with cationic/anionic/neutral surfactants. These

TABLE 5.1
Adsorption Capacities of Different MNMs Doped with Metals or Their Oxides

Adsorbent	Coating/Doping Agent	Pollutant Removed	Adsorption Capacity (mg/g)	Ref.
$CuFe_2O_4$	-	Cd(II)	13.9	[22]
$BiFeO_3$	-	RhB	64.2	[23]
$CoFe_2O_4$	-	Titan yellow	212.8	[24]
		CR	200.0	
$Mn_xCu_{1-x}Fe_2O_4$	Al	As(III)	25.5	[25]
$NiFe_2O_4$-graphene oxide	Graphene oxide	Pb(II)	121.9	[26]
		Cr(III)	126.5	
$Ni_xZn_{1-x}Fe_2O_4$	Ni	Cr(VI)	48.5	[27]
		Mo(VI)	38.8	
		V(V)	28.0	
$CuFe_2O_4$-CeO_2	CeO_2	Pb(II)	972.4	[28]
		Ni(II)	686.1	
		V(V)	798.6	

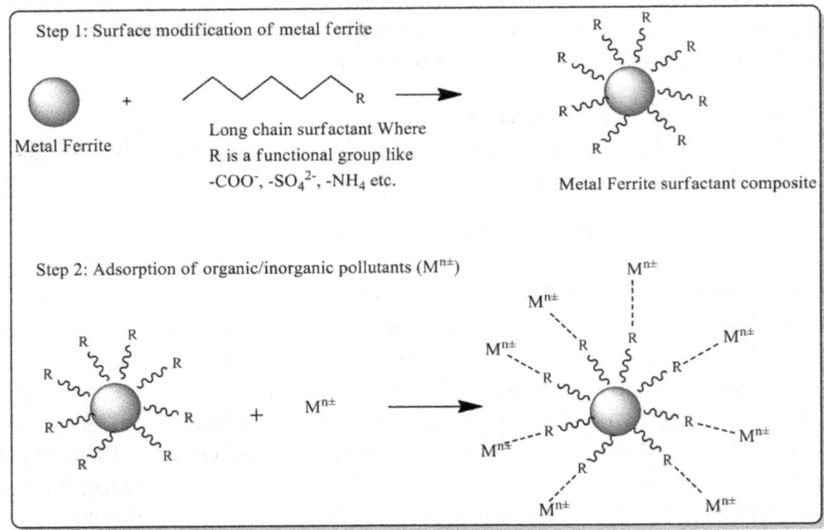

FIGURE 5.3 General mechanism of surfactant-coated nanoferrites. (See Ref. [29].)

TABLE 5.2
Adsorption Capacities of Different MNMs Coated with Surfactants

MNM	Surfactant	Pollutant Removed	Adsorption Capacity (mg/g)	Ref.
$ZnFe_2O_4$/SDS	SDS	Basic Blue 41	42.0	[30]
		Basic Red 18	61.0	
Fe_3O_4/CTAB	CTAB	Nyloset Yellow	136.0	[31]
CoC/SDS	SDS	CV	105.0	[32]
$ZnFe_2O_4$/SDS	SDS	CV	135.5	[33]
$ZnFe_2O_4$/CTAB	CTAB	Direct Red 23	26.1	[34]
		Direct Red 31	55.6	

surfactants due to their different nature provide functionality to the surface of the MNMs, which acts as active sites for the adsorption of the pollutant. The nature of the used surfactant also provides specificity to the magnetic adsorbents. This helps in the adsorption of different pollutants from aqueous systems.

A protective layer is formed on the surface of nanoferrites by the surfactant molecules. The selectivity of the adsorbent can be adjusted using cationic or anionic surfactants. Table 5.2 illustrates the adsorption capacities of different MNMs coated with surfactants.

5.5.1.3 Polymer Coating

Figure 5.4 depicts the general process of polymer coating and the adsorption mechanism of polymer-coated nanoferrites [29]. Various polymers that have been reported to modify the surface of MNMs are polyvinyl alcohol, polyaniline, chitosan and alginate. Polymer coating has the following benefits:

- It helps to protect nanoferrites from oxidation and degradation.
- It enhances the stability of MNMs.

The biopolymers form a cavity-like structure around the metal ferrite. The functional groups present on the surface of biopolymers provide potential active sites for the adsorption of dyes and heavy metals on their surface. Table 5.3 illustrates the adsorption capacities of different MNMs coated with polymers.

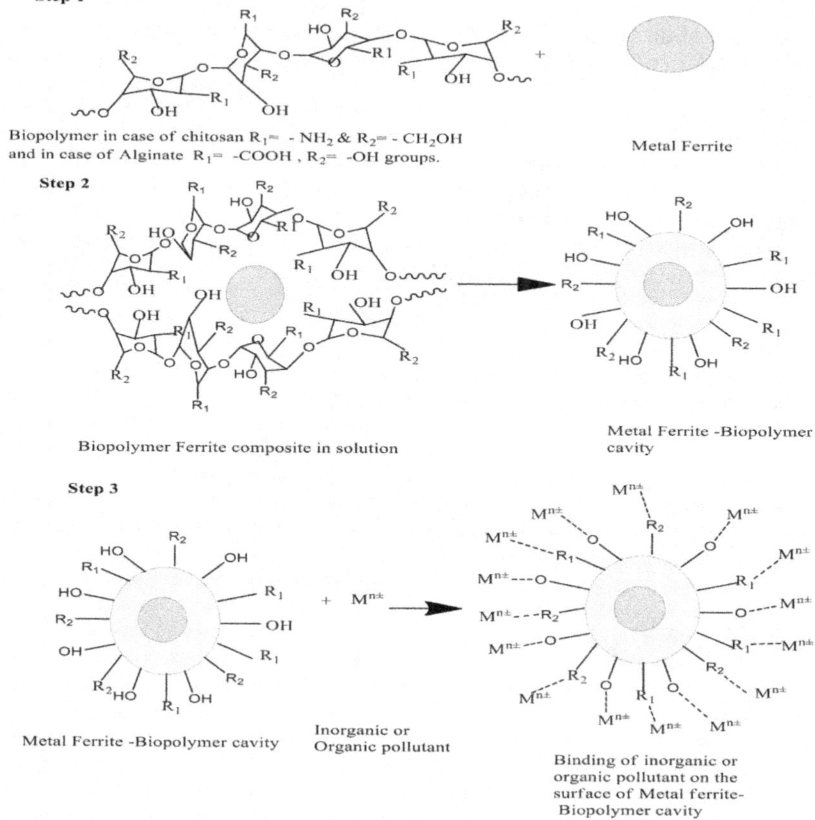

FIGURE 5.4 Preparation and mechanism of removal of pollutants for polymer-coated MNMs. (See Ref. [29].)

TABLE 5.3
Adsorption Capacities of Different MNMs Coated with Polymers

MNMs	Biopolymer Used	Pollutant Removed	Adsorption Capacity (mg/g)	Ref.
$ZnFe_2O_4$/alginic	Sodium alginate	CR	103.2	[35]
		CV	123.5	
$ZnFe_2O_4$/chitosan	Chitosan	Fluoride	6.9	[36]
$NiFe_2O_4$/chitosan			8.3	
$CoFe_2O_4$/chitosan			6.7	
Fe_3O_4/alginate	Sodium alginate	MG	47.8	[37]
Fe_3O_4/chitosan/Al_2O_3	Chitosan/Al_2O_3	MO	417.0	[38]
Fe_3O_4/SiO_2/chitosan	SiO_2/chitosan	MB	43.0	[39]

5.5.2 Oil/Water Separation Using Magnetic Nanomaterials

The presence of oil as fats, lubricants and petroleum products is toxic for aquatic biota. Magnetic nanoparticles have been used to separate oil and water by the technique known as demulsification. Figure 5.5 illustrates the oil/water separation using MNMs. Magnetic nanoparticles easily disperse in continuous phase to target species (dispersed phase). Composites of magnetic nanoparticles with polymers with water absorption capacity, i.e. hydrophilic nature, are when placed in contact with oil/water mixture quickly absorb the water, and the oil is separated. Further, the MNM is easily separated using a magnet. The properties that are required by a material for the retrieval of oil/water mixture are the following:

- Their structure should be anisotropic.
- They should possess magnetic properties.
- They should possess high interfacial and amphiphilic nature.

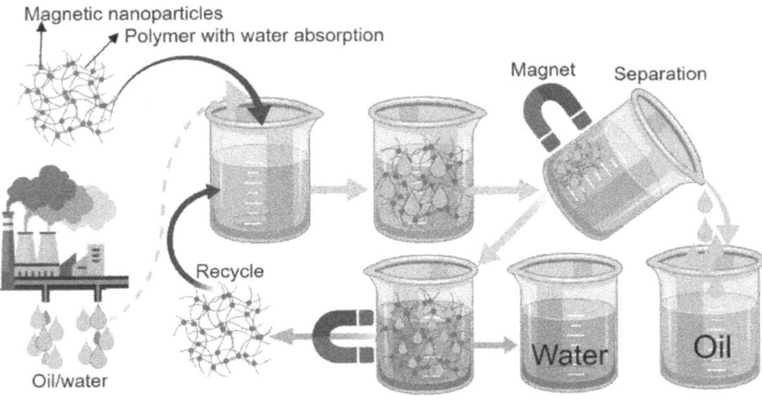

FIGURE 5.5 Representation of separation of oil and water using magnetic nanomaterials. (Adapted with permission from [21].)

FIGURE 5.6 (a) Photograph of separation of water from chloroform; (b) the absorption of water by the MSHO sponge at the bottom of soybean oil. (Adapted with permission from [44].)

For example, P(MMA-AA-DVB)/Fe_3O_4 magnetic composite nanoparticles show pronounced surface properties for the demulsification of stable oil/water emulsions [40–43].

Su et al. [44] explored a novel magnetic superhydrophilic/oleophobic (MSHO) sponge, which has a three-dimensional structure. It is magnetically recoverable in nature. It is used for the selective adsorption of pollutants from wastewater. Figure 5.6a depicts the quick absorption of water by MSHO sponge when it comes in contact with water layer on $CHCl_3$ surface. Because of its low density and oleophobicity, MSHO remained floating on $CHCl_3$ surface. After that, it can be easily recovered using a magnet. Figure 5.6b depicts the quick absorption of water by MSHO when it comes in contact with water droplets (coloured with RhB) at the bottom of soybean oil by an external force.

5.5.3 Magnetic Nanomaterials as Catalysts

MNMs can be effectively used as heterogeneous catalysts due to their following advantages:

- Rapid and easy separation from solution using a magnet.
- MNMs can be dispersed as any nanoparticle in the absence of a magnetic field, whereas they can be selectively precipitated in the presence of a magnetic field.

In this way, they can be easily removed from the reaction vessel by using a magnet and can be reused again. In the field of wastewater treatment, the conventional approaches are photochemical catalysis and Fenton degradation. The other methods include ozonation [45], nitrate reduction [46,47] and oxygen activation degradation [48]. Atony et al. [49] developed novel MNM bimetallic AgNiNPs on Fe_3O_4@chitosan core/shell support as a heterogeneous nanocatalyst. This heterogeneous nanocatalyst can be separated with the help of a magnet and can be reused for several times. Figure 5.7 illustrates the preparation of the Fe_3O_4@CSAgNi nanocomposite and rapid reduction of 4-nitrophenol (4-NP) using $NaBH_4$.

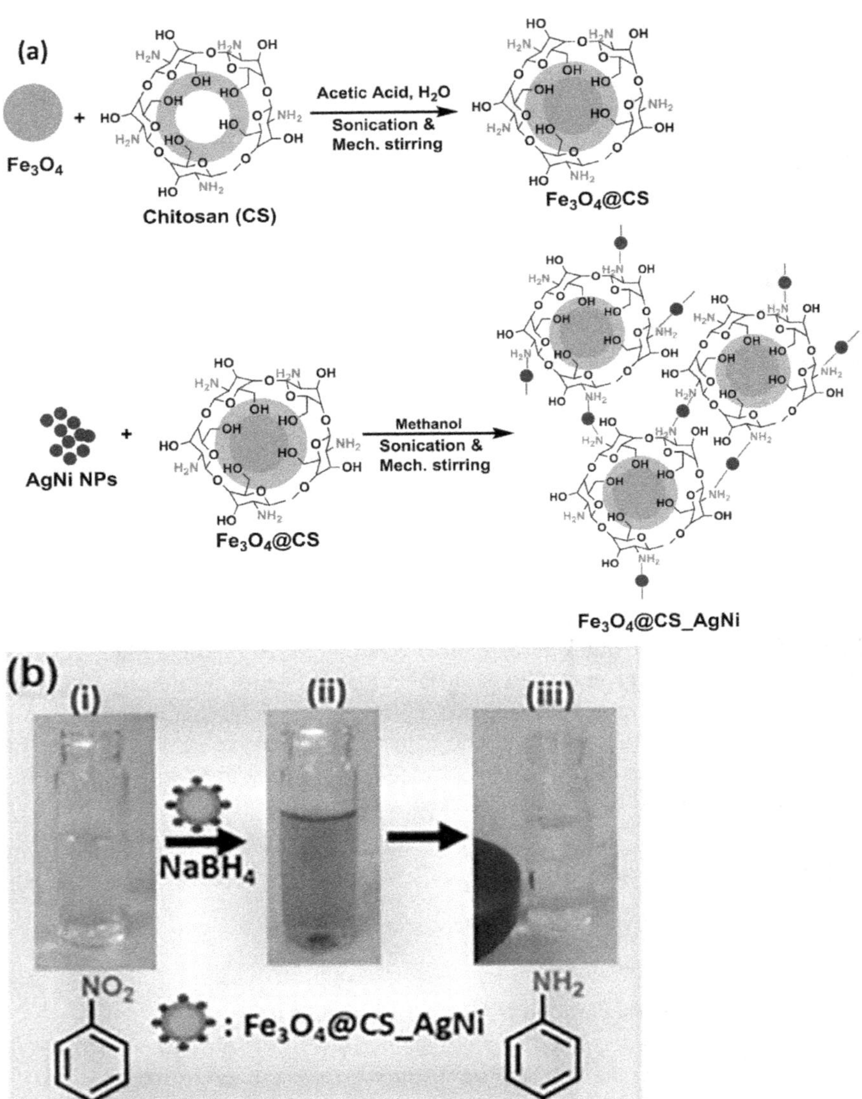

FIGURE 5.7 (a) Synthesis of Fe_3O_4@CSAgNi; (b) change of colour during 4-NP reduction; (i) simply 4-NP, (ii) 4-NP+magnetic nanocomposite+$NaBH_4$; and (iii) separation of MNM by using a magnet after the complete reduction. (Adapted with permission from [49].)

TABLE 5.4
Magnetic Nanocomposites for Catalytic Degradation of Pollutants in Wastewater

Catalytic Method	Pollutant	MNM	Degradation (%)	Ref.
Fenton reaction	Furfural	ZVI	97.5	[50]
Catalytic ozonation	Phenol	$NiFe_2O_4$	97.6	[51]
Photoelectrocatalytic	RhB	$Fe_3O_4@SiO_2@TiO_2$/GO	92.03	[52]
Dechlorination	DDT	ZVI	81–91	[53]
Chemical oxidation	Heavy metals	ZVI	95–99	[54]

Table 5.4 illustrates some examples of MNMs used for catalytic degradation of pollutants in wastewater.

Wang et al. [55] developed mesoporous copper ferrite (meso-$CuFe_2O_4$) MNMs with the following properties:

- High surface area.
- Large pore size.
- Ability of act as a heterogeneous Fenton catalyst.

The pore size and the surface area of meso-$CuFe_2O_4$ MNMs were 9.2 nm and 122 m^2/g, respectively. Meso-$CuFe_2O_4$ MNMs exhibited excellent catalytic activity for the degradation of imidacloprid and achieved complete removal of 10 mg/L imidacloprid after 5 h (reactions conditions of 0.3 g/L catalyst and 40 mM H_2O_2).

Peng et al. [56] developed $TiO_2/SiO_2/Fe_3O_4$ core–shell MNMs, using the sol–gel method. It was comprised of TiO_2 outer shell, an intermediary layer of SiO_2 and core of Fe_3O_4. MB dye was photocatalytically degraded in aqueous solutions using this MNM. It was magnetically recoverable in nature, so it can be reused multiple times, which makes it a cost-effective method for the wastewater remediation. This also reduces the cost of wastewater purification by photocatalytic degradation.

5.5.4 Other Applications

MNMs can also be used to remove microbes and organic pollutants from wastewater.

5.5.4.1 Removal of Microbes

Microalgae present in water have an adverse effect on the quality of water. Zhan et al. [57] synthesized amine-functionalized magnetite (Fe_3O_4-SiO_2-NH_2) nanomaterials for the removal of pathogenic microorganisms. This MNM was effective for the removal of pathogens such as bacteriophage f_2, viruses (poliovirus-1) and bacteria such as *S. aureus*, *E.coli*, *Salmonella* and *B. subtilis*.

5.5.4.2 Organic Pollutants

Organic micropollutants such as bisphenol A (BPA) pose a great threat to water bodies. Kumar et al. [58] synthesized superparamagnetic iron oxide nanoparticles (SPION) and their nanocomposites with β-cyclodextrin for the removal of BPA by the co-precipitation method. During the degradation experiments of BPA, it was found that SPION/β-CD+Solar light+H_2O_2 system performed the best in photodegradation. Moreover, this is magnetically recoverable in nature.

5.6 SEPARATION AND REGENERATION OF MAGNETIC NANOMATERIALS

MNMs can be separated efficiently after wastewater treatment by applying low gradient magnetic field [59]. Figure 5.8 represents the schematic diagram of the removal of magnetic nanoadsorbents by using an external magnetic field.

The regeneration step is important in the field of adsorption/catalysis by MNMs as it make the wastewater treatment process more economical. Various methods of regeneration, such as thermal, electrical, ultrasonic and chemical methods, are being used to make MNMs reusable multiple times [61]. Regeneration depends on the following factors:

- Choice of desorbing agent.
- Temperature.
- pH.
- Concentration.

FIGURE 5.8 Separation of magnetic nanomaterials after use by using an external magnetic field. (Adapted with permission from [60].)

TABLE 5.5
Few Desorption Agents Used in the Regeneration of MNMs

Adsorbent	Adsorbate	Desorption Agent	Desorbing Agent's Conc. (M)	% Removal	Ref.
$CoFe_2O_4$	Titan yellow	NaCl:acetone (2:1)	0.1	99.0	[63]
$ZnFe_2O_4$/alginic	CR	HCl	0.1	91.8	[64]
	CV			92.3	
	BG			90.3	
$ZnFe_2O_4$/chitosan	CV	HCl	0.1	89.5	[65]
	BG			91.7	
Fe_3O_4/chitosan	Cu(II)	EDTA	0.1	96.0	[66]

From the above factors, the efficiency of desorption is mostly affected by the pH value [61,62]. For the regeneration of the MNMs strong acids/bases are usually used as regeneration agents. Few regeneration agents to recover various metal ions and dyes from the adsorbent surfaces are depicted in Table 5.5.

Figure 5.9 represents the systematic process of adsorption and desorption by magnetic nanoadsorbents.

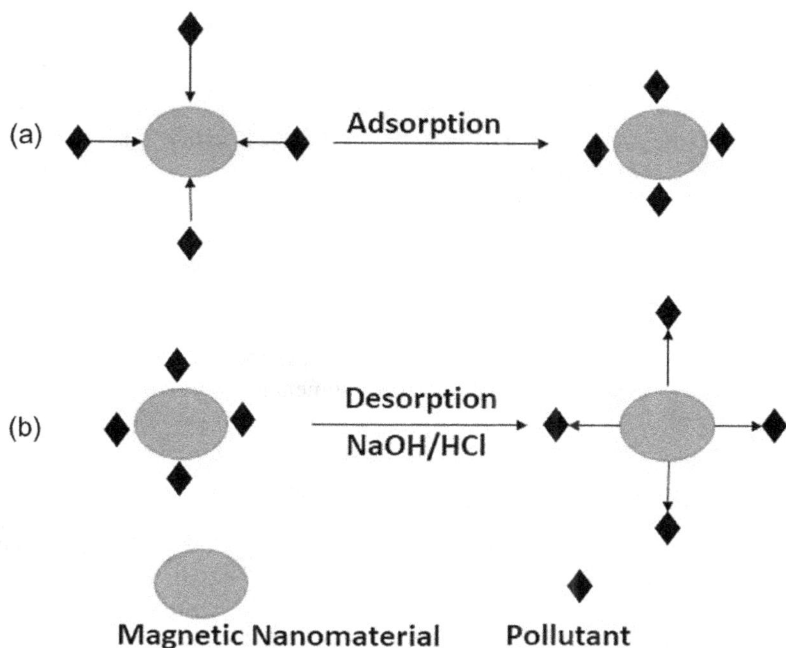

FIGURE 5.9 (a) Adsorption and (b) desorption by magnetic nanomaterials.

TABLE 5.6
Toxicity Studies of MNMs [70]

S. No.	Coating Material	Cell Types	Size (in nm)	Concentration (in mg/mL)	Results
1.	Dextran	Macrophages (human)	100–150	0.1 only	20% Cell viability observed after 7 days
2.	Silica	Human lung Adenocarcinoma epithelial cells	50	4	Dose- and size-dependent damage
3.	Amine surface	Human liver carcinoma cells	61–127	3	Cause severe cytotoxicity due to positive charge
4.	Chitosan	Human hepatocellular carcinoma cells	13.8	0.12	Only 10% cell viability observed after 12 h
5.	Zinc oxide	Bronchial epithelial cells	25	0.2	Ion shedding, oxidative stress due to local ion concentration
6.	Silver	Gill fish cell line	40	0.05	Membrane damage

5.7 TOXICITY OF MAGNETIC NANOMATERIALS

Certain properties such as size, morphology, charge and surface coating affect the interaction between MNMs and biological environments. MNMs can be distributed to various tissues, organs and cells. The bio-distribution of MNMs is 80%–90% in the liver, 5%–8% in the spleen and 1%–2% in the bone marrow [67].

Endothelial cells help in healing of wounds, modulation of vascular bone, blood flow and regulation of immune system [68]. MNMs may get into the blood stream by crossing the epithelial barrier, and consequently, the possibility of the diseases of the liver increases [69]. Table 5.6 illustrates some examples of toxicity studies of MNMs [70].

From the above table, it can be concluded that despite the many advantages MNMS have in wastewater treatment, still there is the need for handling them with care to avoid their escape into the environment so that their harmful effects can be reduced.

5.8 CONCLUSIONS AND FUTURE PROSPECTS

MNMs have attracted the attention of most researchers over past decades. Their potential in wastewater remediation field mainly focuses on the removal of pollutants (dyes/heavy metals), in oil/water separation, in the elimination of microbes and other organic pollutants from wastewater and in catalysis. Their ease of removal with the application of a magnetic field and their regeneration for reuse make them suitable as a low-cost and effective option for wastewater treatment.

In spite of their many advantages, consequent limitations and shortcomings of MNMs in wastewater remediation field are also of great significance to be concerned. Because the lack of comprehensive cognition still exists, their ecotoxic nature should be emphasized, as it is important for ensuring environmental safety. MNMs are still considered as efficient adsorbents and catalysts due to their large surface area, ease of removal by applying an external magnetic field, regeneration and recyclability.

In the last decade, a shift from conventional/non-conventional to MNMs in the field of wastewater remediation has been observed. Although MNM has gained the attention of researchers worldwide for the treatment of wastewater, a gap is still there to effectively transfer a viable technology at industrial scale. Therefore, future researches should focus on the technology development for the implementation of MNMs at industrial scale. Focus is also required to limit the disadvantages of MNMs and ways to cease their escape into the environment. Hence, the development of techniques for wastewater remediation based on MNMs will be an important concern in the future research.

REFERENCES

[1] Y. Anjaneyulu, N. S. Chary, and D. S. Raj, Decolourization of industrial effluents–available methods and emerging technologies–a review, *Rev. Environ. Sci. Bio/Technol.* 2005, 4(4), 245–273.

[2] S. Mondal and S. R. Wickramasinghe, Produced water treatment by nanofiltration and reverse osmosis membranes, *J. Membr. Sci.* 2008, 322(1), 162–170.

[3] F. Sher, A. Malik, and H. Liu, Industrial polymer effluent treatment by chemical coagulation and flocculation, *J. Environ. Chem. Eng.* 2013, 1(4), 684–689.

[4] A. Sonune and R. Ghate, Developments in wastewater treatment methods, *Desalination* 2004, 167, 55–63.

[5] M. M. Matlock, B. S. Howerton, and D. A. Atwood, Chemical precipitation of heavy metals from acid mine drainage, *Water Res.* 2002, 36(19), 4757–4764.

[6] Y. J. Chan, M. F. Chong, C. L. Law, and D. G. Hassell, A review on anaerobic–aerobic treatment of industrial and municipal wastewater, *Chem. Eng.* 2009, 155 (1–2), 1–18.

[7] M. Y. Zhu and G. W. Diao, Review on the progress in synthesis and application of magnetic carbon nanocomposites, *Nanoscale* 2011, 3(7), 2748.

[8] A. Y. Zhang and J. L. Kong, Novel magnetic Fe_3O_4@C nanoparticles as adsorbents for removal of organic dyes from aqueous solution, *J. Hazard. Mater.* 2011, 193, 325–329.

[9] C.C. Lin, Y. S. Lin, and J. M. Ho, Adsorption of reactive red 2 from aqueous solutions using Fe_3O_4 nanoparticles prepared by coprecipitation in a rotating packed bed, *J. Alloys Compd.* 2016, 666, 153–158.

[10] X. H. Guan, Y. K. Sun, H. J. Qin, J. X. Li, I. M. C. Lo, D. He, and H. R. Dong, The limitations of applying zero-valent iron technology in contaminants sequestration and the corresponding countermeasures: the development in zero-valent iron technology in the last two decades (1994–2014), *Water Res.* 2015, 75, 224–248.

[11] M. Kumar, H. S. Dosanjh, and H. Singh, Magnetic zinc ferrite–alginic biopolymer composite: as an alternative adsorbent for the removal of dyes in single and ternary dye system. *J. Inorg. Organomet. Polym. Mater.* 2018, 28(5), 1–18.

[12] M. Kumar, H. S. Dosanjh, and H. Singh, Magnetic zinc ferrite–chitosan bio-composite: synthesis, characterization and adsorption behavior studies for cationic dyes in single and binary systems, *J. Inorg. Organomet. Polym. Mater.* 2018, 28(3), 880–898.

[13] A. H. Lu, E. E. Salabas, and F. Schüth, Magnetic nanoparticles: synthesis, protection, functionalization, and application, *Angew. Chem., Int. Ed.* 2007, 46(8), 1222–1244.

[14] S. Kralj, D. Makovec, S. Čampelj, and M. Drofenik, Producing ultra-thin silica coatings on iron-oxide nanoparticles to improve their surface reactivity, *J. Magn. Magn. Mater.* 2010, 322(13), 1847–1853.

[15] L. Zhang and Y. Wu, Sol–gel synthesized magnetic $MnFe_2O_4$ spinel ferrite nanoparticles as novel catalyst for oxidative degradation of methyl orange, *J. Nanomater.* 2013, 2013, 1–6.

[16] Z. Peng, K. F. Yao, and Z. Liao, The preparation and application of magnetic nanoparticles with core–shell structure, *Int. J. Mod. Phys. B* 2009, 23(6–7), 1523–1528.

[17] A. L. Andrade, D. M. Souza, and M. C. Pereira, Synthesis and characterization of magnetic nanoparticles coated with silica through a sol–gel approach, *Ceramica* 2009, 55, 420–424.

[18] S. S. Ebrahimi and J. Azadmanjiri, Evaluation of $NiFe_2O_4$ ferrite nanocrystalline powder synthesized by a sol–gel auto–combustion method, *J. Non-Cryst. Solids* 2007, 353(8–10), 802–804.

[19] D. K. Yi, S. S. Lee, and J. Y. Ying, Synthesis and applications of magnetic nanocomposite catalysts, *Chem. Mater.* 2006, 18(10):2459–2461.

[20] B. M. Teo, S. K. Suh, and T. A. Hatton, Sonochemical synthesis of magnetic janus nanoparticles, *Langmuir* 2011, 27(1):30–33.

[21] N. Ali, H. Zaman, M. Bilal, H. A. Shah, M. S. Nazir, and H. M. N. Iqbal, Environmental perspectives of interfacially active and magnetically recoverable composite materials – a review, *Sci. Total Environ.* 2019, 670, 523–538.

[22] Y. J. Tu, C. F. You, and C. K. Chang, Chang, kinetics and thermodynamics of adsorption for Cd on green manufactured nano–particles, *J. Hazard. Mater.* 2012, 235, 116–122.

[23] T. Soltani and M. H. Entezari, Sono–synthesis of bismuth ferrite nanoparticles with high photocatalytic activity in degradation of Rhodamine B under solar light irradiation, *Chem. Eng. J.* 2013, 223, 145–154.

[24] M. Ghaemi, G. Absalan, and L. Sheikhian, Adsorption characteristics of Titan yellow and Congo red on $CoFe_2O_4$ magnetic nanoparticles, *J. Iran. Chem. Soc.* 2014, 11(6), 1759–1766.

[25] M. A. Malana, R. B. Qureshi, and M. N. Ashiq, Adsorption studies of arsenic on nano aluminium doped manganese copper ferrite polymer (MA, VA, AA) composite: kinetics and mechanism, *Chem. Eng. J.* 2011, 172(2–3), 721–727.

[26] L. P. Lingamdinne, I. S. Kim, J. H. Ha, Y. Y. Chang, J. R. Koduru, and J. K. Yang, Enhanced adsorption removal of Pb(II) and Cr(III) by using nickel ferrite–reduced graphene oxide nanocomposite, *Metals* 2017, 7(6), 225.

[27] A. Afkhami, S. Aghajani, M. Mohseni, and T. Madrakian, Effectiveness of $Ni0.5Zn0.5Fe_2O_4$ for the removal and preconcentration of Cr(VI), Mo(VI), V(V) and W(VI) oxyanions from water and wastewater samples, *J. Iran. Chem. Soc.* 2015, 12(11), 2007–2013.

[28] F. Talebzadeh, R. Zandipak, and S. Sobhanardakani, CeO2 nanoparticles supported on $CuFe_2O_4$ nanofibers as novel adsorbent for removal of Pb(II), Ni(II), and V(V) ions from petrochemical wastewater, Desalin, *Water Treat.* 2016, 57(58), 28363–28377.

[29] M. Kumar, H. Singh Dosanjh, S. J. Sonika, K. Monir, and H. Singh, Review on magnetic nanoferrites and their composites as alternatives in wastewater treatment: Synthesis, modifications and applications, *Environ. Sci.: Water Res. Technol.* 2020, 6(3), 491–514.

[30] N. M. Mahmoodi, Surface modification of magnetic nanoparticle and dye removal from ternary systems, *J. Ind. Eng. Chem.* 2015, 27, 251–259.

[31] N. Dalali, M. Khoramnezhad, M. Habibizadeh, and M. Faraji, Magnetic removal of acidic dyes from wastewaters using surfactant–coated magnetite nanoparticles: optimization of process by Taguchi method, Proceedings of International Conference on Environmental and Agriculture Engineering IPCBEE, 2011, vol. 15, pp. 89–94.

[32] M. Singh, H. S. Dosanjh, and H. Singh, Surface modified spinel cobalt ferrite nanoparticles for cationic dye removal: Kinetics and thermodynamics studies, *J. Water Process. Eng.* 2016, 11, 152–161.

[33] M. Kumar, H. S. Dosanjh, and H. Singh, Synthesis of spinel $ZnFe_2O_4$ modified with SDS via low temperature combustion method and adsorption behaviour of crystal violet dye, *Asian J. Chem.* 2017, 29(9), 2057–2064.

[34] N. M. Mahmoodi, J. Abdi, and D. Bastani, Direct dyes removal using modified magnetic ferrite nanoparticle, *J. Environ. Health Sci. Eng.* 2014, 12(1), 96.

[35] M. Kumar, H. S. Dosanjh, and H. Singh, Magnetic zinc ferrite–alginic biopolymer composite: as an alternative adsorbent for the removal of dyes in single and ternary dye system, *J. Inorg. Organomet. Polym. Mater.* 2018, 28(5), 1688–1705.

[36] X. Li, Y. Wang, Y. Li, L. Zhou, and X. Jia, Biosorption behaviors of biosorbents based on microorganisms immobilized by Ca–alginate for removing lead (II) from aqueous solution, *Biotechnol. Bioprocess Eng.* 2011, 16(4), 808–820.

[37] A. Mohammadi, H. Daemi, and M. Barikani, Fast removal of malachite green dye using novel superparamagnetic sodium alginate–coated Fe_3O_4 nanoparticles, *Int. J. Biol. Macromol.* 2014, 69, 447–455.

[38] B. Tanhaei, A. Ayati, M. Lahtinen, and M. Sillanpää, Preparation and characterization of a novel chitosan/Al2O3/magnetite nanoparticles composite adsorbent for kinetic, thermodynamic and isotherm studies of Methyl Orange adsorption, *Chem. Eng. J.* 2015, 259, 1–10.

[39] Y. Li, Y. Zhou, W. Nie, L. Song, and P. Chen, Highly efficient methylene blue dyes removal from aqueous systems by chitosan coated magnetic mesoporous silica nanoparticles, *J. Porous Mater.* 2015, 22(5), 1383–1392.

[40] N. Ali, B. Zhang, H. Zhang, W. Zaman, W. Li, and Q. Zhang, 2014. Key synthesis of magnetic Janus nanoparticles using a modified facile method. Particuology 17, 59–65.

[41] N. Ali, B. Zhang, H. Zhang, W. Li, W. Zaman, L. Tian, and Q. Zhang, Novel Janus magnetic micro particle synthesis and its applications as a demulsifier for breaking heavy crude oil and water emulsion. *Fuel* 2015a, 141, 258–267.

[42] N. Ali, Z. Baoliang, H. Zhang, W. Zaman, S. Ali, Z. Ali, and Q. Zhang, Iron oxide-based polymeric magnetic microspheres with a core shell structure: from controlled synthesis to demulsification applications. *J. Polym. Res.* 2015b, 22(11), 219.

[43] N. Ali, B. Zhang, H. Zhang, W. Zaman, S. Ali, Z. Ali, and Q. Zhang, Monodispers and multifunctional magnetic composite core shell microspheres for demulsification applications. *J. Chin. Chem. Soc.* 2015c, 62(8), 695–702.

[44] C. Su, H. Yang, S. Song, B. Lu, and R. Chen, A magnetic superhydrophilic/oleophobic sponge for continuous oil-water separation. *Chem. Eng. J.* 2016, 309, 366–373.

[45] H. Zhao, Y. M. Dong, G. L. Wang, P. P. Jiang, J. J. Zhang, L. N. Wu, and K. Li, Novel magnetically separable nanomaterials for heterogeneous catalytic ozonation of phenol pollutant: $NiFe_2O_4$ and their performances, *Chem. Eng. J.* 2013, 219, 295–302.

[46] G. C. C. Yang and H. L. Lee, (2005) Chemical reduction of nitrate by nanosized iron: kinetics and pathways, *Water Res.* 39, 884–894.

[47] Y. H. Huang and T. C. Zhang, Effects of dissolved oxygen on formation of corrosion products and concomitant oxygen and nitrate reduction in zerovalent iron systems with or without aqueous Fe2+, *Water Res.* 2005, 39, 1751–1760.

[48] C. E. Noradoun and I. F. Cheng, EDTA degradation induced by oxygen activation in a zerovalent iron/air/water system. *Environ. Sci. Technol.* 2005, 39(18), 7158–7163.

[49] R. Antony, R. Marimuthu, and R. Murugavel, Bimetallic nanoparticles anchored on core-shell support as an easily recoverable and reusable catalytic system for efficient nitroarene reduction. *ACS Omega* 2019, 4(5), 9241–9250.

[50] F. Li, J. G. Bao, T. C. Zhang, and Y. T. Lei, A combined process of adsorption and Fenton-like oxidation for furfural removal using zero-valent iron residue. *Environ. Technol.* 2015, 36(24), 3103–3111.

[51] H. Zhao, Y. M. Dong, G. L. Wang, P. P. Jiang, J. J. Zhang, L. N. Wu, and K. Li, Novel magnetically separable nanomaterials for heterogeneous catalytic ozonation of phenol pollutant: $NiFe_2O_4$ and their performances, *Chem. Eng. J.* 2013, 219, 295–302.

[52] F. H. Chen, F. F. Yan, Q. T. Chen, Y. W. Wang, L. F. Han, Z. J. Chen, and S. M. Fang, Fabrication of $Fe_3O_4@SiO_2@TiO_2$ nanoparticles supported by graphene oxide sheets for the repeated adsorption and photocatalytic degradation of rhodamine B under UV irradiation. *Dalton Trans.* 2014, 43, 13537–13544.

[53] J. G. Liu, C. J. Ou, W. Q. Han, F. J. Y. Shen, H. P. Bi, X. Y. Sun, J. S. Li, and L. J. Wang, Selective removal of nitroaromatic compounds from wastewater in an integrated zero valent iron (ZVI) reduction and ZVI/H_2O_2 oxidation process. *RSC Adv.* 2015, 5, 57444–57452.

[54] M. M. Eglal and A. S. Ramamurthy, Competitive adsorption and oxidation behavior of heavy metals on nZVI coated with TEOS, *Water Environ. Res.* 2015, 87(11), 2018–2026.

[55] Y. Wang, H. Zhao, and M. Li, et al., Magnetic ordered mesoporous copper ferrite as a heterogenous Fenton catalyst for the degradation of imidacloprid. *Appl. Catal., B* 2014, 147, 534–545.

[56] Z. Peng, K. F. Yao, and Z. Liao, The preparation and application of magnetic nanoparticles with core–shell structure. *Int. J. Mod. Phys.* 2009, 23(6–7), 1523–1528.

[57] S. Zhan, Y. Yang, Z. Shen, J. Shan, Y. Li, S. Yang, and D. Zhu, Efficient removal of pathogenic bacteria and viruses by multifunctional amine-modified magnetic nanoparticles. *J. Hazard. Mater.* 2014, 274, 115–123.

[58] A. Kumar, G. Sharma, and M. Naushad, SPION/β-cyclodextrin core– shell nanostructures for oil spill remediation and organic pollutant removal from wastewater. *Chem. Eng. J.* 2015, 280, 175–187.

[59] Y. T. Cafer, J. T. Mayo, and W. W. Yu, et al., Low field magnetic separation of monodisperse Fe_3O_4 nanocrystals. *Science* 2006, 314, 964–967.

[60] D. H. K. Reddy and Y. S. Yun, Spinel ferrite magnetic adsorbents: alternative future materials for water purification. *Coord. Chem. Rev.* 2016, 315, 90–111.

[61] S. Kulkarni and J. Kaware, Regeneration and recovery in adsorption–a review, *Int. J. Innov. Res. Sci. Eng. Technol.* 2014, 1(8), 61–64.

[62] I. K. Shah, P. Pre and B. J. Alappat, Steam regeneration of adsorbents: an experimental and technical review, *Chem. Sci. Trans.* 2013, 2(4), 1078–1088.

[63] M. Ghaemi, G. Absalan and L. Sheikhian, Adsorption characteristics of Titan yellow and Congo red on $CoFe_2O_4$ magnetic nanoparticles, *J. Iran. Chem. Soc.* 2014, 11(6), 1759–1766.

[64] M. Kumar, H. S. Dosanjh and H. Singh, Magnetic zinc ferrite–alginic biopolymer composite: as an alternative adsorbent for the removal of dyes in single and ternary dye system, *J. Inorg. Organomet. Polym. Mater.* 2018, 28(5), 1688–1705.
[65] M. Kumar, H. S. Dosanjh, and H. Singh, Magnetic zinc ferrite–chitosan bio–composite: synthesis, characterization and adsorption behavior studies for cationic dyes in single and binary systems, *J. Inorg. Organomet. Polym. Mater.* 2018, 28(3), 880–898.
[66] C. Yuwei and W. Jianlong, Preparation and characterization of magnetic chitosan nanoparticles and its application for Cu(II) removal, *Chem. Eng. J.* 2011, 168(1), 286–292.
[67] A. J. Gupta and M. Gupta, Synthesis and surface engineering of iron oxide nanoparticles for biomedical applications. *Biomaterials* 2005, 26, 3995–4021.
[68] S. S. Luthra, R. R. Narayanan, A. L. Marques, M. M. Chwa, D. W. D. Kim, J. J. Dong, G. M. G. Seigel, A. A. Neekhra, A. L. A. Gramajo, D. J. D. Brown, M. C. M. Kenney, B. D. B. Kuppermann, Evaluation of in vitro effects of bevacizumab (Avastin) on retinal pigment epithelial, neurosensory retinal, and microvascular endothelial cells. *Retina* 2006, 26, 512–518.
[69] K. Donaldson, P. J. A. Borm, V. Castranova, and M. Gulumian, The limits of testing particle-mediated oxidative stress in vitro in predicting diverse pathologies; relevance for testing of nanoparticles. *Part. Fibre Toxicol.* 2009, 6, 1–8.
[70] R. Lakshmanan, Doctoral Thesis. Application of Magnetic nanoparticles and reactive filter materials for wastewater treatment. School of Biotechnology, Royal Institute of Technology, Stockholm, 2013.

6 Alumina-Based Adsorbents

Shaziya H. Siddiqui, Prerna Higgins, and Runit Isaac
Sam Higginbottom University of Agriculture, Technology & Sciences

CONTENTS

6.1 INTRODUCTION

Survival without water is not doable as water is an essential requisite for all the biotic creation. Around 70% of the water covers the surface of the earth, and out of that, only 1% is potable per international standards due to different contaminations (Singh et al. 2018). Toxic pollutant contamination in water will have baleful effect on human being and will effect drinking, irrigation and utilization processes. Any sort of contamination in water prior to living beings will have a baleful effect on them which will further exhibit infelicity in drinking, irrigation and other utilization processes. Water contamination is reported to be an influential aspect of environmental concern. Pollution is increasing at an alarming rate due to the high increase in population rate. The waste is discharged directly into the water bodies without proper treatment. Water contamination is caused mainly by man-made activities such as the discharge of xenobiotic chemicals or alteration

DOI: 10.1201/9781003129042-6

in the water bodies. The industrial activity, agricultural chemicals and improper disposal of wastes cause depletion in the environment.

Heavy metals are important environmental threats, and their effects on the living organisms are devastating as these heavy metal ions are vehemently released from the industries. The pollution of water by the contamination of heavy metals is the one of the most important issues because of the non-biodegradable and carcinogenic nature of the heavy metals. Therefore, heavy metals must be removed from wastewater before its release into the aquatic environment (Ahmad and Haseeb 2017). Heavy metals have toxic effects on the living beings, which include plants, animals and humans.

On the other hand, organic pollutants which include pesticides such as dichlorodiphenyltrichloroethane and lindane; industrial chemicals such as polychlorinated biphenyls, dyes and phenols; and substances such as dioxins are unwanted by-products of manufacturing and combustion processes of oils and greases. Dyes are chemical pollutants that are released from the dyeing and dye manufacturing industries. Several types of dyestuffs have extensively been employed in leathers, textiles, paper and plastic industries (Garg et al. 2004; Liu et al. 2015; Dawood and Sen 2012). The majority of the heavily coloured and toxic dyes are released annually as industrial effluents into the natural water bodies (Gupta et al. 2011). The presence of organic dyes has a great impact on photosynthesis in aquatic plants by preventing the light from penetrating the water, destroying the food chain in water ecosystems and damaging the aesthetic quality of water (Shan et al. 2015; Lei et al. 2017). The highly toxic dyes and their metabolites are carcinogenic in nature and, therefore, harmful to the biota (El Gaini et al. 2009; Lian et al. 2009).

A plethora of pharmaceutical drugs are being used cosmopolitanly for the treatment of disorders (Memmert et al. 2013). A revolution has been brought in human health by the development of the pharmaceuticals only if they serve their intent and not jeopardize the health of the living beings as these are recalcitrant contaminants which inevitably find their way into the food chain at trace concentration through bioaccumulation. As per the World Health Organization (WHO), the pharmaceutical industries contribute to 25% of the world water pollution directly or indirectly mainly through the process of manufacturing, which in fact primarily affects the environment (Haseena et al. 2017). Municipal waste, hospital and industrial effluents that are released into water bodies without proper treatment are also responsible for their contamination.

Therefore, there is a necessity to find cheap and effective ways to maintain a clean and healthy environment for the future generations. To treat serious water pollution problems, especially those involve inorganic pollutants, organic dyes, drugs and heavy metal ions in water bodies, many methods have been developed, such as chemical precipitation (Kurniawan et al. 2006), ozonation (Kadirvelu et al. 2003), photocatalysis (Xu et al. 2015; Ramana et al. 2017; Zhao et al. 2017), ion exchange (Cavaco et al. 2007; Lim and Kim 2015), ultrafiltration (Chakraborty et al. 2014), biodegradation (Gopinath et al. 2009) and membrane separation (Doke and Yadav 2014). Moreover, various water purification technologies such as complexation, chemical oxidation or reduction, solvent extraction, chemical precipitation, reverse osmosis, ion exchange, filtration, membrane process, evaporation and coagulation (Eren et al. 2009; Aksu et al. 2002) have also been employed. Adsorption is one of the most recommended physicochemical treatment processes for the recovery and treatment of pollutants from aqueous solutions based on the utilization of solid adsorbents from either organic, inorganic, biological or low-cost materials (Mahmoud 2006). Adsorption is an eco-friendly and economical method for the removal of contaminants from wastewater (Zhao et al. 2018).

6.2 ALUMINA AS AN ADSORBENT

Semi-crystalline alumina with an increase in surface area over $200\,m^2/g$ is a type of aluminium oxide (Rabia et al. 2018), which is usually involved as a desiccant and sorbent for the elements such as fluoride, arsenic and selenium present in water. Its amphoteric properties with high porosity and high surface area-to-weight ratio (Salvador et al. 2015) permits it to behave as an acid in a basic medium and as a base in an acidic medium. Alumina is hygroscopic in nature, and thus, it is a favoured desiccant for moisture removal from the air as well as a catalyst in natural gas refining operations. Regarding the removal of contaminants, metal oxides are the most frequently used surfaces; they are inexpensive and widely produced as compared to biological surfaces, apart from their high mechanical properties and resistance to thermal decomposition (Yamani et al. 2014). Aluminium oxide is characterized by its non-toxicity, easy use, chemical stability and multiple hydroxide groups, and thus, it is largely used an efficient adsorbent. Aluminium oxide has been used as an adsorbent since 1923 to eliminate pigments, antibiotics, heavy metals and dissolved organics (Hawksworth 2013; Wang et al. 2015) (Figure 6.1).

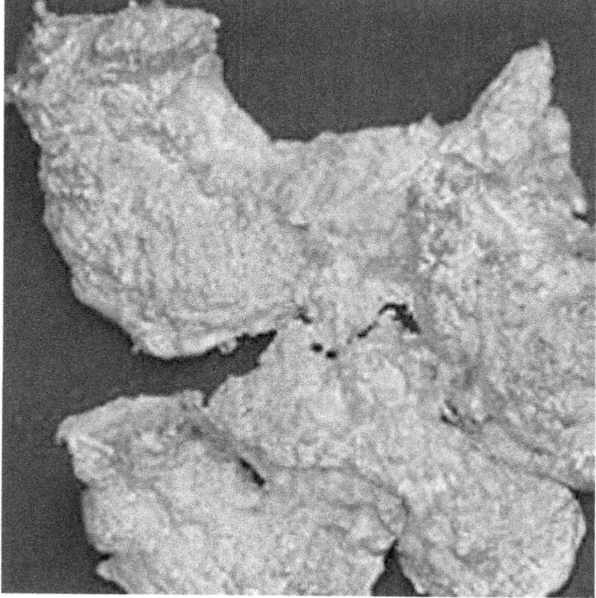

FIGURE 6.1 Image of alumina.

6.2.1 Alumina Activation and Efficiency

Thermal and chemical treatments are the two common treatment methods for the activation of Al_2O_3. The ability of alumina to efficiently adsorb pollutants depends upon two primary factors: the chemisorption adsorption site and the physisorption surface area available. Activation of adsorbent enhances not only the adsorptive sites, but also the surface area (Barakat 2011; Wang et al. 2009). Three driving forces, namely, the van der Waals forces, dipole–dipole interaction and the hydrogen bonding, are responsible for the physisorption where there is no electron exchange (Rabia et al. 2018) between the adsorbent and the adsorbate because it does not involve activation energy. The chemisorption mostly involves electron exchange, which may be either a covalent bond called weak chemisorption, or a heavy chemical adsorption called an ionic bond. An adsorbent's high surface area increases its adsorption performance, thereby making the adsorption process more feasible. Chua et al. (2009) and Cui et al. (2015) found out that the surface area of the activated alumina ranged from 500 to 2000 m^2/L, which greatly helped increase the adsorption rate. In addition, activation severely affected the movement of adsorbate molecules from the bulk fluid to the interface of the fluid and the solid (Cui et al. 2015).

6.3 MODIFICATION OF ALUMINA-BASED ADSORBENTS

Different hydroxides and metal oxides (Ahn et al. 2009; Sun et al. 2014), surfactants, oxidizing agents (Owlad et al. 2010), acids and bases (Fashi et al. 2018) and the other rare earth metals have been found to be promising materials in altering the surface of the adsorbents. Improving the adsorption performance of activated alumina was mainly associated with modifying its surface chemistry. There are several ways to modify the surface of the adsorbent for better adsorption results, such as physical treatment of the adsorbent, chemical treatment, functionalization and impregnation (Girish 2018; Lira et al. 2016). The performance of the modified adsorbent depends to a greater degree on what type of modifying agent has been employed. Fashi et al. (2018) indicated that when Al_2O_3 surface is chemically modified with piperazine, it produces a shift in the concentration of active or binding sites. The modification of Al_2O_3 increased the average pore size with an increase in the specific surface area that enhanced adsorption (Sergey et al. 2019), showing adsorbent stability with good regeneration capacity (Jain et al. 2015). The adsorbent modification results in strong affinity due to the rise in the number of hydroxyl groups (Peng et al. 2017). The modification results in enhanced rate of sorption. This is due to the modifying agent's physical and chemical structure and other experimental conditions.

6.4 ALUMINA-BASED ADSORBENTS FOR ADSORPTIVE REMOVAL OF HEAVY METALS

Suman et al. (2016) employed bentonite/gamma alumina adsorbent for the removal of Mg^{2+}. The porous nature, high feasibility and larger surface area make it a reliable adsorbent for wastewater treatment (i.e. 100% removal capacity was attained at pH 6.0 in 10 mg/L aqueous solution at an equilibrium time of 90 min). The XRD spectra support adsorbent crystalline nature, and the FTIR spectra show a feasible interaction between the adsorbate and the adsorbent by the IR data. The best-suited models are the pseudo-second-order kinetics and the Langmuir adsorption isotherm with R^2 of 0.999 and 0.938, respectively. Saeide et al. (2020) used bentonite/gamma alumina nanocomposite, which proved to be a promising adsorbent for the removal of Mg^{2+} with the removal capacity of 100% in the acidic medium of pH 4, and equilibrium was achieved in 30 min. Laleh et al. (2019) focused on the removal of arsenic by making use of $Fe_3O_4@Al_2O_3@$Zn-Fe LDH as a new magnetic nanoadsorbent based on ultrasound-assisted dispersive magnetic solid-phase extraction. It was adopted because of its high extraction efficiency as a result of high specific surface area, easiness of handling, inexpensiveness and rapidness (Campeli et al. 2009; Lei et al. 2019). The removal capacity was 97.3% at pH 4.0 with a dosage of 28.9 mg, and the contact time was 12.8 min. The adsorption isotherms followed the Langmuir equation with a maximum adsorption capacity of 67.57 mg/g. Moreover, the detailed kinetics studies confirmed that the adsorption kinetics followed pseudo-second-order model. Shikha et al. (2016) discussed the utilization of gamma alumina nanoadsorbent for the removal of Cr^{6+} ions. The adsorbent showed effective removal of Cr^{6+} ions in the acidic medium at pH 2 at 60 min equilibrium time. The governing models for kinetics and isotherm were pseudo-second-order and Langmuir equations. Gupta et al. (2013) carried out a research on the adsorption of fluoride using chloride-infused activated alumina as an adsorbent. A removal capacity of about 60%–70% of fluoride was found. However, a slight decrease in the removal ability of fluoride was observed due to the presence of competing ions, especially chloride ions (Table 6.1).

TABLE 6.1
Comparison of Adsorption Capacity of Alumina-Based Composites for Heavy Metals Removal

Adsorbent	Heavy Metals	Conc. (mg/L)	pH Range	Contact Time (min)	Q_m (mg/g)	Ref.
Bentonite/gamma alumina nanocomposites	Mg^{2+}	100	4	30	3.478	Saeide et al. (2020)
Fe–Al–Ce hydroxide	F^-	10–250	7.0	1440	51.3	Zhang et al. (2011)
Nanoalumina	PO_4^{3-}	10	6	90	0.165	Suman et al. (2016)
Gamma alumina adsorbent	Cr^{6+}	5	2.03	60	1.049	Shikha et al. (2016)
$Fe_3O_4@Al_2O_3@$Zn-Fe LDH	As	6.56	4	12.8	67.57	Laleh et al. (2019)
Nanoalumina	Co^{2+}	52.15	10	35.5	75.78	Dehghani et al. (2020)

6.5 ALUMINA-BASED METAL–ORGANIC FRAMEWORK ADSORBENTS

6.5.1 Characteristics

Metal–organic frameworks (MOFs) have aroused interest due to their interesting characteristics such as large surface area, increased thermal stability, large pore volume and flexibility of their structure (Farha et al. 2012; Furukawa et al. 2011). Aluminium-based MOFs have been found to be a promising adsorbent for wastewater treatment among the MOFs because they are stable in aqueous solutions (Samokhvalov 2018). The aluminium–benzene dicarboxylate MOF (Al-BDC, MIL-53(Al)) is a well-recognized aluminium-based MOF that has successfully been employed in the adsorption of both organic (Li et al. 2015; Xiao et al. 2014; Zhao et al. 2013) and inorganic (Li et al. 2014) compounds from water. It was also used as an adsorbent in the removal of sulphur from liquid fuels (Blanco et al. 2011) (Figure 6.2).

6.5.2 Mechanism Involved in Adsorption

The adsorption performance of pharmaceutical drugs such as naproxen and diclofenac was investigated using MIL-53(Al) from their aqueous solutions, and adsorption of the drugs was done (Ramirez et al. 2021). Furthermore, the possible mechanism associated with the adsorption is shown in Figure 6.3. The MOF demonstrates a great ability to extract the drugs from their aqueous solutions and can therefore be used when purifying waters exposed to such pharmaceuticals.

The mechanisms involved in the adsorption process included the formation of hydrogen bonds and the interaction/stacking of π–π bonds between the adsorbate molecules and the adsorbent. Hydrogen bonds can be feasibly formed with the

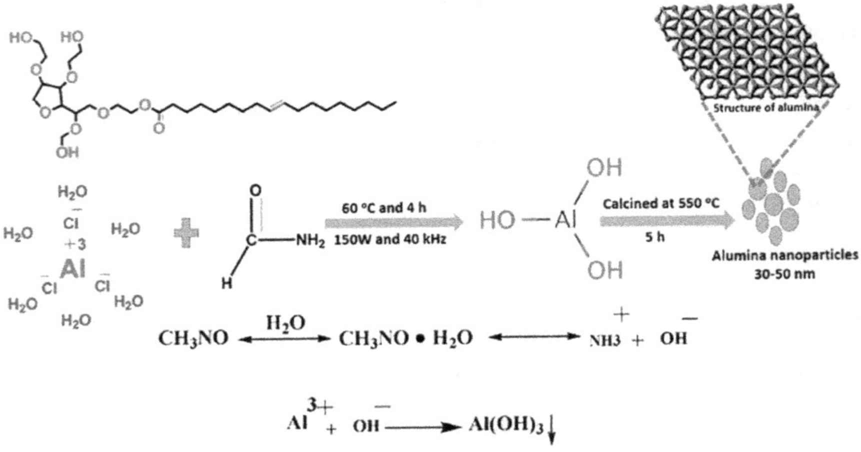

FIGURE 6.2 Synthesis scheme for the formation of alumina nanoparticles. (See Ali et al. 2019).

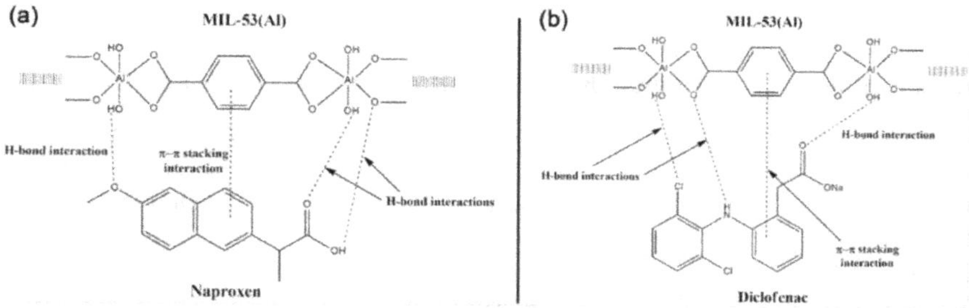

FIGURE 6.3 Chemical structure of (a) naproxen and (b) diclofenac.

pharmaceuticals as the molecules of MIL-53(Al) have hydrogen bond acceptor and donor atoms as well as carboxylic groups and bridging hydroxyl groups, respectively (Bhadra et al. 2017; Lin et al. 2018). Therefore, the drug can attach to the MOF's surface through hydrogen bond formation (Gao et al. 2018). The second possible adsorption mechanism is the π–π interaction/stacking, which occurs between the drug's benzene dimer and the MIL-53(Al) carboxylic ligand (Gao et al. 2018). This type of interaction has previously been reported for the bisphenol A (BPA) adsorption using various MOFs (Czaja et al. 2009; Kitagawa et al. 2004).

The crystallinity of MIL-53(Al) is confirmed by the XRD spectra in which the sample patterns exhibited changes in peak positions and/or intensity due to the breathing effect that occurs by the diffusion of the adsorbate molecules through the MOF pores (Gao et al. 2018). The FTIR spectra give us information about the functional groups present on the adsorbent by the calculated IR spectra of the MOFs and the drugs. Due to the presence of –OH groups obtained from free water molecules are confined into the pores of MOF and forming a bridge connecting the AlO_6 in the framework respectively. The –COO group is co-ordinated to Al atom in the spectra confirming the adsorption mechanism between the adsorbate and adsorbent. (Abid et al. 2016).

Therefore, an effective adsorption of naproxen and diclofenac (pharmaceutical drugs) by MIL-53(Al) was achieved. The efficiency of removal was high, and the adsorption process rate was rapid and feasible. The drug diclofenac in the adsorption experiments also had greater selectivity than the drug naproxen. Hydrogen bonding and stacking/interaction of π–π bonds were the governing reactions involved in the adsorption mechanism. The effect of pH in the adsorption experiments showed electrostatic repulsion between naproxen/diclofenac and MIL-53(Al) when the pH is increased, thereby showing a decrease in adsorption capacity. The pseudo-second-order kinetic model was best fitted with R^2 higher than 0.99. The adsorption isotherm experiments were governed by the Langmuir model. The calculated maximum adsorption capacities were found equal to 297 and 422 mg/g for naproxen and diclofenac, respectively. The adsorption capacity values for MIL-53(Al) with naproxen and diclofenac were higher than those for other pharmaceutical drugs obtained involving the use of other MOFs (Table 6.2).

TABLE 6.2
Different MOF Adsorbents Used for Pharmaceutical Drugs Removal

MOF Adsorbent	Pharmaceutical Drug	Q_m (mg/g)	Ref.
CDM6-K1000	Ibuprofen	408	An et al. (2018)
PCN-222	Chloramphenicol	370	Zhao et al. (2018)
AMSA-MIL-101	Naproxen	93	Hasan et al. (2013)
MIL-101-$(OH)_3$	Ketoprofen	80	Song et al. (2017)
ZIF-8	Oxytetracycline	28.3	Da silva et al. (2015)
UiO-67	Carbamazepine	18.9	Akpinar et al. (2017)
MIL-101/Fe_3O_4	Ciprofloxacin	80.7	Bayazit et al. (2017)
Bio-MOF-1-derived porous carbons	Bisphenol A	390	Bhadra et al. (2018)
MIL-53(Al)	Diclofenac	422	Karamia et al. (2020)
MIL-53(Al)	Naproxen	297	Karamia et al. (2020)

6.6 ALUMINA-BASED NANOADSORBENTS FOR DYE REMOVAL

The use of nanoadsorbents has gained popularity in terms of water treatment in recent years. In general, the application of nanotechnology to wastewater treatment has shown its significant and effective use (Brame et al. 2011). The characteristics of nanoadsorbents or nanoscale materials, such as small size, large surface area, large number of active sites and high capacity of regeneration have shown tremendous results when it comes to water purification (Ali 2012). Among all the inorganic nanomaterials, nanoalumina is one of the most widely used for making membranes for the adsorption of wastewater and its treatment; particularly, aluminium-based nanoadsorbents are used because they are cost-effective (Li et al. 2008).

6.6.1 Dyes and Their Toxic Effects

Textile industries extensively use dyes for colouring and for making their products attractive. These dyes have strong colour and hence affect the water bodies directly or indirectly. They further degrade the quality of life of marine life forms, and their carcinogenic properties (Lasari et al. 2007) cause severe damage to marine life as well as to human on consumption. It becomes really important to remove the dyes from the wastewater since they have catastrophic effects on living life forms.

6.6.2 ϒ-Phase Alumina Adsorbents

ϒ-Phase alumina nanomaterial is synthesized by using the technique of sol–gel precipitation in the presence of ethanol (Wang et al. 2008). The solution of $AlCl_3{\cdot}6H_2O$ is prepared in 95% pure ethanol. Further, liquid ammonia of 0.1 M, which acts as the precipitating agent, is added to the ethanol solution with continuous stirring. A white gelatinous precipitate of $Al(OH)_3$ will be formed and will be further filtered and washed again with ethanol. Dry the resultant precipitate at 90°C for around 6h. The mass of precipitate is calcified in muffle furnace at 600°C for 3h. This will further convert $Al(OH)_3$ into Al_2O_3 powder with a yield of 91% (Banerjee et al. 2017). The γ-phase alumina nanoadsorbent was used to remove the dye Orange G. Orange G, also known as Orange Gelb, is used in textile industries as a staining formulation. Its high carcinogenic behaviour and teratogenic nature makes it a highly toxic mono-azo dye. After the adsorption of the dye by the adsorbent, it was found that the removal was overall dependent on the pH (i.e. around 98.4% of dye removal was achieved at pH 2.5), as shown in Figure 6.3. Rapid removal was seen initially by the adsorbent; the possible reason may be that the sorption sites on the adsorbent are vacant that could further adsorb the dye from the solution phase (Ghaedi et al. 2012). FTIR and X-ray diffraction prove the formation of γ-phase alumina. FTIR showed a peak at 3457 cm^{-1}, proving the stretching vibration of –OH group from AL-OH structure, and a peak at 1629 cm^{-1}, proving the presence of water molecule (Yang et al. 2010). Through the morphology of SEM, it was seen that weak van der Waals' bond existed between the particles. The particles were straight and smooth, and the peaks of aluminium and oxygen showed that the alumina nanoparticles have been formed; this was analysed through EDX spectrum, and also sulphur, nitrogen and sodium ions of the dye were present on the relevant sites. TEM was analysed to know whether the particle size lies in the range of 20–30 nm or not. The alumina adsorbent forms an outer sphere surface complexes on the surface of the metal hydroxide (Hordern 2004), since it has high affinity for SO_4^{2-}, PO_4^{3-} and $C_2O_4^{2-}$.

Solutions such as HCl, $HClO_4$, HNO_3, H_2SO_4 and H_3PO_4 are tested to know their desorption behaviour. Batch mode study was performed and analysed with the help of spectrophotometric analysis to get the concentration of the dye desorbed. It was concluded that the above solution showed desorption in the manner: $H_3PO_4 < H_2SO_4 < HNO_3 < HCl < HClO_4$, where the maximum desorption was around 93%. This further proved that the alumina nanoparticles can be used several times without the loss of adsorption capacity of the adsorbent (Table 6.3).

The reactions in Figure 6.4 suggest that the surface reaction of alumina nanoparticles is governed by pH. Reaction 1 indicates that at pH 7.4, the net surface charge of the adsorbent is zero and so it is in neutral condition. Reaction 2 indicates that

TABLE 6.3
Alumina-Based Adsorbents for Removal of Dyes

Alumina-Based Adsorbent	Dye	Q_m (mg/g)	Ref.
γ-Phase alumina	Orange G	93.3	Banerjee et al. (2017)
Fe_2O_3-Al_2O_3 nanocomposite	Congo red	498	Mahapatra et al. (2013)
MWCNT/alumina composite nodules	Methylene blue	187.5	Kunde et al. (2019)
Immobilized activated alumina	Cibacron reactive yellow	25	Wasti et al. (2016)

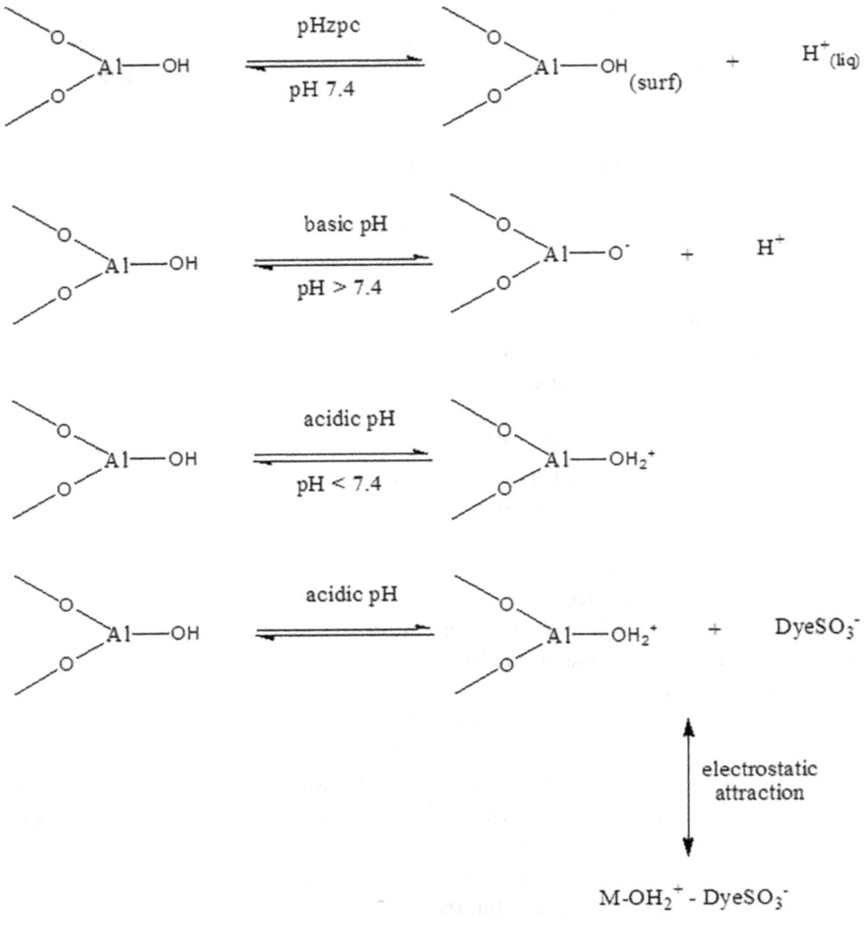

$$(O)_2Al{-}OH \underset{pH\ 7.4}{\overset{pHzpc}{\rightleftharpoons}} (O)_2Al{-}OH_{(surf)} + H^+_{(liq)}$$

$$(O)_2Al{-}OH \underset{pH > 7.4}{\overset{basic\ pH}{\rightleftharpoons}} (O)_2Al{-}O^- + H^+$$

$$(O)_2Al{-}OH \underset{pH < 7.4}{\overset{acidic\ pH}{\rightleftharpoons}} (O)_2Al{-}OH_2^+$$

$$(O)_2Al{-}OH \overset{acidic\ pH}{\rightleftharpoons} (O)_2Al{-}OH_2^+ + DyeSO_3^-$$

electrostatic attraction

$M\text{-}OH_2^+ - DyeSO_3^-$

FIGURE 6.4 Reaction mechanism of gamma alumina nanoparticles for the removal of dye.

pH>pHzc in basic medium, due to which the surface becomes deprotonated and negatively charged, which causes hindrance in the adsorption of anionic dyes due to columbic repulsion. In reaction 3, pH<pHzc, so the surface becomes positively charged, which causes electrostatic attraction of anionic species which leads to higher adsorption of dye anions (Banerjee et al. 2019). So these reactions indicate that the acidic pH range is the optimum for the adsorption of anionic dyes.

6.7 INDUSTRIAL APPLICATIONS OF ALUMINA

Alumina nanoparticles are used as additives due to their positive influence on fuel combustion and emission performance. They are used to improve the spray characteristics of liquid fuel. Alumina nanoparticles also increase the thermophysical properties of diesel fuel, such as thermal conductivity, mass diffusivity, flash point and kinematic viscosity (Hosseini et al. 2017). Alumina is also used in ceramic materials used in aerospace, biomedical and ballistic engineering. Porous alumina ceramics prepared by using agricultural wastes show good tensile strength and porosity. Multi-walled carbon nanotubes filled with alumina show good machinability and can be used for machining (Song et al. 2017). Alumina oxide is also used as a ceramic material in spark plugs, tap washer and abrasion-resistant tiles. High-purity alumina is used in diverse engineering applications, including electronic industry, chemical industry, synthetic gems and oil and gas processing. Activated alumina is also used for its high absorptive properties as a desiccant. Calcined alumina is used in refractory and ceramic industries. They are also used as cosmetic fillers. Zirconia-toughened alumina materials are also used for biomedical applications. Alumina material has effectively been used in the treatment of household water supply. The activated alumina is used in the defluoridation of water (Srimurali and Karthikeyan 2008). In this study, activated alumina candle water filters were used in column packed with cotton for the removal of varying fluoride concentration from 2 to 20 mg/L. Alumina oxide (Al_2O_3)–ceramic membranes are effectively used for the treatment of wastewater by the process of filtration and reverse osmosis.

Banerjee et al. (2019) reported the use of alumina nanoparticles for the treatment of dye water collected from carpet industry. In this study, alumina nanoparticles effectively removed the Orange G dye from the wastewater obtained from industry. The alumina nanoparticles can effectively remove the dye at a controlled by pH up to 50%, indicating the adsorbent is effective for the treatment of wastewater. It was also observed that alumina nanoparticles were effective in reducing the organic loads such as BOD and COD from real wastewater effluent.

6.8 CONCLUSIONS AND FUTURE PROSPECTS

Alumina-based adsorbents show effective and high removal of heavy metals, dyes and drugs because of their high surface area, high porosity, good regeneration ability and amphoteric nature that permits it to behave as acid. It is hygroscopic in nature and thus is a favoured desiccant for the removal of moisture from the air, and it is also used as a catalyst in natural gas and refining operations. In the removal of contaminants, metal oxides are the most frequently used surfaces; they are inexpensive and are widely produced. This chapter covered various alumina-based adsorbents such as Fe_3O_4@Al_2O_3@Zn-Fe LDH and nanoalumina showing higher adsorption capacity for the removal of heavy metals such as arsenic and cobalt. Themodified metal–organic framework alumina based adsorbents are used for the removal of drugs showing higher removal and adsorption efficiency such as MIL-53(Al) towards diclofenac and naproxen drug due to hydrogen bonding and stacking/interaction of π–π bonds. Several other adsorbents such as gamma phase nanoalumina adsorbents showed promising results towards the carcinogenic dye Orange G with adsorption capacity of 93.3 mg/g, and the Fe_2O_3-Al_2O_3 nanocomposite showed 498 mg/g adsorption capacity towards Congo red dye.

Activated alumina is widely used in the development of reverse osmosis (RO) membrane technology. They are widely used as a desiccant and catalyst. Alumina and its products are used effectively in various industries such as water purification, ceramics and biomedical industries. Hence, it is concluded that due to their high surface area, amphoteric nature and good regeneration ability, alumina and alumina-based adsorbents show promising results and can be used for the treatment of wastewater.

REFERENCES

Aazza, M., Moussout, H., Marzouk, R., Ahlafi, H. 2017. Kinetic and thermodynamic studies of malachite green adsorption on alumina. *J. Mater.* 8: 2694–2703.

Abid, H. R., Rada, Z. H., Shang, J., Wang, S. 2016. Synthesis, characterization, and CO_2 adsorption of three metal-organic frameworks (MOFs): MIL-53, MIL-96, and amino-MIL-53, *Polyhedron* 120: 103–111.

Ademola, J. A., Olatunji, M. A. 2013. Evaluation of NORM and dose assessment in an aluminium industry in Nigeria. *World J. Nuclear Sci. Technol.* 3: 150.

Afroze, S., Sen, T. K. 2018. A review on heavy metal ions and dye adsorption from water by agricultural solid waste adsorbents. *Water Air Soil Pollut.* 229: 225–230.

Ahmad, R., Haseeb, S. 2017. Adsorption of Pb(II) on Mentha piperita carbon (MTC) in single and quaternary systems. *Arabian J. Chem.* 10: S412–S421.

Ahn, C. K., Park, D., Woo, S. H., Park, J. M. 2009. Removal of cationic heavy metal from aqueous solution by activated carbon impregnated with anionic surfactants. *J. Hazard. Mater.* 164: 1130–1136.

Akpinar, I., Yazaydin, A.O. 2017. Rapid and efficient removal of carbamazepine from water by UiO-67. *Ind. Eng. Chem. Res.* 56: 15122–15130.

Aksu, Z., Gonen, F., Demircan, Z. 2002. Biosorption of chromium (VI) ions by mowital B3OH resin immobilized activated sludge in a packed bed: comparison with granular activated carbon. *Process Biochem.* 38: 175–186.

Al-Ghouti, M. A., Khraisheh, M. A. M., Allen, S. J., Ahmad, M. N. 2003. The removal of dyes from textile wastewater: a study of the physical characteristics and adsorption mechanisms of diatomaceous earth. *J. Environ. Manage.* 69: 229–238.

Ali, I. 2012. New generation adsorbents for water treatment. *Chem. Revolution* 112: 5073.

Ali, S., Abbas, Y., Zuhra, Z., Butler, I. S. 2019.Synthesis of 'ϒ alumina(Al_2O_3) nanoparticles and their potential use as an adsorbent in the removal of methylene blue from industrial wastewater. *Nanoscale Adv.* 1, 213–218.

An, H. J., Bhadra, B. N., Khan, N. A., Jhung, S. H. 2018. Adsorptive removal of wide range of pharmaceutical and personal care products from water by using metal azolate framework-6-derived porous carbon. *Chem. Eng. J.* 343: 447–454.rsenic Free Water to pave way for a Healthy Life. 2019. World Health Organization.

ATSDR Agency for Toxic Substances and Disease Registry. 2011. U.S. Department of Health and Human Services.

Banerjee, S., Dubey, S., Gautam, R. K., Chattopadhyaya, M. C., Sharma, Y. C. 2019. Adsorption characteristics of alumina nanoparticles for the removal of hazardous dye, Orange G from aqueous solutions. *Arabian J. Chem.* 12: 5339–5354.

Barakat, M. 2011. New trends in removing heavy metals from industrial wastewater. *Arabian J. Chem.* 4: 361–377.

Bayazit, S. S., Danalioglu, S. T., Salam, M. A., Kuyumcu, O. K. 2017. Preparation of magnetic MIL-101(Cr) for efficient removal of ciprofloxacin. *Environ. Sci. Pollut. Res.* 24: 25452–25461.

Bhadra, B. N., Ahmed, I., Kim, S., Jhung, S. H. 2017. Adsorptive removal of ibuprofen and diclofenac from water using metal-organic framework-derived porous carbon. *Chem. Eng. J.* 314: 50–58.

Brieva, B. G., Martin, C. J. M., Al-Zahrani, S. M., Fierro, J. L. G. 2011. Effectiveness of metal–organic frameworks for removal of refractory organo-sulfur compound present in liquid fuels. *Fuel* 90: 190–197.

Čampelj, S., Makovec, D., Drofenik, M. 2009, Functionalization of magnetic nanoparticles with 3-aminopropyl silane. *J. Magn. Mater.* 321: 1346–1350.

Cavaco, S. A., Fernandes, S., Quina, M. M., Ferreira, L. M., 2007. Removal of chromium from electroplating industry effluents by ion exchange resins. *J. Hazard. Mater.* 144: 634–638.

Chakraborty, S., Dasgupta, J., Farooq, U., Sikder, J., Drioli, E., Curcio, S., 2014. Experimental analysis, modeling and optimization of chromium (VI) removal from aqueous solutions by polymer-enhanced ultrafiltration. *J. Memb. Sci.* 456: 139–154.

Chen, L., Wang, T. J., Wu, H. X., Jin, Y., Zhang, Y., Dou, X. M. 2011. Optimization of a Fe–Al–Ce nano-adsorbent granulation process that used spray coating in a fluidized bed for fluoride removal from drinking water. *Powder Technol.* 206: 291–296.

Cherukumilli, K., Delaire, C., Amrose, S., Gadgil, A. 2017. Factors governing the performance of bauxite for fluoride remediation of groundwater. *Environ. Sci. Technol.* 51: 2321–2328.

Chua, J. H., Chee, R.E., Agarwal, A., Wong, S. M., Zhang, G. J. 2009. Label-free electrical detection of cardiac biomarker with complementary metal-oxide semiconductor compatible silicon nanowire sensor arrays. *Anal. Chem.* 81: 6266–6271.

Cui, L., Wu, J., Ju, H. 2015. Electrochemical sensing of heavy metal ions with inorganic, organic and bio-materials. *Biosen. Bioelectron.* 63: 276–286.

Czaja, A. U., Trukhan, N., Muller, U. 2009. Industrial applications of metal-organic frameworks. *Chem. Soc. Revolution* 38: 1284–1293.

Dawood, S., Sen, T. K. 2012. Removal of anionic dye Congo red from aqueous solution by raw pine and acid treated pine cone powder as adsorbent: equilibrium, thermodynamic, kinetics, mechanism and process design. *Water Res.* 46: 1933–1946.

Dehghani, M. H., Yetilmezsoy, K., Salari, M., Hedarinejad, Z., Yousefi, M., Silanpaa, M. 2020. Adsorptive removal of Cobalt(II) from aqueous solutions using multi walled carbon Nanotubes and ϒ-Alumina as novel adsorbents: modelling and optimization based on response surface methodology and Artificial neural network. *J. Mol. Liquid* 299: 112154.

Dhawale, V. P., Khobragade, V., Kulkarni, S. D. 2018. Synthesis and characterization of aluminium oxide (Al_2O_3) nanoparticles and its application in azodye decolourisation. *Chemistry* 27, 31.

Do, X. D., Hoang, V. T., Kaliaguine, S. 2011. MIL-53(Al) mesostructured metal-organic frameworks. *Microporous Mesoporous Mater.* 141: 135–139.

Doke, S. M., Yadav, G. D. 2014. Process efficacy and novelty of Titania membrane prepared by polymeric sol–gel method in removal of chromium (VI) by surfactant enhanced microfiltration. *Chem. Eng. J.* 255: 483–491.

El Gaini, L., Lakraimi, M., Sebbar, E., Meghea, A., Bakasse, M. 2009. Removal of Indigo Carmine dye from water to Mg–Al–CO_3- calcined layered double hydroxides. *J. Hazard. Mater.* 161: 627–632.

Farha, O. K., Eryazici, I., Jeong, N. C., Hauser, B. G., Wilmer, C. E., Sarjeant, A. A., Snurr, R. Q., Nguyen, S. T., Yazaydın, A., Hupp, J. T. 2012. Metal-organic framework materials with ultrahigh surface areas: is the sky the limit? *J. Am. Chem. Soc.* 134: 15016–1502.

Fashi, F., Ghaemi, A., Moradi, P. 2018. Piperazin-modified activated alumni as a novel promising candidate for CO_2 capture; Experimental and modeling. *Greenhouse Gas Sci. Technol.* 9: 37–51.

Furukawa, H., Go, Y. B., Ko, N., Park, Y. K., Uribe-Romo, F. J., Kim, J., O'Keeffe. M., Yaghi, O. M. 2011. Isoreticular expansion of metal–organic frameworks with triangular and square building units and the lowest calculated density for porous crystals. *Inorg. Chem.* 50: 9147–9152.

Gao, Y., Liu, K., Kang, R., Xia, J., Yu, G., Deng, S. 2018. A comparative study of rigid and flexible MOFs for the adsorption of pharmaceuticals: kinetics, isotherms and mechanisms. *J. Hazard. Mater.* 359: 248–257.

Garg, V., Kumar, R., Gupta, R., 2004. Removal of malachite green dye from aqueous solution by adsorption using agro-industry waste: a case study of Prosopis cineraria. Dyes Pigments 62: 1–10.

Ghaedi, M., Tavallali, H., Sharifi, M., Kokhdan, S. N., Asghari, A. 2012. Preparation of low cost activated carbon from Myrtus communis and pomegranate and their efficient application for removal of Congo red from aqueous solution. *Spectrochim. Acta Part A Mol. Biomol. Spectrosc.* 86: 107–114.

Girish, C. R. 2018. Various impregnation methods used for surface modification of the adsorbent: a review. *Int. J. Eng. Technol.* 7: 330–334.

Gopinath, K. P., Murugesan, S., Abraham, J., Muthukumar, K. 2009. Bacillus sp. mutant for improved biodegradation of Congo red: random mutagenesis approach. *Bioresource Technol.* 100: 6295–6300.

Gupta, A. B., George, S. S., Mandal, P. 2013. Effect on activated alumina fluoride removal capacity in the presence of chloride. Proceedings of International Conference on Materials for the Future – in- Innovative Materials, Processes, Products, and Applications-ICMF, pp. 598–600.

Gupta, V. K., Jain, R., Nayak, A., Agarwal, S., Shrivastava, M. 2011. Removal of the hazardous dye—tartrazine by photodegradation on titanium dioxide surface. *Mater. Sci. Eng. C* 31: 1062–1067.

Hasan, Z., Choi, E. J., Jhung, S. H. 2013. Adsorption of naproxen and Clofibric acid over a metal–organic framework MIL-101 functionalized with acidic and basic groups. *Chem. Eng. J.* 219: 537–544.

Hasan, Z., Jeon, J., Jhung, S. H. 2012. Adsorptive removal of naproxen and clofibric acid from water using metal-organic frameworks, *J. Hazardous Mater.* 209: 151–157.

Hasan, Z., Khan, N. A., Jhung, S. H. 2016. Adsorptive removal of Diclofenac sodium from water with Zr-based metal–organic frameworks. *Chem. Eng. J.* 284: 1406–1413.

Haseena, M., Malik, M F., Javed, A., Arshad, S., Asif, N., Zulfiqar, S., Hanif, J 2017. Water pollution and human health. *Environ. Risk Assess. Remed.* 1: 16–19.

Hawksworth, D. K. 2013. *Fluxless Brazing of Aluminium.* Woodhead, Canada, pp. 566–585.

Hordern, B. K. 2004. Chemistry of alumina, reactions in aqueous solution and its application in water treatment. *Adv. Colloid Interface Sci.* 110(1–2), 19–48.

Jain, S., Bansiwal, A., Biniwale, R. B., Milmille, S., Das, S., Tiwari, S., Antony, P. S. 2015. Enhancing adsorption of nitrate using metal impregnated alumina. *J. Environ. Chem. Eng.* 3: 2342–2349.

Jung, C., Boateng, L. K., Flora, J. R. V., Oh J, Braswell, M. C., Son, A., Yoon, Y. 2015. Competitive adsorption of selected non-steroidal anti-inflammatory drugs on activated biochars: experimental and molecular modeling study. *Chem. Eng. J.* 264: 1–9.

Kadirvelu, K., Kavipriya, M., Karthika, C., Radhika, M., Vennilamani, N., Pattabhi, S. 2003. Utilization of various agricultural wastes for activated carbon preparation and application for the removal of dyes and metal ions from aqueous solutions. *Bioresource Technol.* 87: 129–132.

Karamia, A., Sabouni, R., Ghommem, M. 2020. Experimental investigation of competitive co-adsorption of naproxen and diclofenac from water by an aluminum-based metal organic framework. *J. Mol. Liquids* 305: 112808.

Kitagawa, S., Kitaura, R., Noro, S. 2004. Functional porous coordination polymers. *Angew. Chem., Int. Ed.* 43: 2334–2375.

Kunde, G. B., Sehgal, B., Ganguli, A. K. 2019. Synthesis of mesoporous rebar MWCNT/Alumina composite nodules for the effective removal of Cr(VI) and Methylene Blue from an Aqueous medium. *J. Hazardous Mater.* 15: 140–151.

Kurniawan, T. A., Chan, G. Y., Lo, W.-H., Babel, S. 2006. Physico–chemical treatment techniques for wastewater laden with heavy metals. *Chem. Eng. J.* 118: 83–98.

Laasri, L., Elamrani, M. K., Cherkaoui, O. 2007. Removal of two cationic dyes from a textile effluent by filtration-adsorption on wood sawdust. *Environ. Sci. Pollut. Res. Int.* 14: 237–240.

Laleh, A., Nafiseh, S., Akram, M. 2019. Optimization of arsenic removal with Fe_3O_4@Al_2O_3@Zn-Fe LDH as a new magnetic nano adsorbent using Box-Behnken design. *J. Environ. Chem. Eng.* 7: 102974.

Lei, C., Pi, M., Jiang, C., Cheng, B., Yu, J. 2017. Synthesis of hierarchical porous zinc oxide (ZnO) microspheres with highly efficient adsorption of Congo red. *J. Colloid Interface Sci.* 490: 242–251.

Lei, Z., Pang, X., Li, N., Lin, L., Li, Y. 2019. A novel two-step modifying process for preparation of chitosan-coated Fe_3O_4/SiO_2 microspheres. *J. Mater. Process Technol.* 209: 3218–3225.

Li, C., Xiong, Z., Zhang, J., Wu, C. 2015. The strengthening role of the amino group in metal–organic framework MIL-53 (Al) for methylene blue and malachite green dye adsorption. *J. Chem. Eng. Data* 60: 3414–3422.

Li, Z., Wu, Y., Li, J., Zhang, Y., Zou, X., Li, F. 2015. The metal-organic framework MIL-53(Al) constructed from multiple metal sources: alumina, aluminum hydroxide, and boehmite. *Chem. Euro. J.* 21: 6913–6920.

Li, J., Wu, Y., Li, Z., Zhu, M., Li, F. 2014. Characteristics of Arsenate removal from water by metal-organic frameworks (MOFs). *Water Sci. Technol.* 70: 1391–1397.

Lian, L., Guo, L., Guo, C., 2009. Adsorption of Congo red from aqueous solutions onto ca-bentonite. *J. Hazard. Mater.* 161: 126–131.

Lim, S. J., Kim, T. H., 2015. Combined treatment of swine wastewater by electron beam irradiation and ion exchange biological reactor system. *Sep. Purif. Technol.* 146: 42–49.

Lin, S., Zhao, Y., Yun, Y. S. 2018. Highly effective removal of nonsteroidal anti-inflammatory pharmaceuticals from water by Zr(IV)-based metal–organic framework: adsorption performance and mechanisms. *ACS Appl. Mater. Interfaces* 10: 28076–28085.

Lira, M. A., Navarro, R., Saucedo, I., Martinez, M., Guibal, E. 2016. Influence of the textural characteristics of the support on Au (III) sorption from HCl solutions using Cyphos IL101-impregnated Amberlite resins. *Chem. Eng. J.* 302: 426–436.

Liu, M., Xu, J., Cheng, B., Ho, W., Yu, J., 2015. Synthesis and adsorption performance of $Mg(OH)_2$ hexagonal nanosheet–graphene oxide composites. *Appl. Surf. Sci.* 332: 121–129.

Loiseau, T., Serre, C., Huguenard, C., Fink, G., Taulelle, F., Henry, M., Bataille, T., Férey, G. 2004. A rationale for the large breathing of the porous aluminum terephthalate (MIL-53) upon hydration. *Chem. Euro. J.* 10: 1373–1382.

Mahapatra, A., Mishra, B. G., Hota, G. 2013. Adsorptive removal of Congo red Dye from wastewater by mixed Iron oxide-alumina nanocomposites. *Ceram. Int.* 5: 5443–5451

Mahmoud, M. E., 2006. *Encyclopedia of Chromatography*, Second Edition, pp. 1–14, Published on 27 March, 10.1081/E-ECHR-12004279.

Meilikhov, M., Yusenko, K., Fischer, R. A. 2009. The adsorbate structure of ferrocene inside [Al(OH)(bdc)] x (MIL-53): a powder X-ray diffraction study. *Dalton Trans.* 600–602.

Naiya, T. K., Bhattacharya, A. K., Das, S. K. 2009. Adsorption of Cd (II) and Pb (II) from aqueous solutions on activated alumina. *J. Colloid Interface Sci.* 333: 14–26.

Owlad, M., Aroua, M. K., Daud, W. M. A. W. 2010. Hexavalent chromium adsorption on impregnated palm shell activated carbon with polyethyleneimine. *Bioresource Technol.* 101: 5098–5103.

Peng, H., Gao, P., Chu, G., Pan, B., Peng, J., Xing, B. 2017. Enhanced adsorption of Cu (II) and Cd (II) by phosphoric acid-modified biochars. *Environ. Pollut.* 229: 846–853.

Pourshadlou, S., Mobasherpour, I., Majidian, H., Salahi, E., Bidabadi, F. S., Mei, C. T., Ebrahimi, M. 2020, Adsorption system for Mg^{2+} removal from aqueous solutions using bentonite/gamma-alumina nanocomposite. *J. Colloid Interface Sci.* 568: 245–254.

Rabia, A. R. I., Brahim, A. H., Zulkepli, N. N. 2018. Activated alumina preparation and characterization: The review on recent advancement. E3S Web of Conferences, 34: 02049.

Ramana, P. V., Viswadevarayulu, A., Kumar, K., Reddy, S. A., 2017. Synergetic enhanced day-light driven photocatalytic reduction of heavy metal Cr(VI) by graphene supported ZnO nanocubes. *Adv. Mater. Lett.* 8: 303–308.

Ramirez, A. A. C., Garcia, E. R., Medina, R. L., Larios, J. L. C., Parra, R. L., Franco, A. M. M., 2021. Selective adsorption of aqueous diclofenac sodium, naproxen sodium and ibuprofen using a stable Fe_3O_4-FeBTC metal organic framework. *Materials* 14:2293.

Salifu, A., Petrusevski, B., Ghebremichael, K., Modestus, L., Buamah, R., Aubry, C., Amy, G. L. 2013. Aluminum hydroxide-coated pumice for fluoride removal from drinking water: Synthesis, equilibrium, kinetics and mechanism. *Chem. Eng. J.* 228: 63–74.

Salvador, F., Martin-Sanchez, N., Sanchez-Hernandez, R., Sanchez-Montero, M. J., Izquierdo, C. 2015. Regeneration of carbonaceous adsorbents. Part II: Chemical, microbiological and vacuum regeneration. *Microporous Mesoporous Mater.* 202: 277–296.

Samokhvalov, A. 2018. Aluminum metal–organic frameworks for sorption in solution: a review. *Coord. Chem. Revolution* 374: 236–253.

Sergey, R., Irina, K., Alesia, L., Eugene, M., Lynbov, L. 2019. Effects of Li, Na and K modification of alumina on its physical and chemical properties and water adsorption ability. *Materials* 12: 4212.

Shan, R. R., Yan, L. G., Yang, Y. M., Yang, K., Yu, S. J., Yu, H. Q. 2015. Highly efficient removal of three red dyes by adsorption onto Mg–Al-layered double hydroxide. *J. Ind. Eng. Chem.* 21: 561–568.

Sharma, Y. C., Srivastava, V., Mukherjee, A. K. 2010. Synthesis and application of nano-Al_2O_3 powder for the reclamation of hexavalent chromium from aqueous solutions. *J. Chem. Eng. Data* 55: 2390–2398.

Sharma, Y. C., Srivastava, V., Upadhyay, S. N., Weng, C. H. 2008. Alumina nanoparticles for the removal of Ni(II) from aqueous solutions. *Ind. Eng. Chem. Res.* 47: 8095.

Shikha D, Siddh Nath, U., Yogesh, C. S. 2016, Optimization of removal of Cr by gamma-alumina nano-adsorbent using response surface methodology. *Ecol. Eng.* 97: 272–283.

Silva, D. J. D. F., Malo, D. L., Bataglion, G. A., Eberlin, M. N., Ronconi, C. M., Alves, S., de Sa, G. F. 2015. Adsorption in a fixed-bed column and stability of the antibiotic oxytetracycline supported on Zn(II)-2-methylimidazolate frameworks in aqueous media. *PLoS One* 10: e0128436.

Singh, N. B., Nagpal, G., Agrawal, S. 2018. Water purification by using adsorbents: a review. *Environ. Technol. Innov.* 11: 187–240.

Song J. Y., Jhung S. H. 2017. Adsorption of pharmaceuticals and personal care products over metal-organic frameworks functionalized with hydroxyl groups: quantitative analyses of H-bonding in adsorption. *Chem. Eng. J.* 322: 366–374.

Song J. Y., Jhung S. H. 2017. Adsorption of pharmaceuticals and personal care products over metal-organic frameworks functionalized with hydroxyl groups: quantitative analyses of H-bonding in adsorption. *Chem. Eng. J.* 322: 366–374.

Srimurali, M., Karthikeyan, J. 2008. Activated alumina: defluoridation of water and household application-A study. Twelfth International Water Technology conference, IWTC, vol. 12 p. 2008.

Suman, M., Kalzang, C., Pooja, N., Khaiwal, R. 2016. Utilization of nano-alumina and activated charcoal for phosphate removal from wastewater. *Environ. Nanotechnol. Monit. Manage.* 7: 15–23.

Sun, W., Li, H., Li, H., Li, S., Cao, X. 2019. Adsorption mechanisms of ibuprofen and naproxen to UiO-66 and UiO-66-NH2: batch experiment and DFT calculation. *Chem. Eng. J.* 360: 645–653.

Sun, X. D., Lin, W. G., Wang, L. J., Zhou, B., Lv, X. L., Wang, Y., Zheng, S. J., Wang, W. M., Tong, Y. G., Zhu, J. H. 2014. Liquid adsorption of tobacco-specific N-nitrosamines by zeolite and activated carbon. *Microporous Mesoporous Mater.* 200: 260–268.

Tiwari, D., Lalhriatpuia, C., Lee, S. M. 2015. Hybrid materials in the removal of diclofenac sodium from aqueous solutions: Batch and column studies. *J. Ind. Eng. Chem.* 30: 167–173.

United States Department of Health and Human Services 2019. U.S. Department of Health and Human Services.

Vardhan, V., Srimurali, M. 2016. Removal of fluoride from water using novel sorbent lanthanum-impregnated bauxite. *Springer Plus* 5: 1426–1439.

Wang, S. G., Ma, Y., Shi, Y. J., Gong, W. X. 2009. Defluoridation performance and mechanism of nano-scale aluminum oxide hydroxide in aqueous solution. *J. Chem. Technol. Biotechnol.* 84: 1043–1050.

Wang, S. X., Li, S., Wang, Y., Li, Y., Zhai. 2008. Synthesis of γ-alumina via precipitation in ethanol. *Mater. Lett.* 62: 3552.

Wang, X., Zhan, C., Kong, B., Zhu, X., Liu, J., Xu, W., Wang, H. 2015. Self-curled coral-like γ-Al_2O_3 nanoplates for use as an adsorbent. *J. Colloid Interface Sci.* 453: 244–251.

Wasti, A., Awan, M. A. 2016. Adsorption of textile dye onto modified immobilized activated alumina. *J. Assoc. Arab Univ. Basic Appl. Sci.* 20: 26–31.

Xiao, Y., Han, T., Xiao, G., Ying, Y., Huang, H., Yang, Q., Liu, D., Zhong, C. 2014. Highly selective adsorption and separation of aniline/phenol from aqueous solutions by microporous MIL-53(Al): a combined experimental and computational study. *Langmuir* 30: 12229–12235.

Xu, D., Cheng, B., Cao, S., Yu, J. 2015. Enhanced photocatalytic activity and stability of Z-scheme Ag_2CrO_4-GO composite photocatalysts for organic pollutant degradation. *Appl. Catal., B Environ.* 164: 380–388.

Yamani, J. S., Lounsbury, A. W., Zimmerman, J. B. 2014. Adsorption of selenite and selenate by nanocrystalline aluminum oxide, neat and impregnated in chitosan beads. *Water Res.* 50: 373–381.

Yang, D. J., Paul, B., Xu, W. J., Yuan, Y., Liu, E. M., Ke, X. B., Wellard, R. M., Guo, C., Xu, Y., Sun, Y. H., Zhu, H. Y. 2010. Alumina nanofibers grafted with functional groups: a new design in efficient sorbents for removal of toxic contaminants from water. *Water Res.* 44: 741.

Yang, J., Hou, B., Wang, J., Tian, B., Bi, J., Wang, N., Huang, X. 2019. Nanomaterials for the removal of heavy metals from wastewater. *Nanomaterials* 9: 424.

Zhang, Z., Tan, Y., Zhong, M. 2011. Defluorination of wastewater by calcium chloride modified natural zeolite. *Desalination* 276: 246–252.

Zhao, X., Su, Y., Li, S., Bi, Y., Han, X. 2018. A green method to synthesize flowerlike $Fe(OH)_3$ microspheres for enhanced adsorption performance toward organic and heavy metal pollutants. *J. Environ. Sci.* doi: 10.1016/j.jes.2018.01.010.

Zhao, X., Su, Y., Qi, X., Han, X., 2017. A facile method to prepare novel Ag_2O/Ag_2CO_3 three-dimensional hollow hierarchical structures and their water purification function. *ACS Sustainable Chem. Eng.* 5: 6148–6158.

Zhao, X. D., Zhao, H. F., Dai, W. J, Wei, Y. A., Wang, Y. Y., Zhang, Y. Z., Zhi, L. F., Huang, H. L., Gao, Z. Q. 2018. A metal-organic framework with large 1-D channels and rich -OH sites for high-efficiency chloramphenicol removal from water. *J. Colloid Interface Sci.* 526: 28–34.

Zhou, M., Wu, Y., Qiao, J., Zhang, J., McDonald, A., Li, G., Li, F. 2013. The removal of bisphenol a from aqueous solutions byMIL-53(Al) and mesostructured MIL-53(Al). *J. Colloid Interface Sci.* 405: 157–163.

7 Silica Nanomaterials for Water Remediation

Md. Abu Taleb, Rajeev Kumar, and Mohamed Abou El-Fetouh Barakat
King Abdulaziz University

CONTENTS

7.1 INTRODUCTION

The recent increasing trend in urbanization and industrialization has caused the pollution of water by a wide variety of pollutants such as poisonous heavy metals (HMs) and organic contaminants including pesticides, dyes and drugs [1]. Thousands of previous studies have reported that globally, water has been contaminated due to the presence of contaminants in higher amounts than the permissible limits [2,3]. The various human-induced activities such as landfilling, use of pesticides in agriculture, mine drainage, direct release of untreated industrial effluents, dumping of municipal effluents, excessive use of solvents and dumping of radioactive wastes and chemical wastes are directly or indirectly responsible for the increase in the water contamination. The chronic health hazards including cancers, coronary disease, liver damage, neurological and cardiovascular

DOI: 10.1201/9781003129042-7

diseases, damage of central nervous system and sensory disturbances are due to some toxic pollutants such as heavy metals, dyes and antibiotics [4–7].

Therefore, researchers are looking for the decontamination of wastewater by appropriate treatment methods. Numerous conventional methods have been applied for wastewater remediation, such as chemical precipitation [8–10], ion exchange [8,10], adsorption [9,11], membrane filtration [12] and electrochemical treatment technologies [13]. The conventional methods are less efficient due to the resistance and persistence of pollutants in wastewater [14]. Nonetheless, these methods have their own limitations such as discharging of toxic slurry, challenges of sludge management and precarious working environments [15]. Therefore, it is demanding to find environment-friendly and cost-effective water remediation methods. Adsorption is one of the best methods having high competency, simplicity and low cost for water decontamination. Various materials such as alumina (Al_2O_3) [14], activated carbon [16], titanium dioxide (TiO_2) [17] and magnetic graphene oxide [18] have been reported for the adsorption of pollutants. Costly adsorbents are not preferred now for use as an adsorbent material [14]; thus, many researchers are paying to attention on making low-price and efficient adsorbents that are naturally available, such as silica. In the last few decades, researchers have placed increased emphasis on modifying composite adsorbents due to their exclusive characteristics such as enlarged surface area, small size, available reactive sites and high efficiency of regeneration [19].

Recently, nanotechnology for water remediation has received extended attention because of the discovery of exciting new functional materials with excessive efficiencies to entrap numerous contaminants from wastewater [20]. An extensive variety of nanomaterials are reported in the previous researches for the decontamination wastewater [20,21]. Nanomaterials are materials with size in the range of 1–100 nm, and they can be synthesized via top-down and bottom-up techniques [20]. Due to the extremely small size of nanomaterials, their surface area could be very excessive, which will increase the wastewater and the nanomaterial contact [21]. This expanded contact increases the adsorption performance of the nanomaterials towards numerous contaminants. Even though different nanomaterials including metal oxides and magnetic, fabricated nanoparticles are reported for wastewater decontamination, silica and silica-based nanomaterials have gained great attention due to their significant properties. Naturally abundant and low-cost silica having two-dimensional array of channels is a highly ordered material and can efficiently remove pollutants due to extended surface area [20,21]. In the last few years, a substantial number of studies have been performed regarding the design of diverse cost-effective and competent synthetic routes, including hydrothermal, sol–gel, flame and reverse microemulsion, for the synthesis of silica nanomaterials as well as silica composite nanomaterials [21,22]. The synthesis route is an important step for the preparation of efficient and cost effective silica nanomaterials [23]. This advancement in processes has given accessibility to a series of various silica nanomaterials with a variety of shapes, sizes and morphologies for the decontamination of wastewater [24].

In this chapter, the selectivity and possible adsorption mechanisms of silica nanomaterials have been reviewed from very recent studies on wastewater decontamination from different pollutants including organic and inorganic pollutants. The salient factors affecting the adsorption of pollutants on silica nanomaterials are also briefly explained using schemes and figures where appropriate.

7.2 PROPERTIES OF SILICA

Silicon dioxide (SiO_2) or silica is one of the most abundant elements in the nature, since sand is available in the earth crust and subsequently silica is also plentiful in the nature as the primary ore source of silicon [25]. The chemically modified silica is generally fabricated in huge quantities for different commercial applications. Chemically, silica can be synthesized using $SiCl_4$ (silicon tetrachloride) by applying the flame pyrolysis method and the fabricated silica is called fumed silica. Another chemically synthesized silica can be obtained by using the precipitation method from silicate solution, and it is called precipitated silica. The other chemical type of silica named silica gels are synthesized by applying the sol–gel method [26].

The surface of silica is mainly composed of siloxane groups (Si–O–Si) in the inward region and silanol groups (Si–OH) distributed on the surface. The silanols can be classified into three types as isolated (single) silanol, geminal (double) silanol and vicinal silanol [25]. The silica surface having silanol groups enthusiastically respond to numerous reagents. Moreover, the silanol groups (Si–OH) on the surface make silica a better solid support for the immobilization of various organic and inorganic groups [27–31]. Many of the adsorption, adhesion, chemical and catalytic properties of silica depend on the chemistry and geometry of its surface [32]. The content of residual Si–OH groups was measured at high temperature. Above 400°C–450°C, more hydroxyl groups are removed, retaining large siloxane areas [31]. The physically adsorbed water was removed at 115°C, and the bound water was present as hydroxyl group's layer on the surface. According to the literature, silanol groups not only are found on the surface, but may also occur inside the silica skeleton and very fine ultra-micropores (diameter <1 nm) [32]. The silica surface contains both hydrophilic and hydrophobic patches. Silanol groups impart hydrophilicity, while siloxane groups confer hydrophobicity [33]. The silanols can be protonated or deprotonated depending on the pH of the environment [25].

7.3 NANOMATERIALS

Nanocomposites are defined as functional elements that are of nanosize or within the range of 1–100 nm. They are reported in many types, including carbon-based, metal-based, dendrimer and composite nanomaterials. Nanocomposites are exhibits diverse adsorption properties and efficient treatment facilities of wastewater decontamination due to surface chemistry, surface area and selectivity for different pollutants [24]. Nanomaterials can be visualized by using scanning electron microscopy (SEM), and usually, they are synthesized from polymers, metals, glasses and ceramics [25]. Nanocomposites can be made by different modification methods, including surface modification, modification of shapes of materials, multifunctionalization, assembling, size control and composition of materials. Nowadays, nanomaterials are of immense methodical and scientific interest for water decontamination. They are reported to efficiently act as a bridge between molecular structure and massive materials [23].

7.4 SILICA NANOMATERIALS

Silica was reviewed as the excellent supportive material of nanocomposite adsorbent for water remediation due to its surface chemistry as well as other exciting features related to the adsorption properties, including excellent mechanical resistance, thermodynamic stability and smooth mass exchange characteristics. Different silica materials have been used for wastewater decontamination, including silica gel, fumed silica, amorphous silica and mesoporous silica (Table 7.1). Among the other silica materials, mesoporous silica is the best one and can be synthesized using the hydrolysis process along with apposite catalyst and the presence of suitable surfactant (template) to make their siloxanes network (Si–O–Si linkages). The sol–gel method is mainly used for the preparation of different types of ordered mesoporous silica, including SBA-15, SBA-1, MCM-48 and MCM-41, having enlarged surface area, high porosity and abundant silanol groups. The ordered silica is also further hybridized into organic–inorganic composite materials by immobilization of numerous functional groups [34].

The amorphous silica showed high porosity which can be supportive to graft with other molecules as well as functional groups keeping their own properties besides novel entities [24].

7.5 SYNTHESIS ROUTES OF SILICA NANOMATERIALS

Chemical synthesis and biogenic synthesis are the most common advanced synthesis routes for the fabrication of silica nanomaterials. Both of the methods have their own advantages and limitations. Biogenic synthesis of silica nanomaterials

TABLE 7.1
Silica Adsorbents and Their Modification Routes

Silica	Adsorbent	BET Surface Area (m^2/g)	Modification Routes	Ref.
Mesoporous silica	MCM-41	1322	Sol–gel, direct or co-condensation, semi-direct synthesis, post-grafting	[35]
	SBA-15	633		[36]
	SBA-15	757.4		[37]
Amorphous silica	Amorphous SG	311	Sol–gel and hydrothermal	[38]
Silica gel	Aerogel	328.7	Sol–gel and hydrothermal	[39]
	Xerogel	701.6		[39]
Fumed silica	-	153.3172	Thermal process	[40]

is considered environmental friendly but relatively low purification capacity than the chemically synthesized silica [23].

7.5.1 Chemical Synthesis of Silica Nanomaterials

Since long time before silica nanomaterials have been designed applying diverse processes, including the Stober method, reverse microemulsion method, flame synthesis and the very popular sol–gel method. Recently, some novel methods have been utilized for the synthesis of silica nanomaterials. The chemical route fabrication of silica nanomaterials are easiest and widely used methods. In addition, this chemical route has various options to fabricate silica nanomaterials with desired parameters. Surfactant molecules were used in water for producing spherical micelles in reverse microemulsion, and the polar head groups assign themselves to form the reverse micelles (microcavities containing water) [41]. Different functional groups were attached by appropriate coating of silica nanomaterials for sufficient wastewater remediation [24,41]. The chemical vapour condensation (CVC) is a widely practiced method using metalorganic precursors by high-temperature flame decomposition [41]. For the last few decades, the sol–gel method has been considered as the most popular process for the production of silica, glass and ceramic substances due to its propensity to form pristine and homogeneous products at clement form. This technique entails hydrolysis and condensation of metal alkoxides consisting of TEOS, TMOS or inorganic salts such as sodium silicate in the presence of mineral acid (e.g. HCl) or base (e.g. NH_3) as a catalyst [24,41].

7.5.2 Biological Synthesis of Silica Nanomaterials

Recently, the biogenic synthesis of silica nanomaterials using biomass (fungi, bacteria and algae) and plant materials (extract and metabolites) has been found as a promising and successful synthesis route [42]. Sometimes, silica has been fabricated by silicon hydrolysis process using diatoms and fungi.

7.6 SILICA NANOMATERIALS FOR WATER REMEDIATION

Silica nanomaterials are widely used for wastewater decontamination due to their hydrophilic nature, low cost, good biocompatibility, extended surface area, exciting pore volume and controlled particle size. Numerous silica materials have been synthesized with various materials including polymers, inorganic materials (metal oxides), biomaterials (biosorbent), minerals, and industrial wastes to improve the adsorption capacity for water decontamination [34]. Some vital factors related to the improvement of the capacity of the modified silica for wastewater decontamination are structural texture and other surface features including specific area, porosity and availability of functional groups. Another factor to be considered is surface modification; for example, mesoporous silica has been explored different adsorption capacity among SBA-15 and MCM-41 as same type silica upon their surface modification, after modification, SBA-15 showed better mechanical constancy and subsequent adsorption capacity than the MCM-41 due to their good attachment and exchange of smooth hexagonal mesopore channels by thick walled micropores. Therefore, surface modification is another vital factor affecting the fabrication of silica materials and subsequent improvement of adsorption capacity of the materials. Silica surface modification using organic groups can be done by applying post-grafting and hydrolysis synthesis routes. The pre-synthesized mesoporous silica can be functionalized using appropriate organosilane along with the surface silanol. Another route of organically modified silica synthesis by using silica precursors such as tetraethyl orthosilicate (TEOS) and subsequently adding suitable organic groups via co-condensation process is named as one-pot synthesis. Moreover, semi-direct process of ordered mesoporous silica synthesis can provide more functional groups where organosilanes are added before hydrothermal process with TEOS gel [43]. It is worth mentioning that silica surface modification can be done by using not only organic groups, but also inorganic materials such as metal oxides, and in addition, different polymers and biosorbents may also be used for the surface modification of silica materials [44,45].

7.6.1 Surface Modification of Silica Nanomaterials for Improved Adsorption

Different types of pristine silica materials including mesoporous silica, silica gels, and amorphous silica have been used for the removal of metal ions from wastewater. It was explored that the utilized pristine silica adsorbents have low efficiencies due to their poor surface properties. Therefore, in order to enhance the surface properties, functionalization and surface modification of silica materials with appropriate functional groups are considered as imperative tools.

Several silica materials have been fabricated with organic functional moieties including nitrogen-containing organic groups and widely applied for metal ions adsorption. It was found that silica is a reliable material to support withorganic groups subsequently gained high capacities of metal adsorption from wastewater. The silica organic materials were fabricated by direct as well as post-grafting processes. The silica adsorbents having functionalized groups showed high loading of organic contents (C, H and N atoms).

The aliphatic amine having nitrogen-containing functional group as seen in Figure 7.1 was used for Cr(VI) adsorption on (3-aminopropyl)trimethoxysilane (APTMS) with mesoporous silica [46]. It was observed that this organically modified silica material had endured electrostatic attraction ahead in chromate ions in acidic environment. Silica having an amine group was shown to convert into protonated ions in acidic solution, subsequently increasing chromate ions adsorption with increased number of nitrogen atoms. Similar results were observed for metals adsorption on fabricated silica SBA-1 and MCM-41, when synthesized with monoamino (–N), diamino (–NN) and triamino (–NNN) groups as listed in Table 7.2.

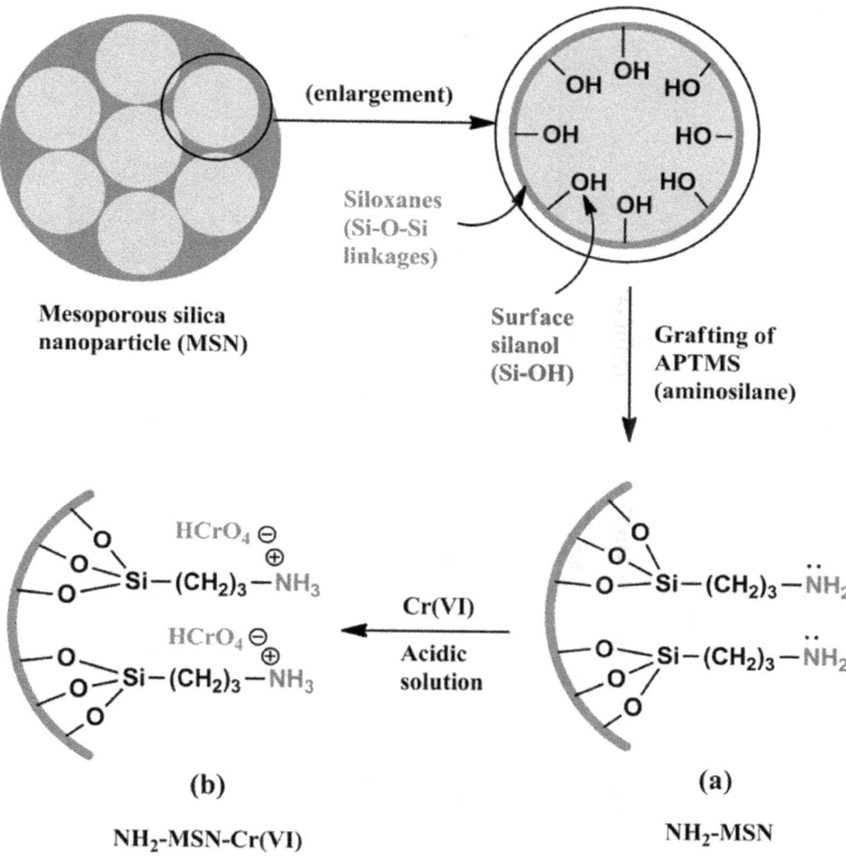

FIGURE 7.1 A schematic illustration of organically modified silica materials with APTMS for Cr(VI) adsorption. (See Ref. [46].)

Wu et al. [47] prepared 2-acetylthiophene-functionalized SBA-15 (A-SBA-15) by using APTMS and TEOS and applied it for sequestration of Cr(III). The adsorption efficiency of A-SBA-15 was 114.2 mg/g under optimum adsorption conditions, which was much higher than that of the pure SBA-15 (15.6 mg/g). The characterization measures confirmed that chemical interaction of organic functional groups through coordination bond was involved in the Cr(III) adsorption on A-SBA-15. Moreover, the presence of nitrogen and sulphur atoms in the structure of the functionalized adsorbent was anticipated to enhance the constancy of A-SBA-15 with chromate ions. The prepared A-SBA-15 was reused for five cycles without significant loss of adsorption efficiency. In contrast to other recently used adsorbents, A-SBA-15 showed better adsorption capacity, reusability and selectivity as a potential organic functional mesoporous silica adsorbent for metal adsorption from wastewater.

TABLE 7.2
Adsorption Capacity, Conditions and Surface Area of Different Surface-Modified Silica Nanomaterials

Surface-Modified Silica Nanomaterials										
				Adsorption Conditions						
Sl. No.	Name of the Composite	Pollutant Adsorption Capacity (mg/g)	BET Surface Area (m^2/g)	Mass (g)	pH	Temp. (°C)	Time (min)	Conc. (mg/L)	Vol. (mL)	Ref.
1	2-Acetylthiophene-functionalized SBA-15 (A-SBA-15)	Cr(III) – 114.2	335.6	0.05	6.0	50	360	200	50	[47]
2	Sodium dodecyl sulphate-functionalized mesoporous silica (SDS-MCM-41)	Cu(II) – 9.51 Cd(II) – 8.56 Zn(II) – 5.78	285	0.02	5-7	25	180	20	20	[48]
3	DANFR-silica composite	W(VI) – 55.32	-	0.04	6.0	25	45	100	50	[49]
4	APTES-grafted silica (GSIL)	Alizarin Red S – 59.8	440	0.025	2.0	30	30	-	25	[50]
5	Silane-coated nickel ferrite nanocomposite ($NiFe_2O_4@SiO_2$)	Acetaminophen – 58	-	0.01	6.0	25	30	12	100	[51]

Kaewprachum et al. [48] functionalized MCM-41 applying organic modification by using sodium dodecyl sulphate (SDS). The novel SDS-MCM-41 was used for the removal of Cu(II), Cd(II) and Zn(II) from contaminated aqueous solution. Adsorption efficiency and reaeration capacity of the new composite were quantified, and the results revealed that adsorption equilibrium was attained within 30 min for Cd(II) and 60 min for Cu(II) and Zn(II) and the material can applied successfully for up to six cycles. The material adsorption efficiency depends on the solution pH; neutral solution was more favourable for better adsorption than the acidic solution. The adsorption capacity of the novel and pristine materials was compared, and it was shows that the adsorption of Cu(II) and Cd(II) ions on novel SDS-MCM-41 was more favourable than on MCM-41, indicating SDS did not favour Zn(II) adsorption.

Dinker et al. [49] modified silica gel by coating 1,8-diaminonaphthalene formaldehyde resin (DANFR) and applied it for W(VI) adsorption. The developed silica material was treated with HCl for the activation of binding sites ($-NH_3^+ Cl^-$) on its surface. The satisfactory sequestration was attained by the anion exchange mechanism of Cl^- with W(VI), while other cations got repelled from the surface ($-NH_3^+$) of the DANFR-silica composite, and 55.32 mg/g of W(VI) was captured from the contaminated water within 45 min of reaction. Concurrent treatment with neat aqueous solution of W carried out 63.27 mg/g of W(VI) removal. Finally, recuperation of WO_4^{2-} ions and renewal of the adsorbent were carried out by using alkaline solution, which demonstrated successful desorption, as investigated by using ion chromatography [49].

Ali et al. [50] fabricated the surface of silica material with grafting by APTES as seen in Figure 7.2. The modified material was applied for the extraction of Alizarin Red S (ARS) dye from wastewater. The synthesized nanomaterial has been found to be a highly promising material for the adsorption of the organic dye

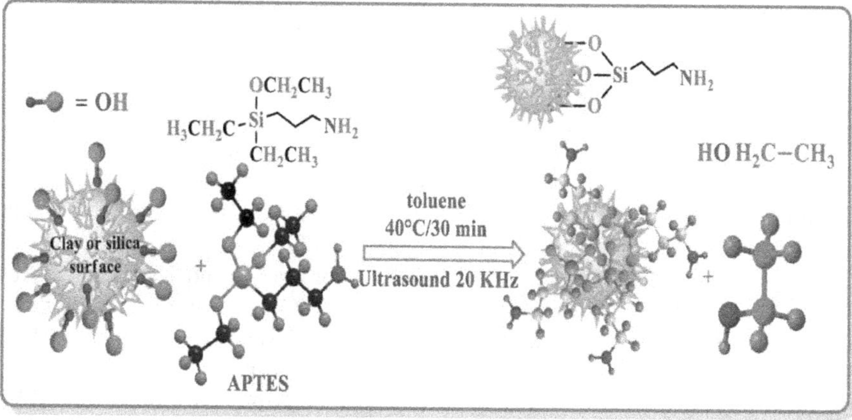

FIGURE 7.2 Proposed mechanism for the modification of APTES-grafted mesoporous silica. (See Ref. [50].)

(ARS) with 59.8 mg/g monolayer adsorption capacity within 30 min of reaction time in acidic condition at optimum pH 2.0 [50].

Kollarahithlu and Balakrishnan [51] synthesized a superparamagnetic silica nanomaterial with nickel ferrite applying co-precipitation using aminosilane. The prepared material has been evaluated for the properties of acetaminophen adsorption from wastewater. The 94% of acetaminophen was removed quickly within 15 min of reaction time with 58 mg/g adsorption capacity. The nanomaterials showed excellent recycling capacity up to four successive regeneration cycles [51].

7.6.2 Silica Nanocomposites for Metal Ions Adsorption

Different silica nanocomposites including polymeric silica nanomaterials, silica metal oxides, carbon-based silica nanomaterials and silica bio-nanominerals have been applied for the removal of heavy metals [23,24,52]. Among the other silica nanomaterials, the polymeric nanomaterials have been considered as promising materials due to some exciting physical, chemical and economical properties, such as easy synthesis, availability, low cost, mechanical and environmental stability, active binding sites along the walls of replicated parts of monomers and functional groups with enlarged surface area [53]. After the integration of polymeric properties into the silica materials, modified silica polymeric adsorbents showed promising adsorption properties for better adsorption of metal ions. Among the other polymers, polypyrrole (PPy), polyaniline (PANI), polyethylenimine (PEI) and polythiophene (PTh) have gained extensive interest for use in the silica modification as a modifier. Some recently reported silica-based polymeric adsorbents are summarized in Table 7.3 for metal ions adsorption. Recent studies have explored that, among the polymers, PANI and its derivatives showed excellent adsorption capacity because of the presence of abundant amine and imine groups in the polymer chain. Moreover, PANI derivatives have the same structure of PANI and other additional pendant functional groups at ortho, meta and para positions may have enhanced the surface properties of the modified adsorbent [44,54]. On the other hand, the N and S atoms of polymers are mainly responsible for metal adsorption and N atoms in polypyrrole (PPy) are more efficient than the S atoms in polythiophene (PTh) [55].

A multifunctional graphene oxide/silica@polyaniline ($GO/SiO_2@PANI$) microsphere was prepared by Kumar et al. [44]. The newly developed composite was utilized for the scavenging of Cr(VI) and Cu(II) ions from contaminated solution, and adsorption equilibrium was attained rapidly at pH 3.0 and 5.3, respectively. The adsorption capacity was 512.47 mg/g for Cr(VI) and 258.275 mg/g for Cu(II), which were very high compared to pristine SiO_2 – 189 mg/g and 106 mg/g correspondingly. The receptors geometry of extracted ions onto $GO/SiO_2@PANI$ was deeply interpreted by applying advanced statistical physics models (ASPM) and adsorption-reduction mechanism. The Cr(VI) ions extraction was showed by vertical and horizontal direction, while Cu(II) uptake was reflecting only horizontal direction. Single- and two-layer dual-energy model was fitted in the adsorption studied metal ions onto fabricated polymeric adsorbent.. $GO/SiO_2@PANI$ was capable up to five successive cycles of reaeration and demonstrated the modified material is a promising one for metal ions adsorption from wastewater [44].

Tighadouine et al. [55] illustrated the capacity of novel supramolecular pyridylpyrazole-β-ketoenolreceptorcovalentlyemergedonsilicasurface(SiNPz/Py) for the removal of some toxic heavy metals from wastewater. The SiNPz/Py was found mechanically and chemically stable, reusable up to five effective cycles, wide range of selectivity with excellent adsorption efficiency.

Impressive uptake capacity and selectivity were shown with the sequence of Pb(II)>Zn(II)>Cu(II)>Cd(II). An outstanding removal capacity of 93 mg/g was reported for Pb(II) at an initial stage within 5 min; after 20 min, it was 110 mg/g at optimum pH and concentration [56].

The experimental data was directed to the pseudo-second-order kinetic model and sorption isotherm was best fitted with the Langmuir model. The chemisorption binding mechanism was assumed for studied metal adsorption.

Thermodynamic features were spontaneous and energetically endothermic in nature.

Choi et al. [56] synthesized a PEI-silica nanoparticle and found it to be consistently well-disseminated PEI polymer in silica particles. It was revealed that Cr(VI) adsorption on PEI-silica nanoparticles was 183.7 mg/g under the controlled condition as monolayer adsorption followed by Langmuir isotherm model.

Metal oxides including ferric oxides (Fe_2O_3), titanium oxides (TiO_2), manganese oxides (MnO) and zinc oxide (ZnO) are capable of removing various metal ions from wastewater. Silica having large surface area, mechanically and chemically stable, malleable tetrahedral structure and tunable porous surface leads to the easy prepared of silica metal oxides adsorbents have been widely used for metal adsorption in the field of environmental remediation. Some recent studies are tabulated in Table 7.3.

Among others, silica adsorbents modified with iron oxides (Fe_3O_4, Fe_2O_3) have been found to be good alternative composite materials with satisfactory metal ions adsorption because of their surface properties including large surface area and magnetic response due to their outer magnetic field [57]. Other metal oxides such manganese oxides (MnO) can be used to modify silica to obtain $MgO–SiO_2$ having good stability and cation exchange capability, and this modified silica was found to be non-toxic and harmless with satisfactory metal adsorption efficiency [57]. Besides that, titanium oxide (TiO_2) and zinc oxide (ZnO) are also utilized for immobilization on silica and are found to have reasonable metal adsorption and regeneration capacity.

Xu et al. [58] developed a mesoporous $MgO–SiO_2$ using state-of-the-art synthesis route and compared the adsorption capacity between metal ions and organic pollutants. The adsorption data were directed into the Langmuir isotherm model, and the maximum monolayer capacities were 753.1 mg/g for Pb(II) ions and 481 mg/g for Cu(II) ions with regeneration capacity of up to three successive cycles. The adsorption efficiencies were better for metal ions than for organic pollutants such as methylene blue (315.6 mg/g) and fulvic acid (24.04 mg/g). The adsorption mechanism was assumed to be ion exchange for metal adsorption.

TABLE 7.3
Metals Adsorption Capacity, Conditions and Surface Area of Different Silica Nanomaterial Adsorbents

Type of Silica Nanomaterial	Name of the Nanomaterial	Metals Adsorption Capacity (mg/g)	BET Surface Area (m^2/g)	Adsorption Conditions: Mass (g)	pH	Temp. (°C)	Time (min)	Conc. (mg/L)	Vol. (mL)	Ref.
Silica polymer nanomaterials	Graphene oxide/silica@polyaniline (GO/SiO_2@PANI)	Cu(II) – 258.275 Cr(VI) – 512.47	150.36	0.02	5.3 3.0	30	300	500	20	[44]
	Pyridylpyrazole-β-ketoenol receptor onto silica surface (SiNPz/Py)	Pb(II) – 122.76 Zn(II) – 95.50 Cu(II) – 66.28 Cd(II) – 50.27	298.51	0.01	6.0	25	60	110	10	[55]
	PEI-silica	Cr(VI) – 183.7	1379.6	0.02	4.0	25	24 h	200	50	[56]
Silica metal oxides nanomaterials	Mesoporous silicon-magnesium oxide ($MgO–SiO_2$)	Pb(II) – 753.1 Cu(II) – 481	142.61	-	9.0-5.0	25	120	100	20	[57]
	Fe_2O_3/SiO_2 monolith	Pb(II) – 690 Cr(III) – 770 Cd(II) – 850	562	0.02	6.0	30	120	100	100	[58]
	SiO_2@ZnO	Cu(II) – 42.17	-	0.19	6.5	25	15	10	10	[59]
Silica carbon nanomaterials	(GO@SiO_2-MSp@SiO_2NH_2)	Pb(II) – 223.5	-	0.02	6.0	35	30	200	50	[60]
	Silica gel/graphene oxide (SG@GO-IIP)	Pd(II) – 154.3	-	0.025	3.0	25	90	100	50	[61]
	Hybrid silicate-hydrochar (Mg-Si-HC)	Cu(II) – 214.7 Zn(II) – 227.3	107.7	0.025	2-5	25	90	300	50	[62]
Silica bio-nanomaterials	Silica sand/anionized starch composite (CMS-SS)	Cu(II) – 383.08	-	0.03	5.0	25	60	1000	30	[63]
	Orange peel waste with silica nanospheres (SiO_2@OPW)	Pb(II) – 200	-	0.02	6.0	27	60	50	50	[64]
	Amino-functionalized mesoporous silica onto sodium alginate (aMSP/SA)	U(VI) – 210	-	0.01	4.0	40	240	600	10	[65]

Singh et al. [56] used the Fe_2O_3/SiO_2 monolith for Cr(III), Pb(II) and Cd(II) ions sequestration, and equilibrium of metal ions adsorption was attained at pH 6 within 120 min. It was observed that the results followed the Langmuir isotherm model with maximum monolayer capacities of 770 mg/g for Cr(III) ions, 850 mg/g for Pb(II) ions and 690 mg/g for Cd(II) ions on Fe_2O_3/SiO_2 monolith. The revealed data showed the kinetics followed pseudo-second-order kinetic model, and thermodynamically, it was spontaneous and exothermic in nature. The adsorption efficiency of Cr(III), Pb(II), and Cd(II) ions on Fe_2O_3/SiO_2 monolith was much better than on the pure silica, and it was 120, 100 and 95.9 mg/g, respectively [56].

Ahmad et al. [59] synthesized amorphous silica nanoparticles (SiO_2@ZnO) using TEOS and applied them for Cu(II) ions adsorption. Experimental results best fitted the Langmuir isotherm having a maximum efficiency of 42.17 mg/g, and the adsorption was highly dependent on pH and the mechanism of adsorption was proposed as ion exchange. Adsorption conditions were optimized, and it was found that 0.19 g of SiO_2@ZnO was required to remove 92% Cu(II) within a very short time (15 min) at pH 6.5 [59].

Carbon-based silica materials have shown immense adaptability for the reason that they have excellent physical, chemical, thermal and electric properties. Different carbon composites, including activated carbon (AC), carbon nanotubes (CNTs), multi-walled carbon nanotubes (MWCNTs) and graphene oxide (GO), were modified with silica for metal ions adsorption as tabulated in Table 7.3.

Silica coated with graphene oxides was functionalized using APTMS (GO@SiO_2-MSp@SiO_2NH_2) by Hassan et al. [60]. Adsorption studies were conducted for Pb(II) ions removal using the novel adsorbent, and the efficiency of 323.5 mg/g was achieved under optimum conditions. Chemical adsorption was suggested by ion exchange and complexation reactions through sharing of electrons between the metal ions and adsorbent surface. The material was highly promising with ten cycles of reusability and highly selective for metal ions adsorption from wastewater.

Li et al. [61] synthesized silica graphene oxide nanocomposites (SG@GO-IIP) and used them for sequestration of Pd(II) ions from contaminated solution. The developed material showed a good affinity for palladium ions with 154.3 mg/g efficiency even in the presence of numerous coexisting ions. The adsorption process was very fast at pH 3.0. The material was found highly selective and recyclable up to five (5) times.

Carbon-based hydrochar-modified silica composite (Mg-Si-HC) was found to be a good alternative for the removal of Cu(II) and Zn(II) ions from aqueous media with an efficiency of 214.7 and 227.3 mg/g, respectively, which were higher than those of the pure silica. The material was prepared by Deng et al. [62] applying one-step process utilizing low-cost silica, magnesium and sawdust. The novel material was compared for the adsorption of metal ions and antibiotic (tetracycline); excellent adsorption capacities were revealed for both the metal ions and antibiotic with very good regeneration capacity of at least five successive recycles.

Biosorbents have been considered as environment-friendly and low-cost adsorbents for the decontamination of metal ions. The immobilization of both living and non-living biosorbents on the perfect surface could be enhanced by the physical and mechanical strength, leading to improved adsorption capacity. Silica is considered as an excellent solid support and widely applied for the removal of metal ions from wastewater. Among others, chitosan, rice husk, starch, lignin, alginate and orange peel have been utilized for the modification of silica due to their outstanding characteristics, including low cost, chemical stability, non-toxicity and hydrophilic nature. Some silica-based biosorbents are presented in Table 7.3. The chitosan as a biosorbent with modification of silica was found their active surface properties as well as excellent adsorption capacity of metal ions especially Cr(VI) adsorption. The mechanism Cr(VI) adsorption onto chitosan based silica material was proposed as seen in Figure 7.3.

Li et al. [63] prepared a silica sand-modified biosorbent using anionized starch composite (CMS-SS) and compared to adsorption behaviour of Cu(II) ions and methyl blue from contaminated environment. The adsorption was performed in both single and binary system. The composite biosorbent was found to be a good adsorbent with an adsorption efficiency of 653.31 mg/g for Cu(II) ions and 1246.40 mg/g for methyl blue, while pure silica sand (SS) showed almost no adsorption efficiency because of weak interaction between metal ions and the material. The achieved adsorption efficiency designated that additional carboxymethyl groups in modified adsorbent might be endorsed for better adsorption due to electrostatic interaction of functional groups and pollutants. When comparing the adsorption capacities of CMS-SS and CMS, the composite CMS-SS exhibited higher performance than the pristine CMS for the studied pollutants; this may be due to the superior structure and surface properties of the modified adsorbent.

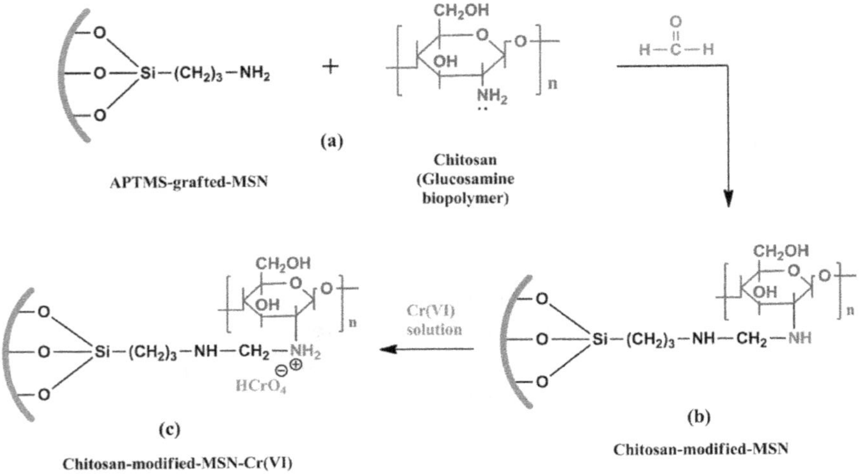

FIGURE 7.3 Proposed schematic diagram of Cr(VI) adsorption on chitosan silica materials. (See Ref. [46].)

Saini et al. [64] chemically confounded silica with orange peel waste (OPW) and developed a novel biosorbent nanocomposite, namely SiO_2@OPW. An adsorption study was conducted for the removal of PB(II) ions from wastewater; under optimum conditions higher adsorption capacity was shown by the composite adsorbent than by the pure OPW, which were 200 and 166.7 mg/g, respectively. The improved adsorption capability of the composite may be ascribed to strong electrostatic attraction among –Si–O–Si–, $Si–O^-$ and Pb^{2+} functional binding sites of SiO_2 nanospheres trapped on the orange peel waste. The prepared material can be reusable up to five regeneration cycles. Amino-functionalized mesoporous silica (aMSP) was modified with sodium alginate (SA) and applied for sequestration of low concentrated U(VI) ions in soluble and mobile forms. Adsorption efficiency was found to be 210 mg/g at pH 4.0 and 40°C, and the U(VI) concentration reduced to 1.0–1.31 ppm. The adsorption of U(VI) on aMSP/SA mainly occurred through ion exchange among carboxyl groups with synchronization of amino and hydroxyl groups [65].

7.6.3 Silica Nanomaterials for Removal of Organic Pollutants (Dyes/Pharmaceuticals)

Since the latter half the last century, the organic pollutants including dyes, pharmaceuticals, pesticides and other volatile organic pollutants have been found to be carcinogenic due to their highly persistent nature and severe toxicity even at very small concentrations in the environment [66, 67]. Among the other organic pollutants, we can highlight dyes and pharmaceuticals as the most common effluents in the wastewater. Dyes come from either natural or industrial effluents as coloured substance. Dyes having complex and large molecular structures are serious toxic compounds, and their products are also carcinogenic [68]. On the other hand, pharmaceutical products such as drugs including antibiotics having severe toxicity contained in the wastewater are of concern for appropriate removal from aquatic environment. Recently, the silica nanomaterials have gained great attention, and vast numbers of previous researches have been reported in the literature for the adsorption organic compounds including drugs and dyes from wastewater. Some of reported studies are summarized in Table 7.4.

Li et al. [69] prepared the SNF/MNP/PS and applied it for the adsorption of methylene blue (MB). The prepared nanomaterial of <10 nm size was evenly distributed on silica and covered the porous structure of silica. The material showed good reusability capacity up to five regeneration cycles with a maximum monolayer capacity of 103.1 mg/g for MB adsorption, which was 81% of the total removal from contaminated environment. Another study by Narayani et al. [70] found 74 mg/g MB adsorption by using a novel core–shell magnetic photocatalyst γFe_2O_3/silica(SiO_2)/TiO_2. The material also showed good magnetic separation and regeneration via non-radiation-driven H_2O_2 activation.

Seyed et al. [71] modified the KIT-6 magnetite mesoporous silica nanoparticles (Fe_3O_4@SiO_2@KIT-6-SO_3H NPs) for the removal of methyl green dye from contaminated water. The material achieved an extraordinary adsorption

TABLE 7.4
Adsorption Capacity for Dyes/Drugs and Adsorption Conditions of Different Silica Nanomaterials

Pollutants	Silica Nanomaterial	Adsorption Capacity (mg/g)	Adsorption Conditions						Ref.
			Mass (g)	pH	Temp. (°C)	Time (min)	Conc. (mg/L)	Vol. (mL)	
Dyes	Silica nanofibre/magnetite nanoparticles/ porous silica (SNF/MNP/PS)	Methylene blue – 103.1	0.3	6.0	25	240	100	100	[69]
	γ-Fe_2O_3/silica (SiO_2)/titania (TiO_2)	Methylene blue – 74.0	0.4	10.0	25	180	300	100	[70]
	Fe_3O_4@SiO_2@KIT-6-SO_3H nanoparticles	Methyl green – 196.0	0.3	3.0	25	120	100	100	[71]
	Silica nanoparticles (porous)	Methylene blue – 347.2	0.1	8.0	25	120	100	20	[72]
	Silica nanoparticles (SiO_2NPs)	Acid Orange 8 – 230	0.1	-	25	120	150	20	[73]
Drugs	N-MCM-41	Ciprofloxacin – 139.25	0.02	7.0	25	120	300	20	[74]
	SBA-NH_2	Cefazolin – 454		5.0	25	120	-	-	[75]
	Fe_3O_4@SiO_2@SBA3- SO_3H	Safranin O	0.12	6.0	25	30	-	25	[76]
	G_1MCM-41	Deferasirox – 295	0.1	4.5	25	-	35	10	[77]
	G_2MCM-41	Deferasirox – 353	0.1	4.5	25	-	35	10	[77]
	G_3MCM-41	Deferasirox – 419	0.1	4.5	25	-	35	10	[77]

efficiency of up to 96.4% of MG with a maximum monolayer capacity of 196 mg/g at acidic pH within 10 min of reaction time. For the adsorption of MB, Yu et al. [72] modified the silica nanomaterial using naturally occurring diatomite. This low-cost material can be reused up to 5 times with a maximum capacity of 347.2 mg/g for MB adsorption [72]. The sugarcane waste was applied as a low-cost biosorbent for the preparation of silica nanoparticles (SiO_2NPs), and the obtained nanoparticles were subsequently evaluated for the removal of Acid Orange 8 (AO8) from wastewater. The economically convenient material exhibited a high recyclability with an adsorption capacity of 230 mg/0g at optimum reaction conditions [73].

Some reported researches were found for the removal of pharmaceutical products including drugs from aqueous solution. Among them, Lu et al. [74] prepared a silica nanomaterial using spherical MCM-41 for the removal of ciprofloxacin (CIP), one of the most popular toxic antibiotics, from wastewater. The CIP elimination capacity was 139.25 mg/g. Szewczyk et al. [75] used the novel SBA-NH_2 for cefazolin (Cef) elimination from wastewater. The silica surface was modified by SBA-15 with APTES applying grafting method (post-synthesis) without changing pore and orderliness of pure silica. The adsorption achieved a maximum monolayer capacity of 454 mg/g at optimum operating conditions. Nanostructured Fe_3O_4@SiO_2@SBA3-SO_3H was used by Danesh et al. [76] for Safranin O extraction from contaminated solution; more than 90% of the contaminant was removed. The material showed high reusability up to nine cycles of regeneration [76]. Taleghani et al. [77] modified silica materials based on MCM-41-type of mesoporous silica nanoparticles with PAMAM dendrimers for deferasirox removal. Drug adsorption capacities were compared with G_1MCM-41, G_2MCM-41 and G_3MCM-41, and a higher capacity of 419 mg/g was shown by G_3MCM-41 due to the electrostatic interactions between the PAMAM-modified MCM-41 surface and deferasirox [77].

7.7 FUTURE PROSPECTS AND CHALLENGES

In this chapter, numerous silica-based modified adsorbents have been reviewed, such as silica polymeric composites, silica metal oxides, silica carbon composites, silica biosorbents and organically modified silica composites. Silica materials showed outstanding alternatives for the fabrication of robust nanoadsorbents due to their excellent chemical, physical and surface properties. The large pore volume of silica materials enabled the immobilization of targeted pollutants, including metals ions and organic compounds. Silica is one of the materials with large area because of its mesoporous surfaces, outer surfaces and mesoporous walls. The mesoporous surfaces are tunable, and their sizes are diversified for the alternatives of different additives and templates.

The metal adsorption capacities of reviewed silica nanomaterails were diverse due to the factors affecting the adsorption process including solution pH, temperature, contact time, metal ions concentration and dose of adsorbent etc.

Considering the economic aspect and in respect of the recycling concept of sustainable environmental management, the reuse and regeneration of adsorbents were extensively investigated. The use of long-chain polymers showed unique adsorption properties compared to the other modified silica-based adsorbents, and they have gained the attention of researchers due to their efficiency and regeneration capacity. A variety of processes were reported for their immobilization in silica matrices. Among the investigated silica polymers, the conduction polymer such as polyaniline played a vital role in the guest–host interactions; such interactions helped in attaining the good selectivity of silica polymeric adsorbents. Particularly, these types of nanocomposites reported as highly potential adsorbents as because of polymeric silica materials are capable to completely removal of metal ions as well as reduction of toxicity of metal ions.. Therefore, more novel polymeric adsorbents should be tested for the adsorption of a wide range of metal ions.

The use of silica metal oxides has shown better selectivity, durability and stability of adsorbent. However, studies were mostly limited to the use of a few metal oxides for the removal of some selected metal ions. Different other metal oxides need to be examined to evaluate their capacity for the removal of comprehensive metal ions, pharmaceuticals and dyes. Silica biosorbents such as chitosan as a biopolymer having amine groups were found to be excellent adsorbents for pollutants adsorption, especially MB and Cr(VI) due to the low cost and environmental compatibility, but their investigation is limited to some selected metal ions adsorption. Therefore, there is a demand for comprehensively examining a wide range of metal ions adsorption. The amalgamation of various organic and inorganic moieties would be promising for material fabrication to using metal ions extractions due to magnificent properties as well as good possibilities of ion exchange mechanism of said materials.

7.8 CONCLUSIONS

In this chapter, it was shown that the adsorption has been considered a promising method for wastewater decontamination because of its feasibility, flexibility and economical convenience. It has also been found that the mechanism of metal ions, dyes and drugs adsorption on different adsorbent materials depends on the nature of adsorbent properties as well as operating factors. A good number of silica-based adsorbents have been reviewed; among them, silica polymers and silica metal oxides showed an excellent capacity to remove all the types of pollutants such as drugs, dyes and metal ions from wastewater. Moreover, silica biosorbents, organically modified silica adsorbents and silica carbon composites are also found to exhibit better selectivity for a wide range of adsorption applications. The findings also recommended the future prospects of water remediation/treatment by silica-based nanocomposites from all kinds of water systems. The collected data of this review would be a useful resource for the modification and preparation of environmentally friendly and low-cost composite adsorbents for wastewater decontamination.

REFERENCES

[1] Isaacson KP, Proctor CR, Wang QE, Edwards EY, Noh Y, Shah AD, Whelton AJ. Drinking water contamination from the thermal degradation of plastics: implications for wildfire and structure fire response. *Environmental Science: Water Research & Technology*, 2021;7(2):274–84.

[2] Hilili JM, Onuora DI, Hilili RU, Annah AF, Onmonya YA, Hilili MH. Ground Water Contamination: Effects and Remedies. *Asian Journal of Environment & Ecology*, 2021;21:39–58.

[3] Okafor UP, Obeta MC, Ayadiuno RU, Onyekwelu AC, Asuoha GC, Eze EJ, Orji-Okafor CE, Igboeli EE. Health implications of stream water contamination by industrial effluents in the Onitsha urban area of Southeastern Nigeria. *Journal of Water and Land Development*, 2021;19.

[4] Al-Makishah NH, Taleb MA, Barakat MA. Arsenic bioaccumulation in arsenic-contaminated soil: a review. *Chemical Papers*, 2020;74(9):2743–57.

[5] Leong, YK, Chang, JS, Bioremediation of heavy metals using microalgae: recent advances and mechanisms. *Bioresource Technology*, 2020,303:122886.

[6] Idrees N, Sarah R, Tabassum B, Abd_Allah, EF, Evaluation of some heavy metals toxicity in Channa punctatus and riverine water of Kosi in Rampur, Uttar Pradesh, India. *Saudi Journal of Biological Sciences*, 2020.

[7] Tu P, Chi L, Bodnar W, Zhang Z, Gao B, Bian X, Stewart J, Fry R, Lu K., Gut microbiome toxicity: connecting the environment and gut microbiome-associated diseases. *Toxics*, 2020;8(1):19.

[8] Adekeye DK, Popoola OK, Asaolu SS, Adebawore AA, Aremu OI, Olabode KO, Adsorption and conventional technologies for environmental remediation and decontamination of heavy metals: an overview. *International Journal of Research and Review*, 2019;6(8):505–16.

[9] Crini G, Lichtfouse E, Advantages and disadvantages of techniques used for wastewater treatment. *Environmental Chemistry Letters*, 2019;17(1):145–55.

[10] You J, Guo Y, Guo R, Liu X, A review of visible light-active photocatalysts for water disinfection: features and prospects. *Chemical Engineering Journal*, 2019.

[11] Shahadat M, Ahmad A, Bushra R, Ismail S, Ahammad SZ, Ali SW, Rafatullah, M, Recent advancement in wastewater decontamination technology. In *Modern Age Waste Water Problems*, 2020, pp. 1–22. Springer, Cham.

[12] Liu T, Chen D, Cao Y, Yang F, Chen J, Kang J, Xu R, Xiang, M. Construction of a composite microporous polyethylene membrane with enhanced fouling resistance for water treatment. *Journal of Membrane Science*, 2021;618:118679.

[13] Sharma K, Kalita S, Sarma NS, Devi A. Treatment of crude oil contaminated wastewater via an electrochemical reaction. *RSC Advances*, 2020;10(4):1925–36.

[14] Barakat MA, Kumar R, Balkhyoura M, Taleb MA, Novel Al_2O_3/GO/halloysite nanotube composite for sequestration of anionic and cationic dyes. *RSC Advances*, 2019;9:13916.

[15] Taleb MA, Kumar R, Al-Rashdi AA, Seliem, MK, Barakat, MA, Fabrication of SiO_2/$CuFe_2O_4$/polyaniline composite: a highly efficient adsorbent for heavy metals removal from aquatic environment. *Arabian Journal of Chemistry*, 2020;13(10):7533–43.

[16] Kavitha G, Subhapriya P, Dhanapal V, Dineshkumar G, Venkateswaran V. Dye removal kinetics and adsorption studies of activated carbon derived from the stems of Phyllanthus reticulatus. *Materials Today: Proceedings*, 2021;8.

[17] Muangmora R, Kemacheevakul P, Chuangchote S. Titanium dioxide and its modified forms as photocatalysts for air treatment. *Current Analytical Chemistry*, 2021;17(2):185–201.

[18] Yang W, Hu W, Zhang J, Wang W, Cai R, Pan M, Huang C, Chen X, Yan B, Zeng H. Tannic acid/Fe^{3+} functionalized magnetic graphene oxide nanocomposite with high loading of silver nanoparticles as ultra-efficient catalyst and disinfectant for wastewater treatment. *Chemical Engineering Journal*, 2021;405:126629.

[19] Youssef AM, El-Naggar ME, Malhat FM, El Sharkawi HM, Efficient removal of pesticides and heavy metals from wastewater and the antimicrobial activity of f-MWCNTs/PVA nanocomposite film, *Journal of Cleaner Production*, 2019;206:315–25.

[20] Gopinath KP, Rajagopal M, Krishnan A, Sreerama SK. A review on recent trends in nanomaterials and nanocomposites for environmental applications. *Current Analytical Chemistry*, 2021;17(2):202–43.

[21] Alizadeh N, Salimi A, Multienzymes activity of metals and metal oxide nanomaterials: applications from biotechnology to medicine and environmental engineering. *Journal of Nanobiotechnology*, 2021;19(1):1–31.

[22] Zheng M, Pan M, Zhang W, Lin H, Wu S, Lu C, Tang S, Liu D, Cai J. Poly (α-l-lysine)-based nanomaterials for versatile biomedical applications: current advances and perspectives. *Bioactive Materials*, 2021;6(7):1878–909.

[23] Ziarani GM, Roshankar S, Mohajer F, Badiei A, The synthesis and application of functionalized mesoporous silica SBA-15 as heterogeneous catalyst in organic synthesis. *Current Organic Chemistry*, 2021;25(3):361–87.

[24] Dev A, Sardoiwala MN, Karmakar S. Silica nanoparticles: methods of fabrication and multidisciplinary applications. *Functionalized Nanomaterials II: Applications*, 2021;12:189.

[25] Singhon R, Adsorption of Cu(II) and Ni(II) ions on functionalized colloidal silica particles model studies for wastewater treatment. Thesis submitted for the degree of Doctor of Science (Chemistry), Université de Franche-Comté, France, 2017.

[26] Yang X, Liu X, Zhang A, Lu D, Li G, Zhang Q, Liu Q, Jiang G, Distinguishing the sources of silica nanoparticles by dual isotopic fingerprinting and machine learning. *Nature Communications*, 2019;10(1):1–9.

[27] Yadav A, Zherebtsov E, Chichkov NB, Gumenyuk R., Melkumo, MA, Yashkov, MV, Dianov EM, Rafailov EU, Optical properties of Ce-doped silica fiber. In The European Conference on Lasers and Electro-Optics (p. ce_p_7). Optical Society of America, 2019.

[28] Kumar SV, Rajkumar R, Umamaheswari N, Study on mechanical and microstructure properties of concrete prepared using metakaolin, silica fume and manufactured sand. *Rasayan Journal of Chemistry*, 2019;12(3):1383–89.

[29] Jal PK, Patel S, Mishra BK, Chemical modification of silica surface by immobilization of functional groups for extractive concentration of metal ions. *Talanta*, 2004;62(5):1005–28.

[30] Lambertz A, Grundler T, Finger F, Hydrogenated amorphous silicon oxide containing a microcrystalline silicon phase and usage as an intermediate reflector in thin-film silicon solar cells. *Journal of Applied Physics*, 2011;109(11):113109.

[31] Coclite AM, Milella A, d'Agostino R, Palumbo F, On the relationship between the structure and the barrier performance of plasma deposited silicon dioxide-like films. *Surface and Coatings Technology*, 2010;204(24):4012–17.

[32] Zhuravlev LT. The surface chemistry of amorphous silica. Zhuravlev model. *Colloids and Surfaces A: Physicochemical and Engineering Aspects*, 2000;173(1–3):1–38.

[33] Wagh PB, Ingale SV. Comparison of some physico-chemical properties of hydrophilic and hydrophobic silica aerogels. *Ceramics International*, 2002;28(1):43–50.

[34] Dinker MK, Kulkarni PS. ChemInform abstract: recent advances in silica-based materials for the removal of hexavalent chromium: a review. *Journal of Chemical & Engineering Data*, 2015a. doi: 10.1021/acs.jced.5b00292.

[35] Mannaa MA, Altass HM, Salama RS. MCM-41 grafted with citric acid: The role of carboxylic groups in enhancing the synthesis of xanthenes and removal of heavy metal ions. *Environmental Nanotechnology, Monitoring & Management*, 2021;15:100410.

[36] Nguyen CH, Fu CC, Chen ZH, Van Tran TT, Liu SH, Juang RS. Enhanced and selective adsorption of urea and creatinine on amine-functionalized mesoporous silica SBA-15 via hydrogen bonding. *Microporous and Mesoporous Materials*, 2021;311:110733.

[37] Wu H, Xiao Y, Guo Y, Miao S, Chen Q, Chen Z. Functionalization of SBA-15 mesoporous materials with 2-acetylthiophene for adsorption of Cr (III) ions. *Microporous and Mesoporous Materials*, 2020;292:109754.

[38] Kot M, Wojcieszak R, Janiszewska E, Pietrowski M, Zieliński M. Effect of modification of amorphous silica with ammonium agents on the physicochemical properties and hydrogenation activity of Ir/SiO_2 catalysts. *Materials*, 2021;14(4):968.

[39] Vareda JP, Durães L, (2017) Functionalized silica xerogels for adsorption of heavy metals from groundwater and soils. *Journal of Sol-Gel Science and Technology*, 2017;84(3):400–8.

[40] Mao J, Ma Y, Zang L, Xue R, Xiao C, Ji D. Efficient adsorption of hydrogen sulfide at room temperature using fumed silica-supported deep eutectic solvents. *Aerosol and Air Quality Research*, 2020;20(1):203–15.

[41] Mitran RA, Ioniţă S, Lincu D, Berger D, Matei C. A review of composite phase change materials based on porous silica nanomaterials for latent heat storage applications. *Molecules*, 2021;26(1):241.

[42] Wallace AK. Engineering diatom peptides for the synthesis of silica nanomaterials, Doctoral Dissertation, Massachusetts Institute of Technology.

[43] Cao X, Zhang Q, Zhang C, Li Z, Zheng W, Liu M, Wang B, Huang S, Li L, Huang X, Kong L, A novel approach to coat silica on quantum dots: forcing decomposition of tetraethyl orthosilicate in toluene at high temperature. *Journal of Alloys and Compounds*, 2020;817:152698.

[44] Kumar R, Barakat MA, Taleb MA, Seliem MK, A recyclable multifunctional graphene oxide/SiO_2@ polyaniline microspheres composite for Cu(II) and Cr(VI) decontamination from wastewater. *Journal of Cleaner Production*, 2020;20:122290.

[45] Ahmad OS, Bedwell TS, Esen C, Garcia-Cruz A, Piletsky SA, Molecularly imprinted polymers in electrochemical and optical sensors. *Trends in Biotechnology*, 2019;37(3):294–309.

[46] Dinker MK, Kulkarni PS. Recent advances in silica-based materials for the removal of hexavalent chromium: a review. *Journal of Chemical & Engineering Data*, 2015;60(9):2521–40.

[47] Wu H, Xiao Y, Guo Y, Miao S, Chen Q, Chen Z, Functionalization of SBA-15 mesoporous materials with 2-acetylthiophene for adsorption of Cr(III) ions. *Microporous and Mesoporous Materials*, 2020;292(15):109754.

[48] Kaewprachum W, Wongsakulphasatch S, Kiatkittipong W, Striolo A, Cheng CK, Assabumrungrat S, SDS modified mesoporous silica MCM-41 for the adsorption of Cu(II), Cd(II), Zn(II) from aqueous systems. *Journal of Environmental Chemical Engineering*, 2020;8(1):102920.

[49] Dinker MK, Patil NV, Kulkarni PS, A diamino based resin modified silica composite for the selective recovery of tungsten from wastewater. *Polymer International*, 2016;65:1387–94.

[50] Ali N, Ali F, Ullah I, Ali Z, Duclaux L, Reinert L, Lévêque JM, Farooq A, Bilal M, Ahmad I. Organically modified micron-sized vermiculite and silica for efficient removal of Alizarin Red S dye pollutant from aqueous solution. *Environmental Technology & Innovation*, 2020;19:101001.

[51] Kollarahithlu SC, Balakrishnan RM. Adsorption of pharmaceuticals pollutants, Ibuprofen, Acetaminophen, and Streptomycin from the Aqueous Phase using Amine Functionalized Superparamagnetic Silica Nanocomposite. *Journal of Cleaner Production*, 2021;29:126155.

[52] Kumar R, Barakat MA, Alseroury FA, Al-Mur BA, Taleb MA. Experimental design and data on the adsorption and photocatalytic properties of boron nitride/cadmium aluminate composite for Cr(VI) and cefoxitin sodium antibiotic. *Data in Brief*, 2020;28:105051.

[53] Saikia J, Adsorption of lead ions from aqueous solution by functionalized polymer aniline–formaldehyde condensate, coated on various support materials. In *Recent Developments in Waste Management*, 2020;271–288. Springer.

[54] Karthik R, Meenakshi S, Removal of hexavalent chromium ions using polyaniline/silica gel composite. *Journal of Water Process Engineering*, 2014;

[55] Tighadouini S, Radi S, Ferbinteanu M, Garcia Y, Highly selective removal of Pb(II) by a Pyridylpyrazole-β-ketoenol receptor covalently bonded onto the silica surface. *ACS Omega*, 2019;4:3954–3964.

[56] Choi K, Lee S, Park JO, Park J, Cho S, Lee SY, Lee JH, Choi J, Chromium removal from aqueous solution by a PEI-silica nanocomposite. *Scientific Report*, 2018;8:1438.

[57] Singh J, Sharma M, S Basu S, Heavy metal ions adsorption and photodegradation of remazol black XP by iron oxide/silica monoliths: kinetic and equilibrium modelling. *Advanced Powder Technology*, 2018;29:2268–79.

[58] Xu C, Shi S, Dong Q, Zhu S, Wang Y, Zhou H, Wang X, Zhu L, Zhang G, Xu D, Citric-acid-assisted sol-gel synthesis of mesoporous silicon-magnesium oxide ceramic fibers and their adsorption characteristics. *Ceramics International*, 2020.

[59] Ahmad I, Siddiqui WA, Ahmad T, Synthesis, characterization of silica nanoparticles and adsorption removal of Cu^{2+} ions in aqueous solution. *International Journal of Emerging Technology and Advanced Engineering*, 2017;7(8):439–45.

[60] Hassan AM, Ibrahim WA, Bakar MB, Sanagi MM, Sutirman ZA, Nodeh HR, Mokhter MA, New effective 3-aminopropyltrimethoxysilane functionalized magnetic sporopollenin-based silica coated graphene oxide adsorbent for removal of Pb(II) from aqueous environment. *Journal of Environmental Management*, 2020;253(1):109658.

[61] Li M, Tang S, Zhao Z, Meng X, Gao F, Jiang S, Chen Y, Feng J, Feng C, (2020) A novel nanocomposite based silica gel/graphene oxide for the selective separation and recovery of palladium from a spent industrial catalyst. *Chemical Engineering Journal*, 2020;386(15):123947.

[62] Deng J, Li X, Wei X, Liu Y, Liang J, Song B, Shao Y, Huang W, Hybrid silicate-hydrochar composite for highly efficient removal of heavy metal and antibiotics: Coadsorption and mechanism. *Chemical Engineering Journal*, 2020;387(1):124097.

[63] Yang W, Ding P, Zhou L, Yu J, Chen X, Jiao F, Preparation of diamine modified mesoporous silica on multi-walled carbon nanotubes for the adsorption of heavy metals in aqueous solution. *Applied Surface Science*, 2013;282:38–45.

[64] Li P, Gao B, Li A, Yang H, Evaluation of the selective adsorption of silica-sand/anionized-starch composite for removal of dyes and Cupper(II) from their aqueous mixtures. *International Journal of Biological Macromolecules*, 2020;149(15):1285–93.

[65] Saini J, Garg VK, Gupta RK, Green synthesized SiO_2@OPW nanocomposites for enhanced Lead(II) removal from water. *Arabian Journal of Chemistry*, 2020;13(1):2496–507.

[66] Dinker MK, Kulkarni PS, Temperature based adsorption studies of Cr(VI) using p-toluidine formaldehyde resin coated silica material. *New Journal of Chemistry*, 2015b;39:3687.

[67] de Paula FD, Effting L, Carbajal Arízaga GG, Giona RM, Tessaro AL, Bezerra FM, Bail A. Spherical mesoporous silica designed for the removal of methylene blue from water under strong acidic conditions. *Environmental Technology*, 2021;5:1.

[68] Alsheheri SZ, Nanocomposites containing titanium dioxide for environmental remediation. *Designed Monomers and Polymers*, 2021;24(1):22–45.

[69] Li Z, Tang X, Liu K, Huang J, Peng Q, Ao M, Huang Z. Fabrication of novel sandwich nanocomposite as an efficient and regenerable adsorbent for methylene blue and Pb(II) ion removal. *Journal of Environmental Management*, 2018;218:363–73.

[70] Narayani H, Jose M, Sriram K, Shukla S. Hydrothermal synthesized magnetically separable mesostructured H_2Ti_3O7/γ-Fe_2O_3 nanocomposite for organic dye removal via adsorption and its regeneration/reuse through synergistic non-radiation driven H_2O_2 activation. *Environmental Science and Pollution Research*, 2018;25(21):20304–19.

[71] Seyed Danesh SM, Faghihian H, Shariati S, Sulfonic acid functionalized magnetite nanoporous-KIT-6 for removal of methyl green from aqueous solutions. *Journal of Nano Research*, 2018;52:54–70.

[72] Yu ZH, Zhai SR, Guo H, Song Y, Zhang F, Ma HC, Removal of methylene blue over low-cost mesoporous silica nanoparticles prepared with naturally occurring diatomite. *Journal of Sol-Gel Science and Technology*, 2018;88(3):541–50.

[73] Rovani S, Santos JJ, Corio P, Fungaro DA. Highly pure silica nanoparticles with high adsorption capacity obtained from sugarcane waste ash. *ACS Omega*, 2018;3(3):2618–27.

[74] Lu D, Xu S, Qiu W, Sun Y, Liu X, Yang J, Ma J, Adsorption and desorption behaviors of antibiotic ciprofloxacin on functionalized spherical MCM-41 for water treatment. *Journal of Cleaner Production*, 2020;10:121644.

[75] Szewczyk A, Prokopowicz M, Sawicki W, Majda D, Walker G. Aminopropyl-functionalized mesoporous silica SBA-15 as drug carrier for cefazolin: adsorption profiles, release studies, and mineralization potential. *Microporous and Mesoporous Materials*, 2019;274:113–26.

[76] Danesh SM, Faghihian H, Shariati S. Sulfonic acid functionalized SBA-3 silica mesoporous magnetite nanocomposite for safranin O dye removal. *Silicon*, 2019;11(4): 1817–27.

[77] Taleghani AS, Ebrahimnejad P, Heydarinasab A, Akbarzadeh A. Adsorption and controlled release of iron-chelating drug from the amino-terminated PAMAM/ordered mesoporous silica hybrid materials. *Journal of Drug Delivery Science and Technology*, 2020;56:101579.

8 Aluminosilicate-Based Adsorbents for Removal of Water Contaminants

Advanced Statistical Physics Models

Moaaz K. Seliem, Mohamed Mobarak, Essam A. Mohamed, and Ali Q. Selim
Beni-Suef University

CONTENTS

DOI: 10.1201/9781003129042-8

8.1 INTRODUCTION

In recent times, rapid growth, industrialization and economic development have resulted in a continuous influx of toxic chemical compounds (i.e. heavy metals, anions and dyes) into aquatic environment, which normally causes water contamination. The continuous discharge of toxic contaminants, even at insignificant concentrations, into aquatic environment is greatly harmful for human beings (Li et al. 2018; Mobarak et al. 2018; Saffari et al. 2020). Furthermore, polluted water is unsuitable for drinking and use in agriculture and industry, and it affects not only humans, but also animals, fish and birds (Sankpal and Naikwade 2012). There are many water contaminants that can enter drinking water naturally or through human activities (Barakat et al., 2020). Based on different advantages including high efficiency, easy regeneration and low cost, the adsorption method is preferred compared to other techniques such as ozonation and coagulation-flocculation in water remediation (Seliem et al. 2020; Barakat et al. 2021). Utilizing adsorbents based on the available and low-cost natural materials is recommended as a favourable technique in wastewater treatment.

Currently, aluminosilicate minerals have received great attention from researchers due to their accessibility, non-toxicity, high stability and low cost (Vimonses et al. 2009; Mobarak et al. 2018). In the area of water remediation, utilizing aluminosilicates such as clays, micas and cancrinite is preferred, especially against heavy metal ions. On the contrary, some aluminosilicates show no perfect attraction for anionic contaminants such as nitrate, perchlorate and hexavalent chromium Cr(VI) due to their negatively charged surfaces (Seliem et al. 2013; Barakat et al. 2020). Moreover, the presence of inorganic cations (i.e. Na^+, K^+ and Ca^{2+}) in their structure results in decreasing their adsorption capacities against organic compounds such as dyes (Mobarak et al. 2018). Consequently, thermal and/or chemical modifications of aluminosilicate minerals are required to improve their affinity for organic pollutants. Furthermore, the interaction between natural aluminosilicate minerals and different modifiers (i.e. cationic surfactants, iron oxide nanoparticles, chitosan and pomegranate peel extract) can fabricate new promising adsorbents against water/wastewater contaminants (Bao et al. 2020). This interaction can cause a perfect change in the surface chemistry, structure and textural properties of aluminosilicate minerals (Xi et al. 2004; Mobarak et al. 2018; Barakat et al. 2020a, 2021).

Explanation of the adsorption results through modelling tools is obligatory to recognize the physicochemical parameters associated with adsorbate–adsorbent interface that could affect the management of water treatment process (Seliem and Mobarak 2019). Based on the former studies (Mobarak et al. 2019; Li et al. 2019, 2020; Barakat et al. 2020; Mohamed et al. 2020; Ramadan et al. 2021), utilizing classical isotherm models such as Langmuir and Freundlich does not offer vital and complete interpretations of the adsorption mechanism (Ramadan et al. 2021). Furthermore, the rules of these traditional adsorption models are not enough to outline the geometry of adsorbate–adsorbent interaction, particularly in composites (Mobarak et al. 2019; Li et al. 2019, 2020). On the other hand, fitting the attained results to the statistical physics theory can offer physico-chemical factors (i.e. steric and energetic parameters). Interpretation of these parameters at varied adsorption temperatures can be used to describe microscopically and macroscopically the interactions between the adsorbates and the employed adsorbents (Li et al. 2019; Barakat et al. 2020; Mohamed et al. 2020; Ramadan et al. 2021).

This chapter explains the importance of the interaction between aluminium silicate minerals (i.e. black clays, cancrinite, serpentine, muscovite and weathered basalt) and different modifiers (i.e. cationic surfactants, iron oxide nanoparticles, chitosan and pomegranate peel extract) in the uptake of water contaminants via experimental and theoretical studies.

8.2 ALUMINOSILICATE MINERALS

8.2.1 Organic Matter-Rich Clay (Black Clay)

Generally, clay is defined as any very fine-grained natural material that becomes plastic (easily shaped) when adding a suitable amount of water and strengthens when either dried or fired. The structure of clay minerals are characterized by the existence of two kinds of sheets with varied chemical composition and coordination. The first one is the tetrahedral sheet [T], which is composed of silicon and oxygen atoms, while the second one is the octahedral sheet [O], which consists of aluminium atoms and hydroxyl groups. When one tetrahedral sheet is connected to one octahedral sheet, clay minerals with 1:1 layered structure are formed (i.e. T:O ratio = 1:1). On the other hand, 2:1 layered structure type is formed by inserting an octahedral sheet between two tetrahedral sheets (i.e. T:O ratio = 2:1) (Figure 8.1). Chemically, clays are hydrous aluminium silicates containing significant amounts of other elements such as potassium, sodium, calcium, magnesium or iron. For instance, the organic matter-rich clay sample has a high percentage of H_2O and organic matter (about 17%), indicating that this clay is rich in organic matter (Mobarak et al. 2018).

In general, clays display low removal efficiency towards anionic water pollutants due to their negatively charged surfaces, and therefore, chemical modification of these raw aluminosilicate materials is obligatory to enhance their affinity to remove different anionic contaminants. Recently, carbon-rich clay sample was modified using a mixture of cetyltrimethylammonium bromide (CTAB) and hydrogen peroxide (H_2O_2). This mixture resulted in increasing the positive charges on the activated clay surface, and therefore, its adsorption capacity for hexavalent chromium (67.05 mg/g) and methyl orange (194.26 mg/g) was greatly improved compared to the original clay sample (Mobarak et al. 2018). Besides, a multifunctional composite through the decoration of CTAB-activated clay by Fe_3O_4 nanoparticles offered a high uptake capacity (447.1 mg/g) for crystal violet dye (Barakat et al. 2020a).

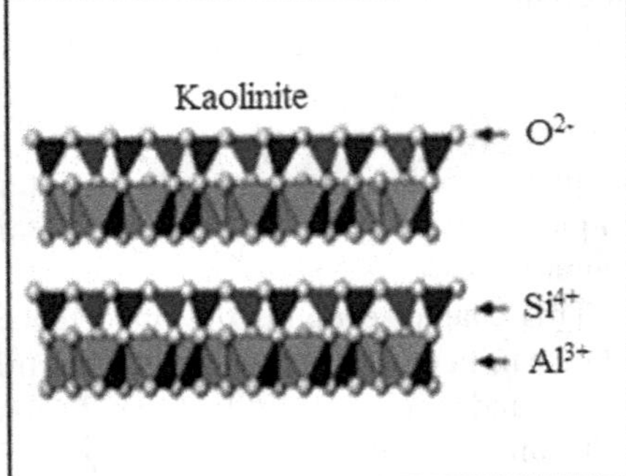

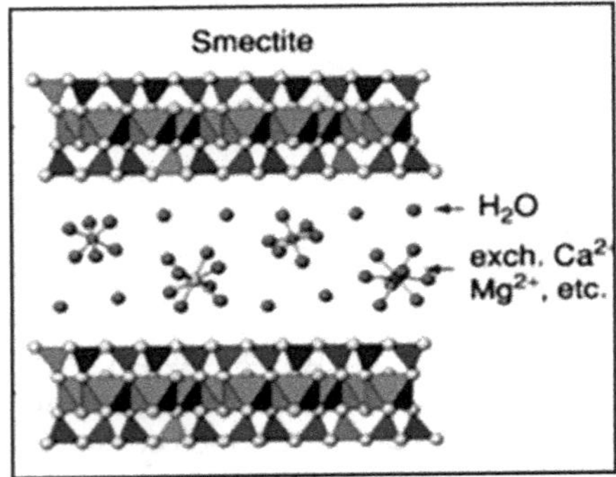

FIGURE 8.1 Structure of clays.

8.2.2 Serpentine

Serpentinization is considered as a common geological low-temperature metamorphic process including water and heat in which low-silica ultramafic and mafic rocks are oxidized. Serpentine with a chemical composition of $Mg_3Si_2O_5(OH)_4$ results from the modification of magnesium-rich silicate minerals such as olivine and pyroxene. The common serpentine minerals are chrysotile, antigorite and lizardite. Structurally, serpentine has an octahedral brucite sheet sandwiched between two silica tetrahedral sheets. Besides magnesium and silicon, aluminium, nickel and iron can be included in the composition of serpentine (Figure 8.2). Exfoliation of serpentine with the 2:1 layered structure was achieved by the interaction between hydrogen peroxide and this aluminosilicate mineral. This separation resulted in simplifying the insertion of the cationic surfactant in the layered structure of serpentine, which commonly enhanced its uptake capacity for hexavalent chromium (54 mg/g) and fluoride (63 mg/g) (Mobarak et al. 2019). In addition, Fe_3O_4 nanoparticles-coated activated serpentine presented the maximum removal of capacities with the values of 201 and 218 mg/g for methylene blue and malachite green dyes, respectively (Seliem et al. 2020).

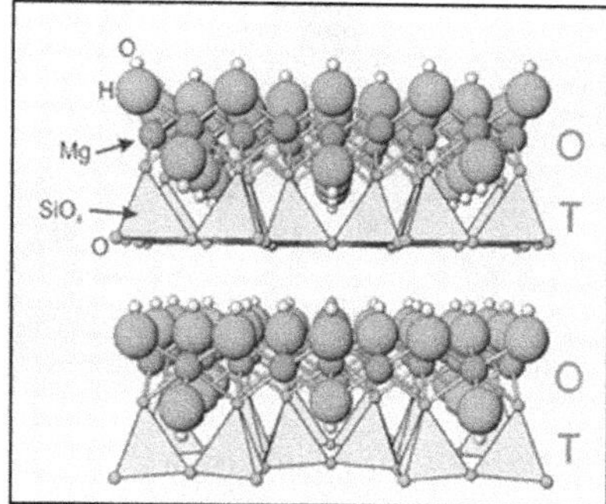

FIGURE 8.2 Structure of serpentine.

8.2.3 Muscovite

Mica, a layered silicate that belongs to phyllosilicate group of minerals, has a 2:1 layered structure of two silica tetrahedral sheets [2T] sandwiching one alumina octahedral sheet [O], i.e. T:O:T structure (Figure 8.3) (Choi et al. 2009). The minerals of layered silicate mica are more common in igneous rocks, particularly the intrusive type, and also, they have varied chemical compositions and physical properties. Muscovite [$KAl_3Si_3O_{10}(OH)_2$] is a dioctahedral Al-rich clay mineral and one of the dominant minerals of the mica family, and phlogopite [$KAlSi_3Mg_3O_{10}(OH)_2$] is a trioctahedral magnesium-rich end member (Choi et al. 2009). Some published articles indicated that muscovite can remove several types of dissolved heavy metals (e.g. arsenic, lead, copper and cadmium) (Yang et al. 2010). A composite of phillipsite and muscovite was utilized for phosphate and ammonium uptake from contaminated solution, and the attained adsorption capacities were 250 and 303 mg/g for phosphate and ammonium, respectively (Abukhadra and Mostafa 2019). Compared to the natural muscovite, the manganese-containing mica resulted in an enhanced uptake of methyl orange by four times and provided the value of 190.44 mg/g (Barakat et al. 2020b).

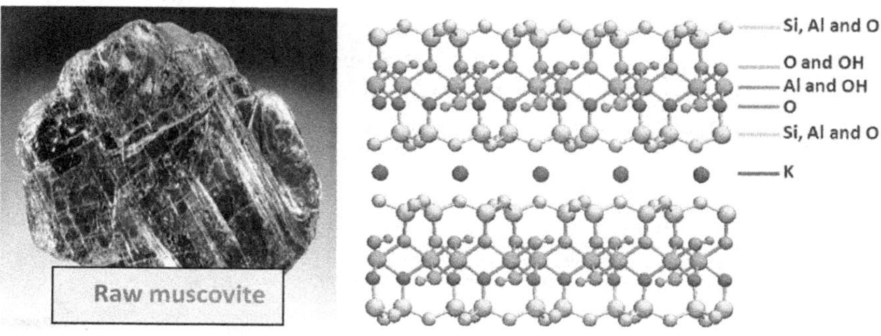

FIGURE 8.3 Structure of muscovite.

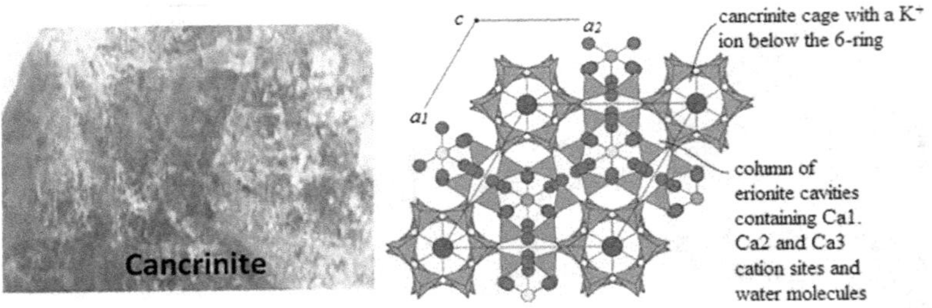

FIGURE 8.4 Structure of cancrinite.

8.2.4 Cancrinite

Cancrinite is a feldspathoid mineral, but it can also be reflected as a zeolite, due to its microporous aluminium silicate framework structure (Selim et al. 2018). This framework structure is composed by stacked layers of silica and alumina tetrahedral in a six-membered rings perpendicular to the C-axis in AB–A′B′ sequence (Gatta and Lotti 2016). The alteration of the stacking sequence results in cancrinite with different types of channels and cages (Selim et al. 2018). The cages and the channels of cancrinite mineral could be occupied by cations, anions, H_2O and CO_2 molecules (Gatta and Lotti 2016) (Figure 8.4). Due to the specific porous structure of cancrinite and its high stability, this mineral has wide applications (Selim et al. 2018). Cancrinite was utilized as an adsorbent for Cu^{2+} (118 mg/g) and Zn^{2+} (67 mg/g) onto a single-compound system. On the other hand, the binary adsorption system gave 93 and 60 mg/g for copper and zinc ions, respectively (Selim et al. 2019a). The competitive impact ascending between Cu^{2+} and Zn^{2+} on the same active adsorption sites of cancrinite resulted in decreasing the adsorption capacity in the binary system.

8.2.5 Weathered Basalt

Basalt, a common basic volcanic rock, results from the rapid cooling of magnesium and iron-rich lava exposed at or near the surface. From the mineralogical point of view, basalt is basically composed of pyroxene, plagioclase and iron-bearing minerals. Weathered basalt is highly susceptible to chemical weathering of the basaltic rock (Figure 8.5). Under the effect of chemical weathering, Ca-plagioclase and pyroxene are transformed to aluminosilicate minerals as zeolites and clays, which are recommended as effective materials in wastewater treatment. The interface between activated weathered basalt and chitosan produced a multifunctional composite with a great active sites number with a high removal efficiency of 115 mg/g (Mohamed et al. 2020).

8.3 MODIFICATION PROCESS

In former studies, modification of aluminosilicate materials created a strong change in the structure, surface chemistry and uptake capacities of these minerals

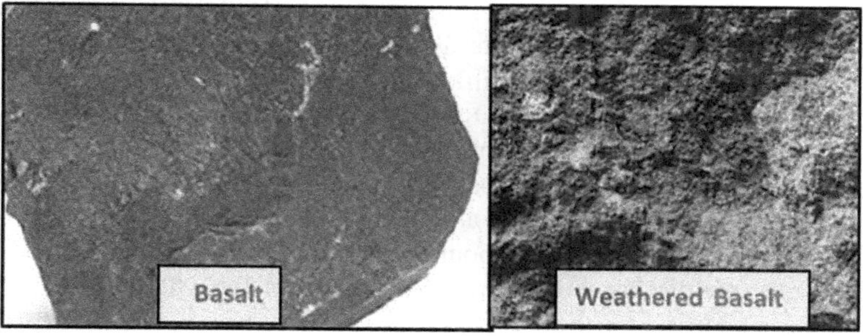

FIGURE 8.5 Photos of basalt and weathered basalt.

(Ramadan et al. 2021). Utilizing a hydrogen peroxide (H_2O_2) in exfoliation of aluminosilicates was found to be an effective approach in enhancing the interaction between aluminosilicate minerals and different modifiers (Mobarak et al. 2018; Seliem et al. 2020; Barakat et al. 2020b, 2021). The activation process via H_2O_2 formed a great number of fissures and holes with dissimilar sizes. The presence of these pores and pits in the structure of the employed raw materials enabled the attachment of different natural and synthetic modifiers to the utilized adsorbent (Barakat et al. 2020b, 2021; Ramadan et al. 2021). Furthermore, composites with varied functional groups were fabricated due to the interaction between H_2O_2-treated aluminosilicates and the tested modifiers. CTAB, magnetic iron oxide nanoparticles, pomegranate peel extract and chitosan were independently used in the modification of these materials.

Surfactants or surface active agents are compounds having the affinity to collect around the interface between two dissimilar materials by changing the properties of this interface. CTAB is characterized by the presence of a positive charge on its polar head group (i.e. cationic surfactant). CTAB is strongly attracted to the negatively charged surface of the H_2O_2-activated carbon-rich clays fabricating a promising adsorbent for wastewater contaminants (Mobarak et al. 2018).

Pomegranate (*Punica granatum* L.) belongs to the Punicaceae family, and its components (i.e. fruits, roots and peels) have attracted great consideration because of their wide-ranging benefits. These elements can be utilized in different applications, including dyeing, drugs and removal of dangerous water compounds (Ramadan et al. 2021). Pomegranate peel, a by-product of the pomegranate juice industry, is composed of numerous constituents including polyphenols and ellagic acids, which have medicinal value for various human infections (Zhang et al. 2007; Salim et al. 2003; Salmani et al. 2017). Pomegranate wastes are considered as low-cost materials that can show adsorption properties to remove metal ions due to the different active sites present on their surfaces (Salmani et al. 2017).

Chitosan as an accessible, non-toxic and low-cost biopolymer is derived from chitin. Chitosan contains different active sites including amino and hydroxyl functional groups, and thus, chitosan-based composites have received great attention in directing the organic and/or inorganic water contaminants (Mohamed et al. 2019).

One of the favourite properties of the employed adsorbents is the probability of their easy separation through external magnetic forces. To fulfil that, natural aluminosilicate minerals can be utilized to form composites with magnetically active phases, such as magnetic Fe_3O_4 nanoparticles. The high surface area, great number of active sites, biocompatibility and easy separation from solutions are important issues in utilizing the fabricated magnetic products in water purification (Pirbazari et al. 2014; Chang et al. 2016).

The used modifiers (i.e. CTAB, pomegranate peel extract, chitosan and Fe_3O_4 nanoparticles) are displayed in Figure 8.6 and listed in Table 8.1.

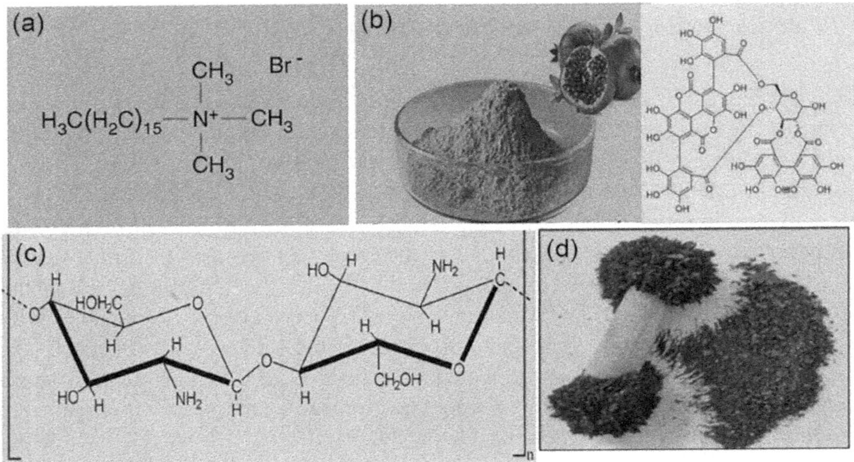

FIGURE 8.6 The employed modifiers: (a) CTAB, (b) pomegranate peel extract, (c) chitosan and (d) Fe_3O_4 nanoparticles.

8.4 ADSORPTION PROCESS

Compared to different advanced techniques utilized in water purification, the adsorption method was recommended to be an effective technique. This is largely because this process is an efficient, easy, simple and low-cost method for the removal of wastewater contaminants (Chen et al. 2017; Sua et al. 2014; Li et al. 2018, 2020; Lafi and Hafiane 2016; Seliem and Komarneni 2016). Overall, the adsorption process can be classified into two major types: (I) physical (physisorption) and (II) chemical (chemisorption). The first type (i.e. physisorption) characterizes the systems that include only comparatively weak intermolecular forces (Sua et al. 2014; Li et al. 2018, 2020). On the other hand, the latter adsorption type basically involves the creation of a new chemical bond between the adsorbent surface and the removed ions (Li et al. 2020; Ramadan et al. 2021). Regeneration and reuse associated with the adsorption process make this technique economical and environmentally friendly (Mobarak et al. 2018).

TABLE 8.1
Aluminosilicate-Based Adsorbents and the Tested Water Contaminants

Adsorbent	The Used Modifier	Water Contaminant	Reference
Qarara clay	H_2O_2/CTAB solution	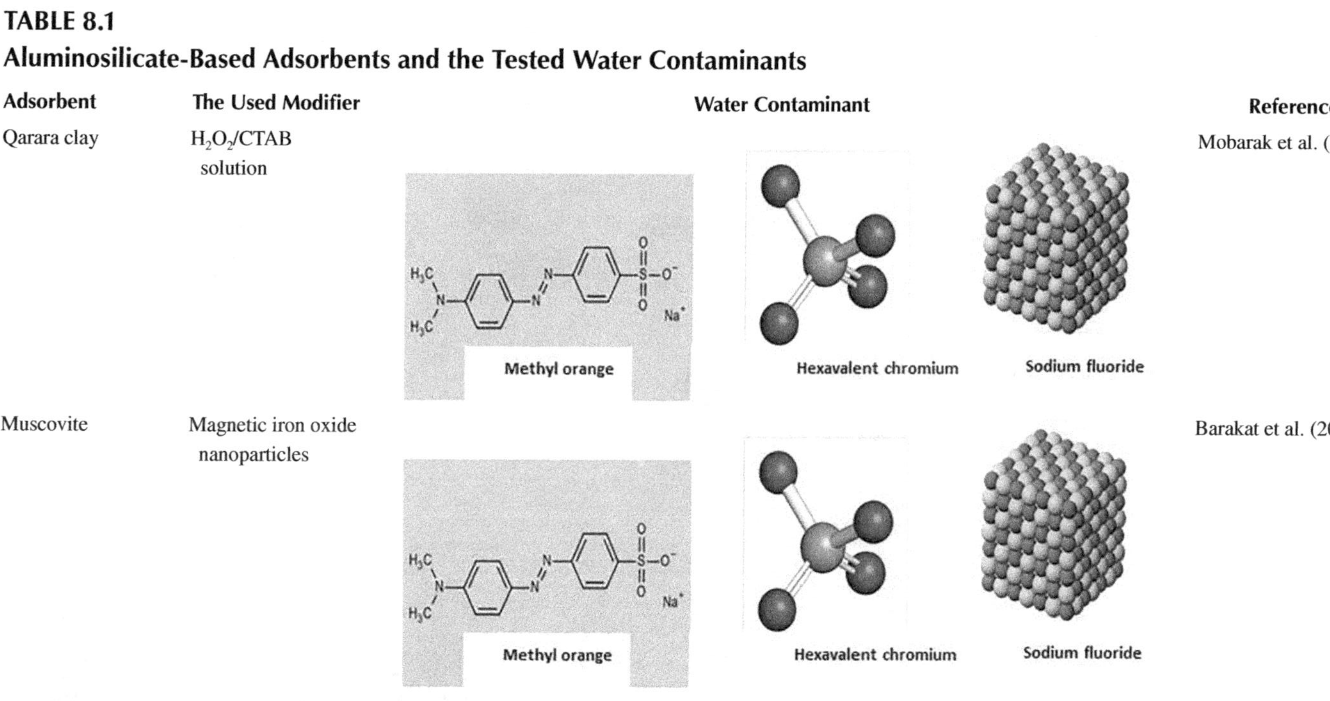	Mobarak et al. (2018)
Muscovite	Magnetic iron oxide nanoparticles		Barakat et al. (2021)

(Continued)

TABLE 8.1 (*Continued*)
Aluminosilicate-Based Adsorbents and the Tested Water Contaminants

Adsorbent	The Used Modifier	Water Contaminant	Reference
Medium clay	Pomegranate peel extract	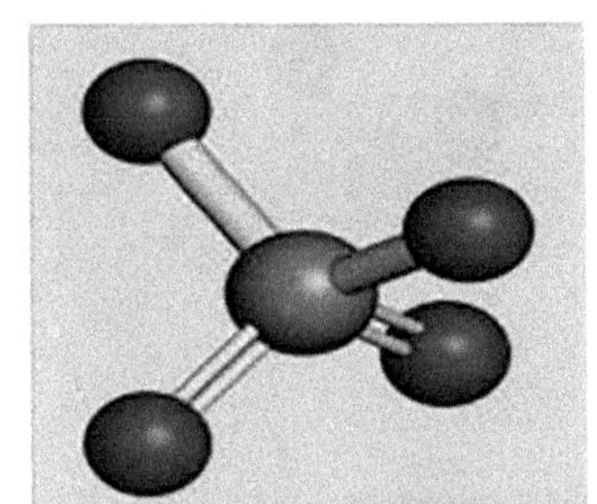 Hexavalent chromium	Ramadan et al. (2021)
Serpentine	Magnetic iron oxide nanoparticles	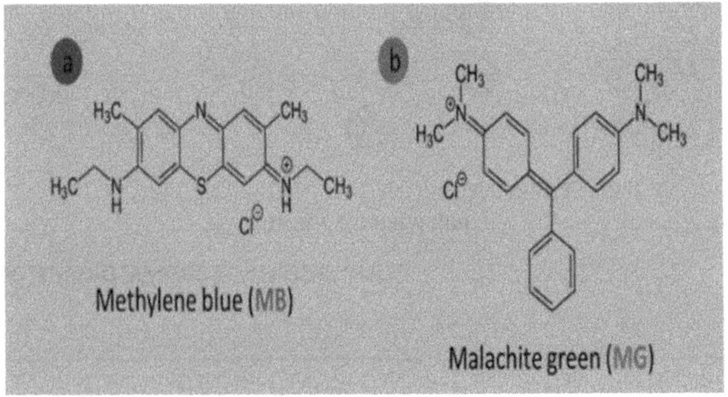 (a) Methylene blue (MB) (b) Malachite green (MG)	Seliem et al. (2020)

(*Continued*)

TABLE 8.1 (*Continued*)
Aluminosilicate-Based Adsorbents and the Tested Water Contaminants

Adsorbent	The Used Modifier	Water Contaminant	Reference
Exfoliated clay	Magnetic iron oxide nanoparticles	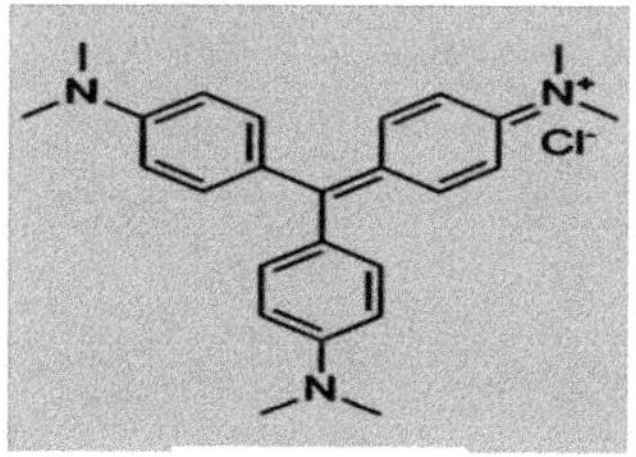 Crystal violet	Barakat et al. (2021)
Weathered basalt	Chitosan	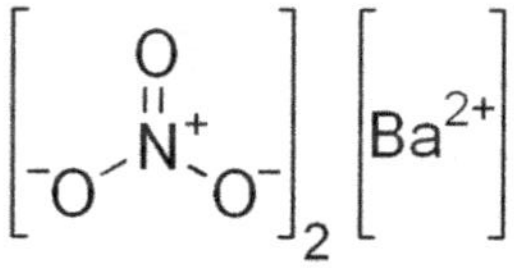 Barium Nitrate	Mohamed et al. (2020)

8.5 EQUILIBRIUM STUDY

Different initial concentrations of each water contaminant were used to perform the isotherm study. The adsorbed amount by aluminosilicate-based adsorbents at equilibrium (q_e) was calculated as follows:

$$q_e\,(\text{mg/g}) = (C_0 - C_e)\frac{V}{m} \tag{8.1}$$

C_e (mg/L) is the concentration of the tested contaminant at equilibrium.

Langmuir (1916) and Freundlich (1906), the most common traditional equilibrium models, were utilized in fitting the attained experimental data at different temperatures as given in Table 8.2 and Figure 8.7.

Most of the stated traditional isotherm models have few important rules that have been expressed to offer simple parameter interpretation for the physicochemical parameters managing the adsorption process (Mobarak et al. 2019; Li et al. 2019, 2020; Barakat et al. 2020b; Mohamed et al. 2020; Ramadan et al. 2021).

TABLE 8.2
Common Classical Adsorption Isotherm Models

Isotherm Model	Formula	Parameters
Langmuir	$\frac{C_e}{q_e} = \frac{1}{q_{max} K_L} + \frac{C_e}{q_{max}}$ $q_e = \frac{q_{max} K_L C_e}{(1 + K_L C_e)}$	C_e (mg/L): equilibrium concentration of the resting contaminant in the solution; q_e (mg/g): removed amount of contaminant at equilibrium; q_{max} (mg/g): maximum adsorption capacity; K_L (L/mg): maximum adsorption capacity; K_L (L/mg): Langmuir constant
Freundlich	$\log q_e = \log K_F + \frac{1}{n} \log C_e$ $q_e = K_F C_e^{1/n}$	K_F: contaminant adsorption capacity; n: heterogeneity factor

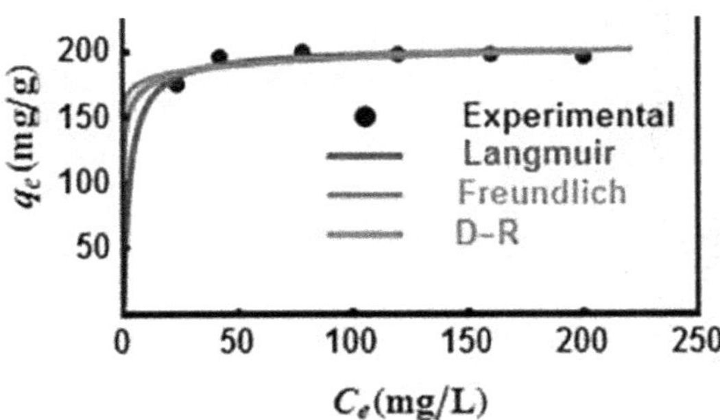

FIGURE 8.7 Application of Langmuir, Freundlich and D–R adsorption models for Cr(VI) uptake by a composite of black clay and pomegranate peel extract. (Reprinted with permission, Ramadan et al. 2021, order ref. https://doi.org/10.1016/j.jece.2021.105352.)

For example, Langmuir model reflects the uniformity of the adsorbed active sites and the maximum uptake capacity without offering the physical meaning of the adsorption process (Seliem and Mobarak 2019). Also, Freundlich isotherm is an empirical equation, and many studies have indicated that this model cannot be related to a multilayer adsorption process (Sellaoui et al. 2016, 2017; Azha et al. 2018; Ramadan et al. 2021).

8.6 STATISTICAL PHYSICS ANALYSIS

In order to get a well and *deeper understanding* of the uptake process, it is required to utilize the analytical models according to the assumptions of the statistical physics theory (Sellaoui et al. 2016, 2017). One of the benefits of this advanced modelling is to give physicochemical meaning of the steric and energetic parameters, which are directly involved in the adsorption process (Barakat et al. 2020a; Mohamed et al. 2020; Ramadan et al. 2021). Besides, this model is appropriate for the determination of important physicochemical parameters such as:

- The number of adsorbed molecules/ions per active site (n).
- The density of receptor sites (N_M).
- The removed amount at saturation (Q_{sat}).
- The concentration at half-saturation ($c_{1/2}$).
- The adsorption energy (ε).

The following advanced statistical physics models were utilized in fitting the adsorption data (Mobarak et al. 2019; Barakat et al. 2020 b; Mohamed et al. 2020; Ramadan et al. 2021).

- **Monolayer model with one energy (Model 1):** Based on this model, the uptake of the tested contaminant occurs onto the adsorbent surface as a single layer with one energy. The average number of the adsorbed molecules/ions is given as follows (Sellaoui et al. 2016):

$$N_o = \frac{N_M}{1+\left(\frac{c_{1/2}}{c}\right)^n} \tag{8.2}$$

where $c_{1/2} = \frac{2\pi m k_B T}{h^2} e^{-\frac{\varepsilon}{k_B T}}$ is the concentration at half-saturation.

The adsorbed amount as a function of concentration is calculated as follows:

$$Q = nN_o = \frac{Q_o}{1+\left(\frac{c_{1/2}}{c}\right)^n} \tag{8.3}$$

where Q_o is the adsorbed quantity at saturation.

- **Monolayer model with two energy sites (Model 2):** The removed molecules or ions were hypothesized to be adsorbed onto two different active sites N_{1M} and N_{2M} on the adsorbent with two energies ε_1 and ε_2, respectively. The total grand canonical partition function and the adsorbed amount are given as given below (Sellaoui et al. 2016; Sellaoui et al. 2017):

$$Z_{\text{gc}} = \left(z_{1\text{gc}}\right)^{N_{1M}} \left(z_{2\text{gc}}\right)^{N_{2M}} \tag{8.4}$$

$$Q = \frac{n_1 N_{1M}}{1+(c_1/c)^{n_1}} + \frac{n_2 N_{2M}}{1+(c_2/c)^{n_2}} \tag{8.5}$$

in which c_1 and c_2 refer to the concentrations at half-saturation intended for the first and the second of receptor sites, respectively.

- **Double-layer model with one energy (Model 3):** According to this model, the adsorption process is expected to occur in two layers. First, the ions or molecules are removed with the similar energy ε. Second, occupancy state N_i can give the values of 0, 1 and 2, and thus, the site is expected to be empty, filled by n molecule and occupied by $2n$ molecule, respectively. The adsorbed quantity against concentration is expressed as follows (Mobarak et al. 2019):

$$Q = Q_o \frac{\left(\frac{c}{c_{1/2}}\right)^n + 2\left(\frac{c}{c_{1/2}}\right)^{2n}}{1+\left(\frac{c}{c_{1/2}}\right)^n + \left(\frac{c}{c_{1/2}}\right)^{2n}} \tag{8.6}$$

- **Double-layer model with two energies (Model 4):** Concerning this model, the first removed layer with an energy level (ε_1) is greater than the second one with an energy level (ε_2). This is mainly related to the direct interaction between the polluted ions and the adsorbent surface. The partition function can be given as follows (Mobarak et al., 2019):

$$Z_{gc} = 1 + e^{\beta(\varepsilon_1+\mu)} + e^{\beta(\varepsilon_1+\varepsilon_2+2\mu)} \tag{8.7}$$

The mathematical equation of the double-layer adsorption model with two energies can be given as follows (Sellaoui et al. 2016, 2017):

$$Q = Q_o \frac{\left(\frac{c}{c_1}\right)^n + 2\left(\frac{c}{c_2}\right)^{2n}}{1+\left(\frac{c}{c_1}\right)^n + \left(\frac{c}{c_2}\right)^{2n}} \tag{8.8}$$

- **Multilayer adsorption model (Model 5):** A multilayer adsorption model, based on the grand canonical ensemble, is used to theoretically describe the adsorption mechanism of ions or molecules onto the utilized adsorbent. This general model suggests that the adsorption process was governed by the formation of a variable number of layers connecting two energies of interactions (Li et al. 2020; Barakat et al. 2020b; Mohamed et al. 2020). The earlier energy resulted from the interaction between the first adsorbed layer of the tested adsorbate (stable number) and the adsorbent surface. On the other hand, the other energy describes the interaction between the layers designed by the adsorbed ions. Consequently, the total number of the removed layers is given by 1+N2.

The mathematical expression of this model is described by the following equation (Li et al. 2020; Barakat et al. 2020b; Mohamed et al. 2020):

$$Q = nN_M \frac{F_1(c)+F_2(c)+F_3(c)+F_4(c)}{G(c)} \tag{8.9}$$

in which

$$F_1(c) = -\frac{2\left(\frac{c}{c_1}\right)^{2n}}{1-\left(\frac{c}{c_1}\right)^{n}} + \frac{\left(\frac{c}{c_1}\right)^{n}\left(1-\left(\frac{c}{c_1}\right)^{2n}\right)}{\left(1-\left(\frac{c}{c_1}\right)^{n}\right)^{2}},$$

$$F_2(c) = \frac{2\left(\frac{c}{c_1}\right)^{n}\left(\frac{c}{c_2}\right)^{n}\left(1-\left(\frac{c}{c_2}\right)^{n N_2}\right)}{1-\left(\frac{c}{c_2}\right)^{n}},$$

$$F_3(c) = -N_2\frac{\left(\frac{c}{c_1}\right)^{n}\left(\frac{c}{c_2}\right)^{n}\left(\frac{c}{c_2}\right)^{n N_2}}{1-\left(\frac{c}{c_2}\right)^{n}},$$

$$F_4(c) = \frac{\left(\frac{c}{c_1}\right)^{n}\left(\frac{c}{c_2}\right)^{2n}\left(1-\left(\frac{c}{c_2}\right)^{n N_2}\right)}{\left(1-\left(\frac{c}{c_2}\right)^{n}\right)^{2}},$$

TABLE 8.3
Advanced Statistical Physics Models for the Adsorption Process

Model	Formula	Parameters
	Statistical Physics Models	
Model 1	$Q = nN_o = \dfrac{nN_M}{1+\left(\dfrac{c_{1/2}}{c}\right)^n} = \dfrac{Q_o}{1+\left(\dfrac{c_{1/2}}{c}\right)^n}$	Q (mg/g): adsorbed quantity, n: ions number per site, N_m (mg/g): receptor sites density, Q_0 (mg/g): adsorbed quantity at saturation, $c_{1/2}$ (mg/L): the concentration at half-saturation
Model 2	$Q = \dfrac{n_1 N_{1M}}{1+(c_1/c)^{n_1}} + \dfrac{n_2 N_{2M}}{1+(c_2/c)^{n_2}}$	c_1 and c_2 (mg/L): concentrations at half-saturation for the first and the second active sites, respectively. n_1 and n_2 (–): ions number per site for the first and the second receptor sites, respectively.
Model 3	$Q = Q_o \dfrac{\left(\dfrac{c}{c_{1/2}}\right)^n + 2\left(\dfrac{c}{c_{1/2}}\right)^{2n}}{1+\left(\dfrac{c}{c_{1/2}}\right)^n + \left(\dfrac{c}{c_{1/2}}\right)^{2n}}$	
Model 4	$Q = Q_o \dfrac{\left(\dfrac{c}{c_1}\right)^n + 2\left(\dfrac{c}{c_2}\right)^{2n}}{1+\left(\dfrac{c}{c_1}\right)^n + \left(\dfrac{c}{c_2}\right)^{2n}}$	
Model 5	$Q = nN_M \dfrac{\left(\dfrac{c}{c_1}\right)^n \left(1-(N_\ell+1)\left(\dfrac{c}{c_2}\right)^{nN_\ell} + N_\ell\left(\dfrac{c}{c_2}\right)^{n(N_\ell+1)}\right)}{\left(1-\left(\dfrac{c}{c_2}\right)^n\right)\left(1-\left(\dfrac{c}{c_2}\right)^n + \left(\dfrac{c}{c_1}\right)^n - \left(\dfrac{c}{c_1}\right)^n \left(\dfrac{c}{c_2}\right)^{nN_\ell}\right)}$	

$$G(c) = \frac{\left(1-\left(\dfrac{c}{c_1}\right)^{2n}\right)}{1-\left(\dfrac{c}{c_1}\right)^n} + \frac{\left(\dfrac{c}{c_1}\right)^n\left(\dfrac{c}{c_2}\right)^n\left(1-\left(\dfrac{c}{c_2}\right)^{n\,N_2}\right)}{\left(1-\left(\dfrac{c}{c_2}\right)^n\right)^2},$$

in which C_1 and C_2 are the concentrations at half-saturation (C_1 is linked to the first layer, while C_2 is related to N_2 layers of the removed ions).

In addition, the N_2 parameter revealed the formed adsorbate layers with a definite adsorption energy.

Table 8.3 summarizes the advanced statistical physics models applied in the adsorption processes.

The best advanced model was determined based on the values of R^2 and the root-mean-square error (RMSE) calculated as follows (Ramadan et al. 2021):

$$R^2 = 1 - \frac{\sum \left(q_{e,\text{exp}} - q_{e,\text{cal}}\right)^2}{\sum \left(q_{e,\text{exp}} - q_{e,\text{mean}}\right)^2} \tag{8.10}$$

$$\text{RMSE} = \sqrt{\frac{\sum_{i=1}^{m} \left(Q_{i\ \text{cal}} - Q_{i\ \text{exp}}\right)^2}{m' - p}} \tag{8.11}$$

in which m' signifies the experimental data and p is the number of adjustable parameters. $Q_{i\ \text{cal}}$ and $Q_{i\ \text{exp}}$ are the values of the calculated and the experimental removed amounts of the adsorbed ions, respectively.

8.7 PHYSICOCHEMICAL PARAMETERS

8.7.1 Steric Parameters

- ***The n parameter:*** The n parameter, as a stoichiometric coefficient, explains the number of adsorbate molecules or ions of per active site of the adsorbent (Sellaoui et al. 2016). Determination of this steric parameter plays a main role in clarification the geometry of the adsorbed molecules or ions onto the adsorbent surface. Based on the values of the n parameter, three probable situations can occur to outline the adsorption process (Sellaoui et al. 2016; Li et al. 2020; Barakat et al. 2020b; Mohamed et al. 2020; Ramadan et al. 2021).
- **Situation 1, n<0.5:** This reveals that one active site of the adsorbent can accept a portion of the removed ion, suggesting a parallel geometry (i.e. a multi-docking removal mechanism).
- **Situation 2, 0.5<n<1:** This suggests that the removed ions can be adsorbed onto the adsorbent via parallel and non-parallel positions (i.e. a mixed geometry).
- **Situation 3, $n \geq 1$:** In this case, the receptor site (functional group) of the applied adsorbent can receive one or more ions of the adsorbate, indicating a non-parallel adsorption location (i.e. a multi-molecular uptake mechanism).
- **The N_M parameter:** This physicochemical parameter represents the number of molecules or ions per receptor site (N_M). Overall, it can be observed that, when the N_M increased with temperature, the corresponding n parameter decreased, and the conflicting condition was also recognized. For instance, the increase in the n parameter is usually related to the aggregation phenomenon, while the reduction in the n value could be attributed to the thermal agitation (Sellaoui et al. 2016; Li et al. 2020; Barakat et al. 2020a; Mohamed et al. 2020; Ramadan et al. 2021). Therefore, aggregation of the removed ions can lead to a reduction in the N_M parameter, but thermal agitation results in enhancing the value of this parameter.

- **The Q_{sat} parameter:** The determination of the adsorption capacity at saturation ($Q_{sat} = n.\ N_M$), for altered temperatures, is a significant factor to designate the efficiency of the tested adsorbent. Due to the opposite trend between the n and N_M parameters, the Q_{sat} style (i.e. increase or decrease with temperature) is mainly related to one of the two steric parameters.

8.7.2 The Energetic Parameters (ε)

The interaction between the removed ions and the adsorbent surface can be energetically evaluated as follows (Sellaoui et al. 2016; Li et al. 2020; Barakat et al. 2020 b):

$$C_1 = C_s e^{-\frac{\Delta E_1}{RT}} \tag{8.12}$$

$$C_2 = C_s e^{-\frac{\Delta E_2}{RT}} \tag{8.13}$$

in which c_1 and c_2 denote the concentrations at half-saturation and c_s is the adsorbate solubility.

The positive (+) and negative (–) energy values establish the endothermic and exothermic interactions between the removed ions and the adsorbent active sites, respectively. Physical adsorption energies can be associated with different interactions as follows (Sellaoui et al. 2016):

- The van der Waals forces (4–10 kJ/mol).
- Hydrophobic bond forces energies (about 5 kJ/mol).
- Dipole bond forces energies (2–29 kJ/mol).
- Electrostatic interactions (10–50 kJ/mol).

On the other hand, the chemical bond forces energies usually give values more than 80 kJ/mol (Sellaoui et al. 2016; Li et al. 2020; Barakat et al. 2020).

8.8 THERMODYNAMIC FUNCTIONS

The determination of the thermodynamic parameters (i.e. entropy, free enthalpy and internal energy) is obligatory to interpret the adsorption process (Sellaoui et al. 2016; Barakat et al. 2021).

- **Entropy:** Entropy has an essential part in measuring the order and disorder of the removed ions from the adsorbent surface. Entropy is associated with the grand potential (J) as a function of the grand canonical partition function (Z_{gc}) as given below (Ramadan et al. 2021):

$$J = -k_B T \ln Z_{gc} = -\frac{\partial \ln Z_{gc}}{\partial \beta} - T\ S_a$$

Consequently, the entropy can be expressed as

$$\frac{S_a}{k_B} = -\beta \frac{\partial \ln Z_{gc}}{\partial \beta} + \ln Z_{gc}$$

where k_B refers to the Boltzmann constant, T is the absolute temperature, and thermodynamic beta (β) = $1/k_BT$.

- **Free enthalpy:** The free enthalpy, which is well defined by the chemical potential (μ), is written as follows (Ramadan et al. 2021):

$$G = \mu\ Q \tag{8.14}$$

where μ is the chemical potential of the adsorbed ion.

- **Internal energy:** The adsorption internal energy is expressed as follows:

$$E_{int} = -\frac{\partial \ln Z_{gc}}{\partial \beta} + \frac{\mu}{\beta}\left(\frac{\partial \ln Z_{gc}}{\partial \mu}\right) \tag{8.15}$$

8.9 INTERPRETATION OF STERIC AND ENERGETIC PARAMETERS

Herein, the advanced statistical physics models are applied to evaluate the adsorption data resulted from the interaction between phosphate (adsorbate) and cationic surfactant-modified black clay as an adsorbent (Selim et al. 2019). The attained values of *RMSE* for the used statistical adsorption models at varied temperatures (i.e. 298, 308 and 318°K) are listed in Table 8.4.

From Table 8.4, we can conclude that monolayer with two energy sites (Model 2) and double-layer model with two energy sites (Model 4) fit well the phosphate adsorption data at (298°K) and (308 and 318°K). Thus, the steric and energetic parameters are considered according to the two adsorption models.

TABLE 8.4
Values of Coefficients of R^2 and RMSE for the Tested Isotherm Models of Phosphate Adsorption by the Modified Black Clay

	298		308		318	
T (°K)	R^2	RMSE	R^2	RMSE	R^2	RMSE
Model 1	0.9983	6.80	0.9993	4.314	0.9981	7.832
Model 2	0.9984	6.399	0.9992	4.313	0.9985	6.746
Model 3	0.9983	6.81	0.9993	4.32	0.9986	6.60
Model 4	0.9983	6.865	0.9992	4.216	0.9986	6.594
Model 5	0.9982	6.489	0.9993	4.266	0.9985	6.879

8.9.1 Steric Parameters

For Model 2, the adsorbed ions number per site (n) resulted to be 1.11 (for n_1) and 1.999 (for n_2). In this case, a vertical adsorption location and a multi-molecular uptake mechanism were involved in the uptake process. The percentage of the adsorbate onto the used adsorbent at 298°K for $n_1 = 1.11$ could be anchored either by one active site (x) or by two adsorption sites ($1 - x$). By the application of this equation $x.1+(1-x)2=1.11$, 89% and 11% of the removed phosphate ions singly and doubly anchored to the adsorption sites, respectively. Correspondingly, in the situation of $n_2 = 1.99$, the removed phosphate ions percentage by activated black clay was 99% by two ions per site and 1% by one ion per active site. The densities of active sites (N_M) of the removed phosphate ions confirmed the opposite trend between the n and N_M parameters. Regarding Model 4, the n values were <1.0 at temperatures of 308 and 318°K. Therefore, a horizontal adsorption location and a multi-docking uptake mechanism were detected in the uptake process. With increasing solution temperature, the N_M value increased. The Q_{sat} value was improved with increasing the adsorption temperature from 308 to 318°K (i.e. this interaction is endothermic). Furthermore, the steric N_M parameter played the main role in managing the adsorption process (i.e. the Q_{sat} and N_M parameters have the same trend with temperature).

8.9.2 Energetic Parameters

The adsorption energies were between 18.0 and 22.0 kJ/mol, and therefore, the phosphate adsorption was physical in nature (Sellaoui et al. 2016; Barakat et al. 2021). Also, the first adsorption energy (ε_1) was greater than the second one (ε_2), confirming the strong direct interaction between the adsorbate ions and the adsorbent surface as compared to the phosphate–phosphate interactions. In addition, the adsorption energies have positive values, indicating that the phosphate adsorption process was endothermic with a physical nature.

8.10 THERMODYNAMIC FUNCTIONS

The expression of the adsorption entropy for Model 2 is (Mobarak et al. 2019):

$$\frac{S_a}{k_B} = -\frac{\begin{array}{c} n_1 N_{1M}\left(\frac{c}{c_1}\right)^{n_1}\ln\left[\frac{c}{c_1}\right]+n_2 N_{2M}\left(\frac{c}{c_2}\right)^{n_2}\ln\left[\frac{c}{c_2}\right] \\ +\left(\frac{c}{c_1}\right)^{n_1}\left(\frac{c}{c_2}\right)^{n_2}\left(n_1 N_{1M}\ln\left[\frac{c}{c_1}\right]+n_2 N_{2M}\ln\left[\frac{c}{c_2}\right]\right)\end{array}}{\left(1+\left(\frac{c}{c_1}\right)^{n_1}\right)\left(1+\left(\frac{c}{c_2}\right)^{n_2}\right)} + \ln\left[\left(1+\left(\frac{c}{c_1}\right)^{n_1}\right)^{N_{1M}}\left(1+\left(\frac{c}{c_2}\right)^{n_2}\right)^{N_{2M}}\right] \quad (8.16)$$

The adsorption entropy for Model 4 is expressed as follows (Mobarak et al. 2019):

$$\frac{S_a}{k_B} = -N_M \left(\frac{\left(\frac{c}{c_1}\right)^n \ln\left(\frac{c}{c_1}\right)^n + \left(\frac{c}{c_2}\right)^{2n} \ln\left(\frac{c}{c_2}\right)^{2n}}{1+\left(\frac{c}{c_1}\right)^n + \left(\frac{c}{c_2}\right)^{2n}} + \ln\left[1+\left(\frac{c}{c_1}\right)^n + \left(\frac{c}{c_2}\right)^{2n}\right] \right) \tag{8.17}$$

The free enthalpy of Model 2 is expressed as follows:

$$\frac{G}{k_B T} = \log\left[\frac{\mu}{z_{tr}}\right] \left(\frac{n_1 N_{1M}}{1+\left(\frac{c_1}{c}\right)^{n_1}} + \frac{n_2 N_{2M}}{1+\left(\frac{c_2}{c}\right)^{n_2}} \right) \tag{8.18}$$

The free enthalpy of Model 4 is given as follows:

$$\frac{G}{k_B T} = \ln\left[\frac{\mu}{z_{tr}}\right] \left(n\, N_M \frac{\left(\frac{c}{c_1}\right)^n + 2\left(\frac{c}{c_2}\right)^{2n}}{1+\left(\frac{c}{c_1}\right)^n + \left(\frac{c}{c_2}\right)^{2n}} \right) \tag{8.19}$$

The internal energy of Model 2 is given as follows:

$$\frac{E_{int}}{k_B T} = -\frac{N_{1M}\left(\frac{c}{c_1}\right)^{n_1}\left(n_1 \ln\left[\frac{c}{c_1}\right] - \mu\right) + N_{2M}\left(\frac{c}{c_2}\right)^{n_2}\left(n_2 \ln\left[\frac{c}{c_2}\right] - \mu\right) + \left(\frac{c}{c_1}\right)^{n_1}\left(\frac{c}{c_2}\right)^{n_2}\left(N_{1M}\left(n_1 \ln\left[\frac{c}{c_1}\right] - \mu\right) + N_{2M}\left(n_2 \ln\left[\frac{c}{c_2}\right] - \mu\right)\right)}{\left(1+\left(\frac{c}{c_1}\right)^{n_1}\right)\left(1+\left(\frac{c}{c_2}\right)^{n_2}\right)} \tag{8.20}$$

The internal energy of Model 4 is given as follows:

$$\frac{E_{int}}{k_B T} = -N_M \frac{\left(\frac{c}{c_1}\right)^n \ln\left(\frac{c}{c_1}\right)^n + 2\left(\frac{c}{c_2}\right)^{2n} \ln\left(\frac{c}{c_2}\right)^{2n} + \mu\left(\left(\frac{c}{c_1}\right)^n + \left(\frac{c}{c_2}\right)^{2n}\right)}{1+\left(\frac{c}{c_1}\right)^n + \left(\frac{c}{c_2}\right)^{2n}} \tag{8.21}$$

The evolution of the phosphate adsorption entropy as a function of its concentration is given in Figure 8.8.

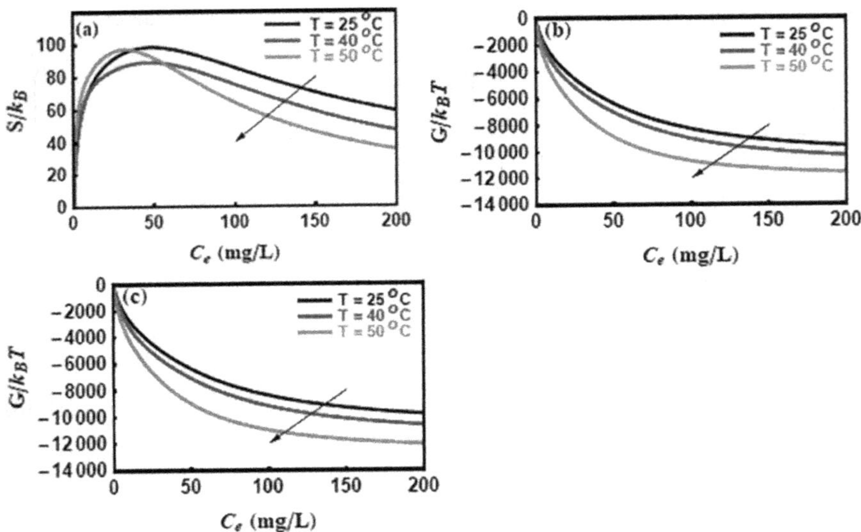

FIGURE 8.8 Evaluation entropy, free enthalpy and Gibbs free energy as a function of Cr(VI) concentrations at 25°C, 40°C and 50°C. (Reprinted with permission, Ramadan et al. 2021, order ref. https://doi.org/10.1016/j.jece.2021.105352.)

Concerning the multilayer adsorption model, several operating scenarios can be reflected to understand the experimental data as follows (Barakat et al. 2020b):

- **Scenario 1:** n and N_2 are free adjustable parameters (i.e. multilayer).
- **Scenario 2:** n is a free adjustable parameter and N_2 = zero (fixed) (i.e. a monolayer).
- **Scenario 3:** n is a free adjustable parameter and N_2 = 1 (fixed) (i.e. double-layer).
- **Scenario 4:** n is a free adjustable parameter and N_2 = 2 (fixed) (i.e. triple-layer).
- **Scenario 5:** n = unity (fixed) and N_2 = zero (fixed) (i.e. Langmuir model).

Surfactant-modified exfoliated clay was further decorated with magnetic Fe_3O_4 nanoparticles. The multilayer model fitted well the interactions between the crystal violet and the developed adsorbent, and the attained parameters are listed in Table 8.5. The n was more than unity at all temperatures, demonstrating a vertical adsorption orientation and a multi-interaction mechanism. The slight variation of N_t ($N_t = 1+N_2$) indicated that the influence of the N_2 on the removal mechanism could be ignored. The increase in crystal violet uptake with increasing temperature was mainly controlled by the number of active sites (N_M) (i.e. the $Q_{\text{sat}} = n.\, N_M.\, N_t$ and the N_M having the same trend) (Figure 8.9). The proposed crystal violet adsorption mechanism through the steric and energetic parameters of the multilayer model is displayed in Figure 8.10.

TABLE 8.5
Steric and Energetic Parameters of the Multilayer Model for the Adsorption of Crystal Violet onto Fe_3O_4/Exfoliated Clay Adsorbent (Barakat et al. 2020)

T	*n*	N_M	$1+N_2$	ΔE_1	ΔE_2	Q_{sat}
(°C)	(-)	(mg/g)	(mg/g)	(KJ/mol)	(KJ/mol)	(mg/g)
25	2.56	162.24	2.19	21.06	12.69	909.57
40	2.23	198.83	2.26	22.99	13.47	1002.06
55	1.8	256.5	2.33	27.22	14.24	1075.76

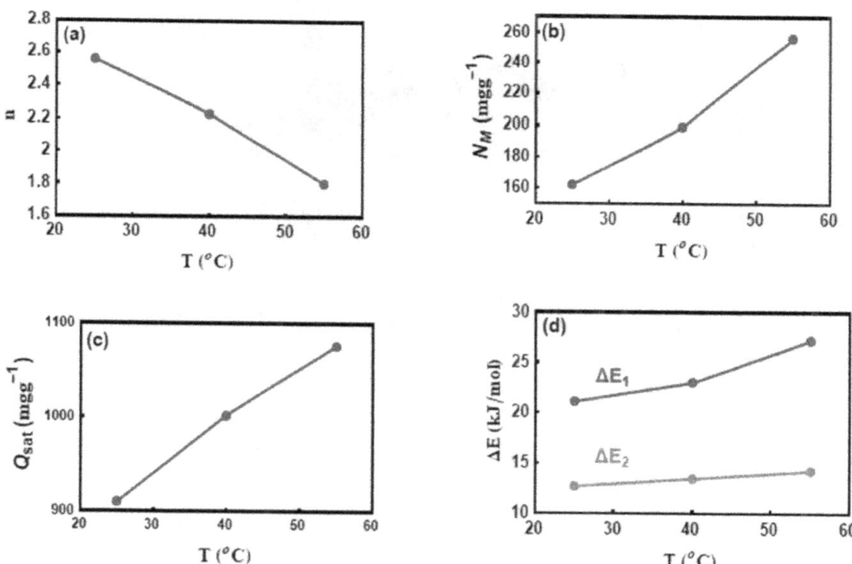

FIGURE 8.9 Evolution of statistical physics parameters: (a) n, (b) N_M, (c) Q_{sat} and (d) ΔE as a function of temperature for the adsorption of crystal violet onto Fe_3O_4/exfoliated clay adsorbent. (See Barakat et al. 2020.)

8.11 CONCLUSIONS AND FUTURE RESEARCH

This chapter covered many studies about the application of aluminosilicate materials (e.g. carbonaceous clays, mica, zeolites, weathered basalt and serpentine) to the adsorption of wastewater contaminants. CTAB, iron oxide (Fe_3O_4) magnetic nanoparticles, pomegranate peel extract and chitosan were individually used in the modification of these materials. The availability, non-toxicity, high stability and low cost of natural aluminosilicates make these materials appropriate for the fabrication of promising adsorbents against different water contaminants. Besides,

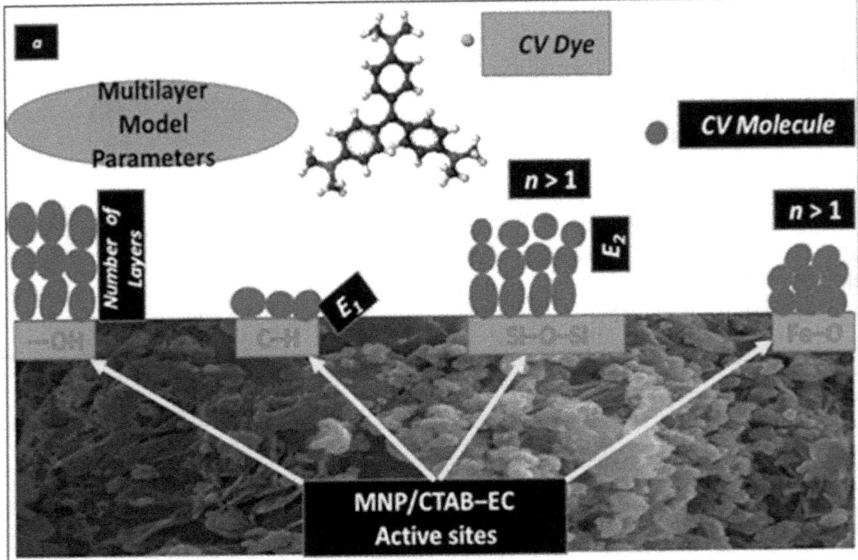

FIGURE 8.10 The proposed crystal violet adsorption mechanism based on the steric and energetic parameters of the multilayer model. (See Barakat et al. 2020.)

a discussion of the advanced modelling via the statistical physics theory was also achieved. Interpretation of the physicochemical parameters based on the statistical physics theory offered deep insights and a proper understanding of the adsorption mechanism. Overall, aluminosilicate minerals were found to be promising in the decontamination of water and wastewater through the adsorption technique.

Several issues that should be considered in the future studies are as follows:

- Evolution of the adsorption process in the multicomponent systems (e.g. binary uptake method).
- Using different modifiers in the activation of aluminosilicates to prepare new composites with varied adsorption sites.
- Utilizing the density functional theory (DFT) to recognize the role of functional groups in the adsorption process.
- Application of the aluminosilicate-based materials in the degradation of organic water contaminants.

ACKNOWLEDGEMENTS

The authors of this study acknowledge the funding from Beni-Suef University, the National Plan for Science, Technology and Innovation (MAARIFAH), King Abdulaziz City and STDF.

REFERENCES

M.R. Abukhadra, M. Mostafa, Effective decontamination of phosphate and ammonium utilising novel muscovite/phillipsite composite: Equilibrium investigation and realistic application, *Science of the Total Environment* (2019) 667, 101–111.

S.F. Azha, L. Sellaoui, M.S. Shamsudin, S. Ismail, A. Bonilla-Petriciolet, A. Ben Lamine, A. Erto, Synthesis and characterization of a novel amphoteric adsorbent coating for anionic and cationic dyes adsorption: experimental investigation and statistical physics modelling, *Chemical Engineering Journal* 351 (2018) 221–229.

T. Bao, M. Mezemir, A. Hosseinzadeh, R.L. Frost, Z.M. Yu, J. Jin, K. Wu, Catalytic degradation of P-chlorophenol by muscovite-supported nano zero-valent iron composite: Synthesis, characterisation, and mechanism studies, *Applied Clay Science* 195 (2020) 105735.

M.A. Barakat, R. Kumar, E.C. Lima, M.K. Seliem, Facile synthesis of muscovite–supported Fe3O4 nanoparticles as an adsorbent and heterogeneous catalyst for effective removal of methyl orange: Characterisation, modelling, and mechanism, *Journal of the Taiwan Institute of Chemical Engineers* 119, (2021) 146–157.

M.A. Barakat, R. Kumar, M.K. Seliem, A.Q. Selim, M. Mobarak, I. Anastopoulos, D. Giannakoudakis, M. Barczak, A. Bonilla-Petriciolet, E.A. Mohamed, Exfoliated clay decorated with magnetic iron nanoparticles for crystal violet adsorption: Modeling and physicochemical interpretation, Nanomaterials, 1454 (2020a) doi: 10.3390/nano10081454.

M.A. Barakat, A.Q. Selim, M. Mobarak, R. Kumar, I. Anastopoulos, D.A. Giannakoudakis, A. Bonilla-Petriciolet, E.A. Mohamed, M.K. Seliem, S. Komarneni. Experimental and theoretical studies of methyl orange uptake by Mn–rich synthetic mica: Insights into manganese role in adsorption and selectivity. *Nanomaterials* 10 (2020b) 1–19.

J. Chang, J. Ma, Q. Ma, D. Zhang, N. Qiaoa, M. Hu, H. Ma, Adsorption of methylene blue onto Fe3O4/activated montmorillonite nanocomposite, *Applied Clay Science* 119 (2016) 132–140.

F. Chen, X. Wu, X. Ran R. Bu, F. Yang. Co–Fe hydrotalcites for efficient removal of dye pollutants via synergistic adsorption and degradation, *RSC Advances* 7 (2017), 41945–41954.

J. Choi, S. Komarneni, K. Grover, H. Katasuki, M. Park, Hydrothermal synthesis of Mn–mica, *Applied Clay Science* 46 (2009) 69–72.

G. Diego Gatta, P. Lotti, Cancrinite-group minerals: crystal-chemical description and properties under non-ambient conditions—a review, *American Mineralogist* 101 (2016) 253–265.

H.M.F. Freundlich, Over the adsorption in solution, *Journal of Physical Chemistry* 57 (1906) 385–471.

R. Lafi, A. Hafiane, Removal of methyl orange (MO) from aqueous solution using cationic surfactants modified coffee waste (MCWs), *Journal of Taiwan Institute Chemical Engineering*, 58 (2016) 424–433.

I. Langmuir, The constitution and fundamental properties of solids and liquids, *Journal of American Chemical Society* 38 (1916) 2221–2295.

Z. Li, H. Hanafy, L. Zhang, L. Sellaoui, M.S. Netto, M.L.S. Oliveira, M.K. Seliem, G.L. Dotto, A. Bonilla-Petricioleti, Q. Li, Adsorption of Congo red and methylene blue dyes on an ashitaba waste and a walnut shell-based activated carbon from aqueous solutions: experiments, characterization, and physical interpretations, *Chemical Engineering Journal* 388 (2020) 1–10.

Z. Li, L. Sellaoui, G.L. Dotto, A.B. Lamine, A.B. Petriciolet, H. Hanafy, H. Belmabrouk, S.N. Matias, A. Erto, Interpretation of the adsorption mechanism of Reactive Black 5 and Ponceau 4R dyes on chitosan/polyamide nanofibers via advanced statistical physics model, *Journal of Molecular Liquids* 285 (2019) 165–170.

Z. Li, G. Wang, K. Zhai, C. He, Q. Li, P. Guo, Methylene blue adsorption from aqueous solution by loofah sponge-based porous carbons, *Colloids and Surfaces A: Physicochemical and Engineering Aspects*, 538 (2018) 28–35.

M. Mobarak, E.A. Mohamed, A.Q. Selim, F.M. Mohamed, L. Sellaoui, A.B. Petriciolet, M.K. Seliem, Statistical physics modeling and interpretation of methyl orange adsorption on high–order mesoporous composite of MCM–48 silica with treated rice husk, *Journal of Molecular Liquids* 285 (2019) 678–687.

M. Mobarak, A.Q. Selim, E. Mohamed, M.K. Seliem, A superior adsorbent of CTAB/H2O2 solution–modified organic carbon rich-clay for hexavalent chromium and methyl orange uptake from solutions, *Molecular Liquids Journal* 259 (2018) 384–397.

E.A. Mohamed, A.Q. Selim, S.A. Ahmed, L. Sellaou, A. Bonilla-Petriciolet, A. Ertoe, Z. Li, Y. Li, M.K. Seliem, H2O2–activated anthracite impregnated with chitosan as a novel composite for Cr(VI) and methyl orange adsorption in single–compound and binary systems: Modeling and mechanism interpretation, *Chemical Engineering Journal* 380 (2020) 122445.

E.A. Mohamed, A.Q. Selim, A.M. Zayed, S. Komarneni, M. Mobarak, M.K. Seliem, Enhancing adsorption capacity of Egyptian diatomaceous earth by thermo-chemical purification: Methylene blue uptake, *Journal of Colloid and Interface Science* 534 (2019) 408–419.

A.E. Pirbazari, E. Saberikhaha, S.S.H. Kozani, Fe3O4–wheat straw: Preparation, characterization and its application for methylene blue adsorption, *Water Resources and Industry* 7–8 (2014) 23–37.

H.S. Ramadan, M. Mobarak, E.C. Lima, A. Bonilla-Petriciolet, Z. Li, M.K. Seliem, Cr(VI) adsorption onto a new composite prepared from Meidum black clay and pomegranate peel extract: experiments and physicochemical interpretations, *Journal of Environmental Chemical Engineering* 9 (2021) 105352.

R. Saffari, Z., Shariatinia, M. Jourshabani, Synthesis and photocatalytic degradation activities of phosphorus-containing ZnO microparticles under visible light irradiation for water treatment applications, *Environmental Pollution* (2020) 259, 113902.

R. Salim, M. Al-Subu, I. Abu-Shqair, H. Braik, Removal of zinc from aqueous solutions by dry plant leaves, *Transactions of the Institution of Chemical Engineers* 81 (2003) 236–242.

M.H. Salmani, M. Abedi, S.A. Mozaffari, H.A. Sadeghian, Modification of pomegranate waste with iron ions a green composite for removal of Pb from aqueous solution: equilibrium, thermodynamic and kinetic studies, *AMP* (2017) 1–8.

S.T. Sankpal, P.V. Naikwade, Physicochemical analysis of effluent discharge of fish processing industries in Ratnagiri India, *Bioscience Discovery* 3(1) (2012) 107–111.

M.K. Seliem, S. Komarneni, Equilibrium and kinetic studies for adsorption of iron from aqueous solution by synthetic Na-A zeolites: statistical modeling and optimization, *Microporous and Mesoporous Materials* 228 (2016) 266–274.

M.K. Seliem, S. Komarneni, T. Byrne, F.S. Cannon, M.G. Shahien, A.A. Khalil, I.M. AbdEl-Gaid, Removal of perchlorate by synthetic organosilicas and organoclay: kinetics and isotherm studies, *Applied Clay Science* 71 (2013) 21–26.

M.K. Seliem, M. Mobarak, Cr(VI) uptake by a new adsorbent of CTAB–modified carbonized coal: Experimental and advanced statistical physics studies, *Molecular Liquids Journal* 294 (2019) 111676.

M.K. Seliem, M. Mobarak, A.Q. Selim, E.A. Mohamed, R.A. Halfaya, H.K. Gomaa, I. Anastopoulos, D.A. Giannakoudakis, E.C. Lima, A. Bonilla-Petriciolet, G. Luiz Dotto, A novel multifunctional adsorbent of pomegranate peel extract and activated anthracite for Mn(VII) and Cr(VI) uptake from solutions: Experiments and theoretical treatment, *Journal of Molecular Liquids* 311 (2020) 113169.

A.Q. Selim, E.A. Mohamed, M.K. Seliem, A.M. Zayed, Synthesis of sole cancrinite phase from raw muscovite: Characterization and optimization, *Journal of Alloys and Compounds* 762 (2018) 653–667.

A.Q. Selim, L. Sellaoui, S.A. Ahmed, M. Mobarak, E.A. Mohamed, A. Ben Lamine, A. Erto, A.B. Petriciolet, M.K. Seliem, Statistical physics-based analysis of the adsorption of Cu2+and Zn2+ontosynthetic cancrinite in single-compound and binary systems, *Journal of Environmental Chemical Engineering* 7 (2019) 103217.

A.Q. Selim, L. Sellaoui, M. Mobarak, Statistical physics modeling of phosphate adsorption onto chemically modified carbonaceous clay, *Journal of Molecular Liquids* 279 (2019) 94–107.

L. Sellaoui, G.L. Dotto, A.B. Lamine, A. Erto, Interpretation of single and competitive adsorption of cadmium and zinc on activated carbon using monolayer and exclusive extended monolayer models, *Environmental Science and Pollution Research* 24 (2017) 19902–19908.

L. Sellaoui, H. Guedidi, S. Masson, L. Reinert, J.M. Levêque, S. Knani, A. Ben Lamine, M. Khalfaoui, L. Duclaux, Steric and energetic interpretations of the equilibrium adsorption of two new pyridinium ionic liquids and ibuprofen on a microporous activated carbon cloth: Statistical and COSMO-RS models, *Fluid Phase Equilibria* 414 (2016) 156–163.

Y. Sua, Y. Jiaoa, C. Doua, R. Hana. Biosorption of methyl orange from aqueous solutions using cationic surfactant-modified wheat straw in batch mode, *Desalination and Water Treatment* 52 (2014) 6145–6155.

V. Vimonses, S. Lei, B. Jina, C.W.K. Chow, C. Saint, Kinetic study and equilibrium isotherm analysis of Congo Red adsorption by clay materials, *Chemical Engineering Journal* 148 (2009) 354–364.

Y. Xi, Z. Ding, H. He, R.L. Frost, Structure of organoclays—an X-ray diffraction and thermogravimetric analysis study, *Journal of Colloid and Interface Science* 277 (2004) 116–120.

J.-S. Yang, J.Y. Lee, Y.-T. Park, K. Baek, J. Choi, Adsorption of As(III), As(V), Cd(II), Cu(II), and Pb(II) from aqueous solutions by natural muscovite, *Separation Science and Technology* 45 (2010) 814–823.

Q. Zhang, D. Jia, K. Yao, Antiliperoxidant activity of pomegranate peel extracts on lard, *Natural Product Research* 21 (2007) 211–216.

9 Antimicrobial Nanomaterials for Water Disinfection

Muzammil Anjum
PMAS Arid Agriculture University

Samia Qadeer
University of Narowal

Muhammad Waqas
Kohat University of Science and Technology

Azeem Khalid
PMAS Arid Agriculture University

Rashid Miandad
University of Peshawar

Mohamed Abou El-Fetouh Barakat
King Abdulaziz University

CONTENTS

DOI: 10.1201/9781003129042-9

9.1 INTRODUCTION

There is no doubt that water is one of the most vital elements on earth needed for life. Like other organisms, humans also completely rely on water resources for their life to continue. Due to human activities and improper management, the pressure is being exerted on water resources due to escalated pollution caused by industrial, commercial, domestic and agricultural activities in both developing and developed worlds (Zhao et al. 2019). This pollution includes toxic and hazardous chemicals, and biological waste components particularly from manufacturing industries and municipal discharge are the main cause of spread of diseases (Mostafiz et al. 2020; Cheriyamundath and Vavilala 2021). About 2.2 million people die every year out of 1 billion consumers of unhealthy water (Zhao et al. 2019).

The wastewater treatment and management systems have largely been studied and established, but they still possess several disadvantages and deficiencies (Crini and Lichtfouse 2019). Typically, in a municipal scenery, a water treatment plant is a systematic project in which water is processed through multiple steps such as primary, secondary and tertiary treatment (Figure 9.1). Briefly described, the primary treatment collects the wastewater in a quiescent reservoir responsible for the settlement of suspended heavy particles to the bottom and lighter particles to the surface, such as oil and grease. In case of secondary treatment, also known as mineralization process, a chemical and biochemical oxidation of organic matter takes place, which is converted into inorganic non-toxic compounds such as water, salts and carbon dioxide. The BOD is reduced by 85% to the value less than

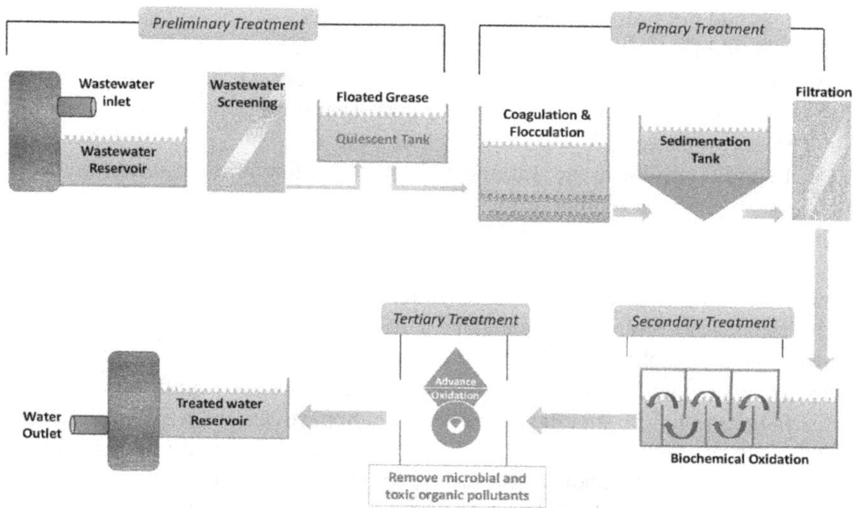

FIGURE 9.1 Steps of wastewater treatment process. Tertiary treatment involves wastewater disinfection.

30 mg/L. Tertiary treatment, also known as advance treatment, thereon eliminates the biological nutrients (N and P) and pathogens (microorganisms). After this step, the wastewater can be reused and is indistinguishable from original resource; thus, it can be released into the environment (river, lake, sea, ground or waste lands, etc.) (Zhao et al. 2019).

For the removal of the microorganisms and pathogens, there are different techniques with their own advantages and disadvantages. For wastewater disinfection, two methods are followed, which include inactivating the pathogens or completely removing them. Reportedly, chlorine disinfection, size exclusion, UV disinfection and ozone disinfection are common treatment methods to serve the purpose (Liu et al. 2012). The use of chlorine is the most recognized way for water disinfection. However, the intensive use of chemicals also increases the amount of disinfection by-products causing secondary harmful effects. The disinfectant such as chlorine and chloramines can react with different contents of water and produce carcinogenic by-products (Li et al. 2008). The use of size exclusion to remove bacteria is also considered as a highly effective technology and can be applied against a variety of pathogens. This filtration is effective even if the size of pathogens is very small like some viruses. However, the high energy consumption and low treatment speed limit its application. Another technology that has drawn high attention is the UV irradiation of water for the removal of microorganisms; however, the UV set-up is very expensive and requires high energy for operation. Hence, keeping in view these limitations there is need to establish and exploit effective and low-cost disinfection methods. On a broader perspective, an ideal disinfectant must show the following properties (Hossain et al. 2014).

1. It should not produce any harmful by-products.
2. It should be easily applicable and economical for the intended use.
3. The disinfectant should have no negative impact on human health.
4. Disinfectant product should not be corrosive for equipment and should be easy to store.
5. It should exhibit broad antimicrobial activity for multiple microorganisms at ambient temperature.
6. It should be safely disposable and environmentally friendly.

Of recent, water disinfection using antimicrobial material such as nanoparticles has obtained the interest of scientists across the globe (Sangari et al. 2015). Nowadays, nanotechnology is dealing with various emerging environmental issues associated with wastewater treatment, where the most important concern is water disinfection. The specific properties of nanoparticles, such as a high surface area-to-volume ratio (Kumar et al. 2014), dissociation reaction initiated over time due to environmental response, and ability to remove a broad spectrum of microorganisms (Singh et al. 2012), have sparked interest in their effective utilization. (Singh et al. 2012). The literature shows that various types of nanomaterials such as zinc, magnesium, copper, titanium, alginate, gold and silver have widely been used due to their antimicrobial property (Mahadevan et al. 2017). In this chapter, various types of nanomaterials and the processes involved in the water disinfection during water treatment have been discussed in detail.

9.2 ANTIMICROBIAL NANOMATERIALS

It has progressively been understood that nanomaterials may add to cutting-edge water treatment advances. Nanomaterials are recognized as materials with a size less than or equal to 100nm (Cheriyamundath and Vavilala 2021). However, nanomaterials with a size of 1–10nm have shown effective antimicrobial properties because of the higher contact area with microbial cells (Ogunsona et al. 2020). Nanomaterials with smaller sizes have orders of magnitude more complex surface area and surface reactive sites than their bulk counterparts (Zhao et al. 2019). These nanoparticles use a variety of pathways to interact with and kill microbes. They may directly interfere with microbes by disrupting their transmembrane or damage microbial cells directly by breaking their cell membrane, or they can damage them indirectly via oxidizing cell components, producing reactive oxygen species (ROS) or producing secondary products that can harm the cells of microorganism (Li et al. 2008). The surface charge on the bacterial cell assists nanomaterials having the opposite surface charge in binding with the cell, causing antimicrobial activity. Various nanomaterials have tunable chemical, physical and electronic properties that are not found in bulk materials (Zhao et al. 2019). Nanoparticles with antimicrobial properties, such as copper, silver, TiO_2, ZnO, zero-valent iron, carbon nanotubes, fullerenes and bio-nanoparticles including chitosan nanocomposites, are all included in the list of efficient nanoparticles that being extensively studied (Ogunsona et al. 2020). For decades, silver has been recognized to have

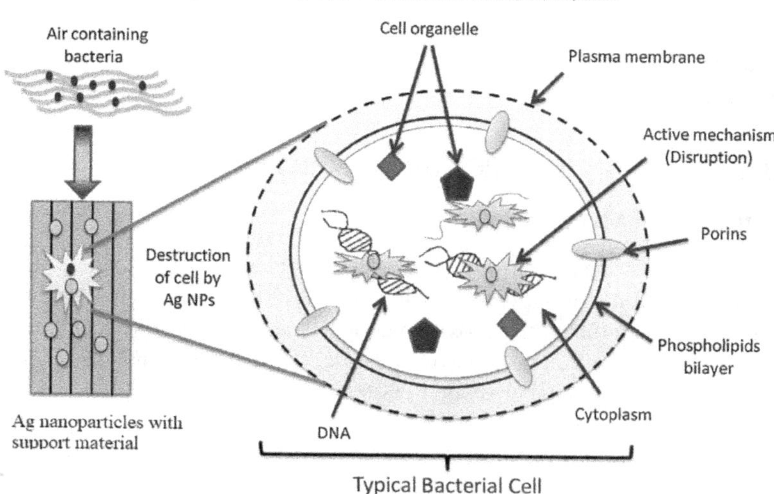

FIGURE 9.2 General mechanism of destruction of a bacterial cell: an example of Ag nanoparticles. (Adapted from Ogunsona et al. (2020) with permission from Elsevier © 2020.)

antibacterial properties. In bacterial protein, the silver ions released by this nanomaterial bind to the thiol groups, causing it to be damaged. These ions have the ability to damage the cell envelope as well as inhibit DNA replication (Qu et al. 2013). Figure 9.2 represents the general mechanism of bacterial cell disruption using Ag nanoparticles over a support material. The figure reveals that the Ag nanoparticles may bind to and penetrate the cell membranes of both gram-positive and gram-negative bacteria, disrupting cell activity by releasing silver ions, making them ideal for microbe inhibition. They can easily pass through the plasma membrane and the lipid bilayer to reach the cytoplasm. The antimicrobial properties of TiO_2 are primarily due to its photocatalytic potential, which is characterized by the formation of free hydroxyl radicals and peroxide when exposed to sunlight or UV light (Li et al. 2008). ZnO is also very appealing. Due to its high chemical stability, low cost, mass production and photocatalytic action, ZnO is also very appealing in photocatalytic performance for the removal of pathogens. It has been demonstrated that doping ZnO with Mg increases its photocatalytic performance, resulting in increased antimicrobial activity (Sadaiyandi et al. 2018). Carbon nanotubes, on the other hand, are also cytotoxic to bacteria; they induce physical disruption of the cell membrane and oxidative stress as well as disrupt the cellular assembly (Qu et al. 2013). Concerning the suitability of nanomaterials for application as disinfectant, they must comply with the following other properties (Hossain et al. 2014):

1. Chemical stability in liquid phase.
2. Physical stability, i.e. no settling or aggregation.

3. If the nanomaterial requires photo-excitation, it must be vigorous under visible or solar light illumination.
4. Capability to eliminate inorganic and organic contamination while disinfection of wastewater.

9.3 CLASSES OF ANTIMICROBIAL NANOMATERIALS

9.3.1 Metals and Metal Oxides

Antimicrobial activity has been observed in a variety of nanoparticulate metals, metal oxides and metal halides (Snigdha et al. 2020). Antimicrobial activity has been discovered in metal/metal oxide nanoparticles such as silicon (Si), titanium dioxide (TiO_2), silver (Ag), silver oxide (Ag_2O), zinc oxide (ZnO), gold (Au), magnesium oxide (MgO), copper oxide (CuO) and calcium oxide (CaO) (Durán et al. 2016). They target several biomolecules and use modes of action that are completely different from those specified for conventional antibiotics, exhibiting efficacy against bacteria that have already established resistance, posing a threat to the growth of resistant strains (Slavin et al. 2017). Numerous methods can be used to classify metal-based nanoparticles. These methods reveal important details about their physicochemical and electrical properties and morphology, all of which are important for their in vivo behaviour. The shape, size, surface energy and roughness are some of the most important properties of nanoparticles (Wang et al. 2017).

9.3.1.1 General Mechanisms

Bacteria have a unique set of traits that describe their actions when they come in contact with metal nanoparticles (Sánchez-López et al. 2020). Understanding the distinctions between cell walls of gram-positive bacteria and gram-negative bacteria is important because the principal toxicological effect of antimicrobial substances in bacteria is produced by direct contact with the surface of the cell (Wang et al. 2017). The surface of both gram-positive and gram-negative bacteria is negatively charged. A thick coating of peptidoglycan is formed in gram-positive bacteria by linear chains of alternate N-acetylglucosamine and N-acetylmuramic acid residues connected by a sequence of 3–5 amino acids that are cross-linked. Gram-negative bacteria, on the other hand, have a more complicated structure than gram-positive bacteria. Gram-negative bacteria have partially phosphorylated lipopolysaccharides in outer membrane that contribute to the negative charge of the cell envelope. On the contrary, gram-negative bacteria have a more complex structure (Bruslind 2017).

A schematic image of the antimicrobial mechanism of metal ions is shown in Figure 9.3. The nanomaterials having positive charge are attracted to the surface of negatively charged bacterial cell walls through electrostatic interactions. Positively charged metal-based nanoparticles, on the other hand, form a strong bond with membranes, disrupting cell walls and thereby penetrating into the cells. Furthermore, metal ions released by nanoparticles from the extracellular space

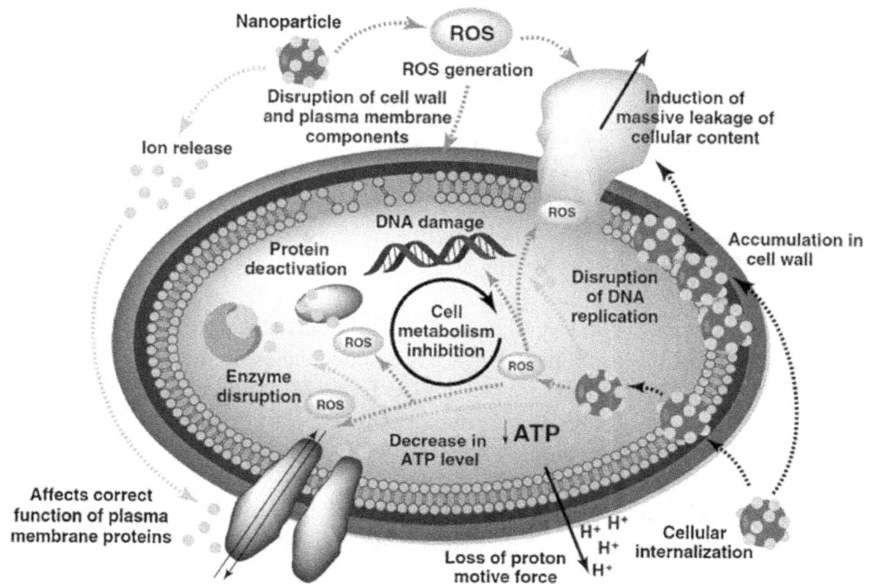

FIGURE 9.3 Schematic image of metal ions antimicrobial mechanisms. (Adapted from Ogunsona et al. (2020) with permission from Elsevier © 2020.)

have the capacity to infiltrate the cell and disturb biological processes (Stensberg et al. 2011). Metal ions or nanoparticles may also cause the release of ROS within the cell. Glutathione is oxidized as a result of oxidative stress, suppressing bacteria's antioxidant defence mechanism towards ROS. Metal ions will then interact with cellular structures (proteins, membranes and DNA), disrupting cell functions (Sánchez-López et al. 2020). Metal ions have the ability to form tight coordination bonds with organic compounds and N, O and S atoms of biomolecules. Metal-based nanoparticles have a broad spectrum of activity due to the non-specific interaction between metal ions and biomolecules (Yuan et al. 2018).

9.3.1.2 Types of Metals and Metal Oxides

9.3.1.2.1 Ag and Ag_2O Nanoparticles

Several studies have suggested that antimicrobial action relies heavily on silver ions (Dizaj et al. 2014; Durán et al. 2016; Snigdha et al. 2020). The surface area of the nanomaterial is a critical parameter in the antimicrobial toxicity. The highest concentration of released silver ions was found in silver nanoparticles with the largest surface area. Silver nanoparticles with a smaller surface area emitted least amount of silver ions, resulting in poor antimicrobial properties (Zawadzka et al. 2014). According to the literature, Ag nanoparticles' antibacterial activity is due to the damage to the bacteria's outer membrane (Lok et al. 2006). According to some researchers, Ag nanoparticles may cause pits and gaps in the bacterial membrane, causing the cell to fragment (Yun et al. 2013). Ag ions have also been shown to interact with the disulphide or sulphhydryl groups of enzymes, causing metabolic damage and, as a result, cell death (Egger et al. 2009).

The effect of size reduction on the antimicrobial effect of Ag nanoparticles was investigated by Jo et al. (2009). *Bipolaris sorokiniana* and *Magnaporthe grisea* were regulated using Ag nanoparticles. They also tested the effectiveness of Ag nanoparticles on a variety of pathogens, including soilborne fungi that rarely produce spores. Ag nanoparticles (20–30 nm) could better penetrate and colonize plant tissue, according to their findings. It was suggested that Ag nanoparticles have a lot of potential for regulating spore-producing fungal plant pathogens. In another study, Mie et al. (2014) used the disc diffusion method to measure the antibacterial activity of the synthesized Ag nanoparticles against eight microorganisms. According to their findings, Ag nanoparticles demonstrated possible antibacterial activity against gram-negative bacteria. As a result, the authors speculated that these synthesized Ag nanoparticles could be used in pharmaceutical and biomedical applications. Besides Ag nanoparticles, the antimicrobial activity of Ag_2O nanoparticles has also been discovered (Allahverdiyev et al. 2011). Metal oxide nanoparticles are thought to be a novel alternative to most antibiotics (Sathyanarayanan et al. 2013), having a great antimicrobial activity (Allahverdiyev et al. 2011). It is believed that metal oxide nanoparticles might be considered as a novel alternative to the most antibiotics (Sathyanarayanan et al. 2013).

9.3.1.2.2 ZnO Nanoparticles

The ZnO-based nanoparticles have shown potential bactericidal performance towards bacteria (gram-positive and gram-negative), as well as spores that are resistant to high strain and temperature. The antibacterial activity of ZnO nanoparticles over microparticles was attributed to the nanoparticles' increased surface area (Xie et al. 2011). Decreasing the particle size increased the bactericidal efficacy of ZnO nanoparticles. The antimicrobial activity of CuO, ZnO and Fe_2O_3 nanoparticles towards gram-negative bacteria (*E. coli* and *Pseudomonas aeruginosa*) and gram-positive bacteria (*S. aureus* and *Bacillus subtilis*) bacteria was examined by Azam et al. (2012). According to their observations, ZnO nanoparticles had the most bactericidal activity, whereas Fe_2O_3 nanoparticles had the least antimicrobial activity.

ZnO decreases the viability of bacteria; however, the exact mechanism of its antibacterial activity is still unknown. One hypothesis is that the production of hydrogen peroxide is a major element in antibacterial activity. Another cause of the antibacterial effect of ZnO particles is thought to be the aggregation of particles on the bacteria surface due to electrostatic forces (Zhang et al. 2008). In addition, ROS produced on particle surfaces, zinc ion release, membrane dysfunction and nanoparticle internalization could all be factors in cell damage (Ravindranadh and Mary 2013). The antibacterial activity of Ag nanoparticles was observed at ultra-low concentrations (Rai et al. 2012); however, the antibacterial activity of ZnO nanoparticles was dependent on concentration and surface area (Rai et al. 2012). As a result, higher concentrations of ZnO nanoparticles with a greater surface area showed better antibacterial activity. The antibacterial properties of ZnO nanoparticles against *Shigella dysenteriae* were investigated by Hosseinkhani et al. (2011). According to their findings, particle size reduction resulted in a significant reduction in the number of bacteria.

9.3.1.2.3 TiO_2 Nanoparticles

Antimicrobial property of TiO_2 is mainly associated with its shape, size and crystal morphology (Dizaj et al. 2014). It is suggested that oxidative stress caused by the production of ROS may be a particularly significant mechanism for TiO_2 nanoparticles (anatase). The ROS then induce DNA damage at a specific location (Haghighi et al. 2013). In a study, it was discovered that in the presence of TiO_2 nanoparticles, the antimicrobial resistance of MRSA (methicillin-resistant *Staphylococcus aureus*) to different antibiotics reduced (Dizaj et al. 2014). The antifungal effect of TiO_2 nanoparticles on fungal biofilms was investigated by Haghighi et al. (2013). According to their findings, the use of TiO_2 nanoparticles showed better antifungal efficiency on *C. albicans* biofilms immune to fluconazole. Furthermore, Haghighi et al. (2013) also suggested that TiO_2 could efficiently constrain the growth of fungal biofilms, especially those formed on the surface of medical devices. TiO_2 nanoparticles' photocatalytic properties aid in the effective eradication of bacteria. In fact, when exposed to UV light, TiO_2 nanoparticles produce ROS. Carré et al. (2014) hypothesized that the antibacterial activity was followed by lipid peroxidation, leading to an increased membrane fluidity and disrupted cell integrity.

The use of TiO_2 nanoparticles in UV light is limited because of their chance to cause cellular mutations in human cells and tissues (Dizaj et al. 2014). It has been demonstrated that doping TiO_2 nanoparticles with metal ions is a viable solution to this problem. Furthermore, doping TiO_2 nanoparticles with metal ions improves their antibacterial and photocatalytic activities substantially. Doping TiO_2 nanoparticles with metal ions changes their light absorption range to visible light, eliminating the need to irradiate them (Allahverdiyev et al. 2011). Furthermore, doping TiO_2 nanoparticles with metal ions improves their antibacterial and photocatalytic properties greatly in visible light, thus, eliminating the need to irradiate them with UV light.

9.3.1.2.4 Au Nanoparticles

The gold-based nanoparticles are regarded as highly beneficial in the development of antibacterial agents due to their non-toxicity, strong functionalization ability, polyvalent effects, easy detection and photothermal activity (Lokina and Narayanan 2013). Despite the fact that most antibiotics and antibacterial nanomaterials cause cellular death due to the formation of ROS, the antibacterial effect of Au nanoparticles does not involve any ROS-related activities (Cui et al. 2012). They observed that tRNA binding to the ribosome is hindered, causing a drop in ATP levels, and that tRNA binding to the ribosome is altered.

Tiwari et al. (2011) tested the antifungal and antibacterial properties of Au nanoparticles with 5-fluorouracil over *S. aureus*, *Micrococcus luteus*, *E. coli*,

Pseudomonas aeruginosa, *Aspergillus niger* and *Aspergillus fumigatus*. These nanoparticles showed more activity on gram-negative bacteria than on gram-positive bacteria. These nanoparticles were found to have antifungal activity against *A. niger* and *A. fumigatus*. Antibacterial activities of Au nanoparticles were tested against *E. coli* and *Bacillus Calmette-Guérin*. It has been claimed that Au nanoparticles had antibacterial activity against both gram-negative (*E. coli*) and gram-positive (*S. aureus*) bacteria. Kundu and Liang (2008) found that the self-assembly of Au nanoparticles functionalized with a tightly bound capping agent resulted in 45-micron-long chains. According to Zhou et al. (2014), these two methods enable delivering a huge number of Au nanoparticles to the bacterium cell wall much easier. Extra-aggregation of weakly bound capping agents such as citrate, on the other hand, lowers surface area and hence nanoparticle interactions.

9.3.1.2.5 Si and SiO_2 Nanoparticles

In recent years, there has been a lot of study into the usage of Si nanoparticles in conjunction with other biocidal metals such as Ag. Egger et al. (2009) focused on the creation of a new Ag-Si nanocomposite with improved antibacterial properties. The nanocomposite had a greater antibacterial impact against a wide spectrum of microbes than traditional materials such as silver nitrate and silver zeolite, according to their findings. In another study, Mukha et al. (2010) developed Au/SiO_2 and Ag/SiO_2 nanostructures and examined their antimicrobial activity. Their findings revealed that Ag/SiO_2 nanocomposites had enhanced antimicrobial capabilities against *E. coli*, *Candida albicans* and *S. aureus*, but Au/SiO_2 nanocomposites had no antibacterial activity against the pathogens mentioned. These nanocomposites may be used to disinfect water. It has been found elsewhere that Ag-decorated Si nanowires were biocompatible to the human lung adenocarcinoma epithelial cell line A549, however, Cu-decorated Si nanowires have shown cytotoxic effect. These findings demonstrate that combining Si compounds particularly nanocomposites, with metals such as Cu offers a lot of potential for the creation of antimicrobial agents. Si nanoparticles can also be utilized as antibacterial agents in biological applications since they are non-toxic.

9.3.1.2.6 Cu and CuO Nanoparticles

Cu nanoparticles are of great interest to scientists due to their special chemical, biological and physical properties, low cost of preparation and antimicrobial activities (Ahamed et al. 2014). Cu-chitosan nanoparticles (2–350 nm) were studied for their antimicrobial properties by Usman et al. (2013). These nanoparticles were tested for antibacterial and antifungal activities against methicillin-resistant *B. subtilis*, *S. aureus*, *Pseudomonas aeruginosa*, *Candida albicans* and *Salmonella choleraesuis*. Their findings suggested that these nanoparticles have a lot of potential as antimicrobial agents (Usman et al. 2013).

Cu nanoparticles, on the other hand, oxidize quickly when exposed to oxygen, limiting their use. CuO nanoparticles were screened for antibacterial activity against *Pseudomonas aeruginosa*, *Salmonella paratyphi*, *Klebsiella pneumoniae* and *Shigella strains* (Mahapatra et al. 2008). According to their findings, these nanoparticles had good antibacterial activity against the bacteria listed. The authors theorized that nanoparticles passing through the cell membrane of bacteria and then destroying the essential enzymes of the bacteria are the crucial factors that caused cell death. CuO nanoparticles have antibacterial activity that is size dependent. CuO nanoparticles were tested for antibacterial activity against two gram-positive bacteria (*S. aureus* and *B. subtilis*) and two gram-negative bacteria (*E. coli* and *Pseudomonas aeruginosa*). CuO nanoparticles were found to have inhibitory effects against both classes of bacteria, according to their findings. The authors concluded that the size, stability and concentration of these nanoparticles added to the growth medium influenced their bactericidal activity.

9.3.2 Semiconductor Photocatalysts

The basic mechanism of photocatalysis involves generation of electron–hole pairs by the induction of photons with energy either equal to or higher than the semiconductor's bandgap (Robertson et al. 2012). These two reactive entities are involved in redox reaction for the continuous production of suitable species in the process. The exciton photogeneration in direct and indirect semiconductors is a well-established process (Robertson et al. 2012). Once a photogenerated electron is produced, it requires an acceptor for further process. In environmental photocatalysis O_2 is assumed to produce a superoxide radical with exited electron while photogenerated holes generated OH when react with a water molecule. If the oxidation ends up with the generation of H_2O and CO_2 and eventually ionization of halides, then its compounds are said to be mineralized. Furthermore, inorganic compounds can be converted into less toxic or less dangerous compounds via photocatalysis (Litter et al. 2010) either by reduction (reduction with electron), or by oxidation (reaction with holes). The biological entities such as bacteria, viruses and macromolecules can be completely removed by disinfection (Benabbou et al. 2011; Robertson et al. 2012). Initially, in valence band a photogenerated hole is created, i.e. h+VB, while in conduction band, a photogenerated electron e-CB is formed.

$$\text{Photocatalyst} + h\nu \rightarrow h_{VB^+} + e_{CB^-}$$

Once h^+ VB reaches the targeted surface, then one of the two mechanisms can occur: (1) immediate oxidation of pollutants or (2) generation of $^{\bullet}OH$, at first, and then followed by the oxidation of pollutants. These two mechanisms are called direct and indirect oxidation of photocatalysis, respectively. In the first mechanism, a contact is established between the photocatalyst and the pollutant. The second mechanism is dominant in systems with weak pollutants.

Several studies were conducted to examine the role of experimental variables in photocatalytic treatment of microorganisms. There are several variables that may affect the process. These variables are the chemical nature of bacterial suspension medium, photocatalyst type, photocatalyst concentration, treatment duration and light intensity (Cushnie et al. 2009). The study concluded that the aforementioned parameters effect the microbial response to photocatalysis. The retention time of the process may depend upon the complexity of the cell wall of the organism. Microorganisms with a complex cell wall are more resistant to the treatment than simple cell wall microorganisms. Protozoa was found to be the most resistant microorganism to photocatalytic treatment, followed by bacterial spores, then mycobacteria, viruses, fungi and bacteria (Malato et al. 2009).

An extensive work was conducted to examine the exact mechanism of photocatalysis of microorganism; still, the exact mode of action has not been identified yet. Generally, in photocatalysis the generation of ROS plays an important role, especially the hydroxyl radical. The parent photocatalyst material, i.e. TiO_2, when reaching the microorganism, initially damages the cell wall, followed by the cytoplasmic membrane, which leads to the intercellular attack (Pablos et al. 2011). Goulhen-Chollet et al. (2009) assessed the mechanism of TiO_2 photocatalysts by adopting the biochemical approaches. The use of 1D and 2D sodium dodecyl sulphate–polyacrylamide gel electrophoresis illustrated that degradation of model and bacterial protein increase with the photocatalysis, decreases the cell activity and eventually end up with the death of cell. Nevertheless, the targeted protein was unable to be identified; thus, bacterial resistance to such treatment is impossible. Wu et al. (2009) investigated the morphology of *E. coli* treated with PdO/TiON photocatalyst and concluded that the photocatalysis damaged the cell membrane by generating holes and pits in cell walls.

The use of photocatalysis for the disinfection of microorganism in real wastewater will have different results as compared to laboratory experiments (Robertson et al. 2012). The presence of different kinds of pollutants such as organic and inorganic compounds, suspended solids and dissolved oxygen in wastewater may affect the efficiency of photocatalysis. Alrousan et al. (2009) utilized the immobilized TiO_2 nanoparticles film for the disinfection of *E. coli* in both distilled water and real wastewater. The study concluded that the presence of pollutants, especially humic acids, in real wastewater decreased the rate of photocatalytic disinfection of *E. coli.* Furthermore, bacteria found in the real wastewater were at different growth stages. Thus, bacterial growth factor should be given due importance during experimental design to make sure the maximum efficiency of the photocatalytic process can be achieved.

9.3.3 Carbon Nanostructures

Several studies reported the successful usage of carbon nanostructures (CNSs) in biological activities such as biomaterials, drug delivery, sensing and antibacterial agents (Bitounis et al. 2013).

9.3.3.1 Fullerene

Fullerene particles were used in biological activities by various researchers (Moor et al. 2016; Dostalova et al. 2016). A number of bactericidal mechanisms based on fullerene particles were observed; however, the powerful antibacterial activities of fullerenes and their derivatives against a wide range of microorganisms were observed in the presence of light (Chen et al. 2016), which may be due to the presence of unique structure of fullerene particles. Basically, fullerene nanoparticles have closed-cage structure in which conjugation is extended by π-electrons. The unique closed-cage structure may absorb the light, which may generate ROS (Kleandrova et al. 2015). The exposure of fullerenes particles to light may excite the C_{60} from ground state to a short-lived (~1.3 ns) state followed by quick decay to lower triplet state that may have longer lifetime (50–100 μs) (Sharma et al. 2011). After that, the fullerenes particles in the presence of molecular oxygen ($3O_2$) may generate ROS along with singlet ($1O_2$) and superoxide anion (O_2^-) via energy transfer photochemical and electron transfer pathways, respectively. Generally, the peroxidation of eukaryotic lipid and interruption of cell membrane occurred due to ROS (Ishaq et al. 2015; Chen et al. 2016). The availability of ROS at high level may lead to acute death of microorganism (Zhou et al. 2016) by enhancing cellular damage such as proteins, nucleic acid and lipids (Prasad et al. 2017). However, the fullerenes particles may act as antioxidants in the absence of light, avoiding peroxidation of lipids caused by superoxide and hydroxyl radicals (Navarro et al. 2008).

The fullerene NPs can disrupt and/or cleave the DNA while interacting with the outer surface of the microbial membrane. This may be due to NPs' high surface hydrophobicity, which can lead to easy interaction with membrane lipids (Grinholc et al. 2015). However, various bacterial species have different cell wall components; thus, the interaction of fullerene particles with cells may vary from species to species. It was observed that these NPs are biologically more active against gram-positive rather than to gram-negative species, suggesting that bactericidal activity of these NPs is mainly based on penetration in to bacterial cell wall. (Markovic and Trajkovic 2008). The study on *P. putida* demonstrated the increase in cyclopropane with the reduction in unsaturated fatty acid on exposure to fullerene NPs. The findings show that fullerene bioactivity may degrade in microorganisms through cell wall degradation, specifically alter the membrane permeability and lipid structure. (Fang et al. 2007).

9.3.3.2 Carbon Nanotubes (CNTs)

Kang et al. (2007) conducted the preliminary study and reported the antimicrobial performance of single-walled carbon nanotubes (SWCNTs) against *E. coli* pathogen. Later on, various researches were carried out to examine the antimicrobial performance of multi-walled carbon nanotubes (MWCNTs) and SWCNTs against microorganisms (Chung et al. 2011). The experimental data show that toxicologically SWCNTs are more effective as compared to MWCNTs and even the furellene-C_{60} (Oyelami and Semple 2015). CNTs are an incredible material and have high antibacterial activity, but still their bactericidal mechanism is yet to be explored.

Several factors are involved in the toxicity mechanism of CNTs. These factors are length, diameter, catalyst residue, surface chemistry and functional groups, electronic structure and CNT coating (Jackson et al. 2013). Nanotube length, in particular, has a significant role in the interaction with cell membrane. The tube with shorter length showed more bactericidal activities as compared to longer ones (Kang et al. 2008). Shorter-length tubes may enhance the chance of interaction between microorganisms and nanotubes, which resulted in the extensive damage of the cell membrane (Aslan et al. 2010). The study illustrated that once the length of MWCNTs reaches 50 µm, they wrapped around the microorganism and resulted in the constant promotion of osmotic lysis of microorganisms (Chen et al. 2013). However, in liquid medium longer-length nanotubes are more effective as compared to shorter ones. The shorter-length nanotubes prefer self-aggregation without involving a large number of molecules, while larger-length nanotubes aggregate more molecules and are thus more bio-effective than shorter-length nanotubes (Yang et al. 2010). Nevertheless, the agglomeration/aggregation of CNTs is inevitable due to their potential van der Waals interactions and unique configuration (Saifuddin et al. 2013).

Furthermore, CNTs' toxic effect during biofilm formation was studied to examine their potential to inhibit the microorganism proliferation and attachment at various stages of the bacterial colonization. Biofilm structure may enhance the bacterial cell protection and increase their resistance against the harmful NPs and physical forces (Bazaka and Bazaka 2014). The microscopic examination of *B. subtilis* and *E. coli* biofilms demonstrated that ~80%–90% microbes were found dead, which were in contact with the coated SWCNTs (Ahmed et al. 2012). However, the CNTs' biofilm interaction is highly dependent on the stage of the biofilm and CNTs' efficiency is high at the early stage of biofilm formation (Dong and Yang 2014), while the microorganism's susceptibility to CNTs reduces as the biofilm gets mature (Rodrigues and Elimelech 2010).

9.3.3.3 Graphene

Graphene as an antibactericidal agent has extensively been examined by various researchers. It includes graphite (Gt), graphene nanosheets (GNS), graphene oxide (GO), pristine graphene, multilayer graphene (MLG) and reduced graphene oxide (RGO). The results show that under the similar experimental conditions, GO has more effective antibacterial activities towards *P. aeruginosa*, followed by RGO and Gt (Gurunathan et al. 2012). However, the diverse properties of various graphene materials, such as the number of layers, sheet size, shape, defect density, hydrophilicity and corrugation, make the examination of actual antimicrobial mechanisms more challenging (Zhu et al. 2017). To understand the interaction mechanism of nanomaterials with microorganisms, various methods have been observed.

Experimental approaches and theoretical simulations both revealed that graphene may physically damage the microorganism by two possible ways: by intensive insertion followed by cell membrane cutting and phospholipids extraction from lipid membrane (Zhou and Gao 2014). The molecular dynamics simulation defined the insertion of graphene sheet, which started when the sheet vibrates at the rate of 10–100 ns by adopting the back and forth motion (Tu et al. 2013). The insertion may occur into both inner and outer membranes. The end result will be the damage of cell membrane due to the strong van der Waals interactions along with hydrophobic and lipids effects. The phospholipids extraction may generate thin lipid bilayer and alteration of membrane due to the strong dragging force of graphene sheet, resulting in the severe damage of living system.

9.3.4 Polymeric Materials

Among polymeric materials, the peptides (AMPs) are a type of polymeric substance that can be used to combat multidrug-resistant infections while also reducing the risk of generating new drug-resistant pathogen strains owing to the physical nature of membrane rupture (Boman et al. 1983). Synthetic macromolecules are emerging antimicrobial components with the potential to combat with multidrug-resistant microbes (Engler et al. 2012). AMPs were used to explore the first macromolecular system. Boman et al. (1983) explored the AMPs and described how multicellular organisms defend themselves naturally from the infectious attack of pathogens. Basically, AMPs were discovered from the immune system of silk moths (*Hyalophora cecropia*). The study concluded that these AMPs are able to kill a variety of microbial pathogens, which may include fungi and gram-positive and gram-negative bacteria. As the pioneer work was carried out on these AMPs, an extensive antimicrobial peptide database has been developed on the basis of the extensive literature (Wang and Wang 2004). AMPs are commonly amphiphilic and share mutual features such as nonpolar amino acid, which create the hydrophobic environment (tryptophan) along with lysine, a cationic amino acid. The establishment of such structures led to the electrostatic interaction of AMPs' cationic charge with the anionic charge of the microbial membrane. Upon interaction, AMPs generate the secondary structure, allowing hydrophobic component insertion into membrane lipid domains, which may disrupt the membrane structure. The possible modes action of polymeric materials towards microbes are as follows.

9.3.4.1 Polymers with Antimicrobial Activity

The antimicrobial activities were displayed by their name themselves. Generally, the chemical structure of these materials was used as the key characteristic for their categories. Polymers were found in various categories such as guanidine-containing polymers, polymers mimicking natural peptides (arylamide, phenylene ethynylene backbone, polynorbornene derivatives, halogen polymers and synthetic peptides) and organometallic polymers.

9.3.4.2 Polymers that Undergo Chemical Modifications to Achieve Antimicrobial Activity

Antimicrobial activity can be incorporated into polymers in a variety of ways. However, it is preferable that chemical alteration does not deteriorate the final polymeric material properties. If chemical alteration is involved, different scenarios can be possible:

1. A polymer is covalently bound to a small molecule having antimicrobial activity.
2. Antimicrobial peptides are attached to a non-functional polymer.
3. Standard polymers are grafted with antimicrobial polymers.
4. Polymers associated with antimicrobial organic compounds: Antimicrobial activity is caused by noncovalent bonds between antimicrobial agents, either natural or synthetic, and polymers, resulting in the release of the corresponding compound.
5. Polymers with antimicrobial inorganic compounds: The biocidal action of inorganic systems, such as metals, metallic oxides or modified clays, is transferred into the polymers to produce antimicrobial activity in the final product.

9.3.5 Bio-Based Nanomaterials

The use of biomass such as the use of plant extract for the synthesis of antimicrobial nanoparticles, is referred to as bio-based materials and process is called green synthesis. . Due to its high strength, availability and sustainability, agricultural waste has a lot of potential to be used in nanocomposites. Agricultural waste's potential has triggered a lot of research into how to use it as a raw material for biocomposite polymer formulation. Agriculture waste biomass obtained from rice, pineapple, oil palm, coconut, banana, and other vegetables and fruits can be used in biocomposites. Lignocellulose is made up of cellulose, hemicellulose and lignin and is a major component of agricultural waste. It also serves as a major source of natural fibres, which can be used as an alternate raw material in biocomposites (Dungani et al. 2017). These bio-nanocomposites can be used in variety of applications including disinfection of water from pathogens. For instance, nanocrystalline cellulose is a new bio-nanomaterial made from cellulosic materials such as agro-food waste. Due to its special properties such as large surface area ($250\,m^2/g$), high stiffness (167.5 GPa), tensile strength (7500 MPa) and large amount of surface hydroxyl groups, it can be used in wastewater disinfection (Teh et al. 2019).

Polysaccharide-based bioflocculants have received a lot of attention in recent years because they are biodegradable, are non-toxic and cause very little secondary pollution. Manivasagan et al. (2015) reported the development of a polysaccharide-based bioflocculant through green synthesis of Ag nanoparticles by *Streptomyces* sp. MBRC-91. The biosynthesized Ag nanoparticles demonstrated high antibacterial activity in water, which could open up new possibilities for wastewater treatment. As a result, the introduction of innovative bactericidal bio-nanomaterials for water treatment and biotechnological applications can be extended using biosynthesized Ag nanoparticles.

Seaweeds and marine algae can also be used for the synthesis of antimicrobial biocomposites. They are considered as a constant source of chemical compounds, including a variety of biologically active secondary metabolites (Vikneshan et al. 2020). *Ulva lactuca* is a green marine algae (Chlorophyta) that is nutritious and has antibacterial properties against various pathogens (Sujatha et al. 2012). It is non-toxic and has a high level of activity. Ulvan is referred to as a sulphated polysaccharide obtained from algae *Ulva spp.* and is responsible for the antibacterial activity of the algae.

9.3.6 Aerogels

Aerogels are materials made up of 95%–99% air, have a large surface area and have a very low density (De France et al. 2017). Aerogels are porous materials with a low density and a large specific surface area and are fabricated through supercritical and freeze-drying techniques (Darabitabar et al. 2020). Lin et al. (2020) prepared a composite film structure of chitosan/okra powder/nanosilicon aerogel and investigated its structural characterization and physicochemical properties. The composite film had excellent mechanical, barrier and optical properties, according to the findings. The composite film is extremely flexible. In terms of antibacterial activity, the composite films were effective against gram-negative (*E. coli*) and gram-positive (*S. aureus*) bacteria, with an inhibition zone of 551.96 mm^2 for *E. coli* and 350.29 mm^2 for *S. aureus*, respectively.

Other studies have shown that the chitosan–gelatin combinations can produce aerogels with a variety of properties, including high surface area availability, porosity, density and structure (Alessandro Kovtun et al. 2020). The reaction between carboxylic and amino groups and cross-linked structure has shown the establishment of permanent covalent bonds. Chitosan enhances mechanical properties and prevents degradation in water (Campodoni et al. 2019). Furthermore, because of the exposed amino groups, chitosan–gelatin aerogels can react with carboxylic and epoxide groups of graphene (Poletti et al. 2020), resulting in stable composites that are attracting a huge interest.

Bingjun et al. (2015) developed a chitosan–nanosilicon aerogel film that inhibited *E. coli* and *S. aureus*, with S. aureus experiencing a superior inhibitory effect. The antibacterial effect of the composite film grew as the chitosan–nanosilicon aerogel content grew, possibly due to an electrostatic interaction between the positively charged chitosan molecule and the negatively charged bacterial cell membrane, causing changes and damage to the bacterial surface. This led to the inhibition of bacterial metabolism and death of bacterial cells. Kadam et al. (2019) used chitosan and silver nanoparticles to develop a composite film that showed a maximum inhibition zone of 225 and 400 mm^2 against *E. coli* and *S. aureus*, respectively. Wang et al. (2015) found the greatest inhibitory zone after adding nanoscale silver to the chitosan–gelatin composite film.

9.4 NANOMATERIAL-BASED PROCESSES FOR WATER DISINFECTION

In the purification of water, nanomaterials can be used in a variety of ways. There are three major processes through which nanoparticles have shown promising results at a laboratory scale in terms of antimicrobial applications. These are as follows:

i. Photocatalysis: nanomaterial disinfection of water activated by light sources.
ii. Direct application: ion leaching nanomaterials for direct antimicrobial activity.
iii. Membrane filtration: development of filtration membranes of antimicrobial nanoparticles.

These applications differ in mechanisms of action using different types of nanoparticles. In the following sections, the different approaches will be described in relation to the reported literature (Ogunsona et al. 2020).

9.4.1 Photocatalysis

Photocatalysis is a promising technology that generates highly oxidative species that kill bacteria and removes a wide spectrum of chemical contaminants by interacting light with solid semiconductor particles. A few of the most used photocatalysts for water disinfection include TiO_2, SiO_2, Cr_2S_3, NiS, Fe_2O_3, ZnO, WO_3, NiS and CdSe. TiO_2 is the widely used material for photocatalysis due to its higher surface activity, low cost, non-toxicity, high stability, wide availability and high photocatalytic performance (Benhabiles et al., 2019). The photocatalytic antimicrobial activity of TiO_2 has extensively been studied in a variety of species, including viruses, bacteria, fungi, algae and cancer cells. This process, which can be used in both fresh and salt waters, involves adding nanoparticles directly to polluted water while exposing it to a light source. Although TiO_2 nanoparticles have extensively been studied in this field and are a leading alternative for photocatalytic reactions, with numerous forms commercially available, they, however, have a number of drawbacks considering their immense potential. Low-energy absorption from visible light sources (they are predominantly active under UV irradiation), difficulty in uniform dispersion and challenges in retrieving nanoparticles after water treatment are all issues that limit their industrial use for water disinfection (Dong et al. 2015). Improvement in antimicrobial efficiency under visible/solar light can be achieved by improving photocatalytic properties with effectively doped TiO_2 particles. The use of solar light in the disinfection phase can be a cost-effective way to save energy while maintaining high disinfection standards. For instance, Wang et al. (2014) developed Zn/TiO_2 nanomaterials using sol–gel method at 500°C and tested their antibacterial activity under visible light. They found a strong *C. albicans* antibacterial effect, and *E. coli* and *S. aureus* were also removed to some extent.

9.4.1.1 Mechanisms

The nanocatalyst is applied directly to the water solution while photocatalysis. After being triggered by light, the nanoparticles may catalyse oxidation–reduction processes directly on biomolecules or generate ROS. Biological macromolecules, such as cellular membranes and DNA, can be oxidized by ROS, which can lead to cell death. When photons are absorbed by the electronic band structure of nanoparticles, electrons on the surface of nanoparticles are excited, allowing them to drive oxidation–reduction processes in the surrounding molecules. Metals are chemical elements having electronic band structures that have been employed in water treatment as photoactivated catalysts with biocidal action.

9.4.2 Nanomembranes

Nanofiltration is another leading method of water purification and microbial removal (Ogunsona et al. 2020). Water with high microbial concentration can be purified using filtration membranes. Since the 1960s, the use of polymer-based membranes has been a dominant water treatment technology producing high-quality water at a low cost (Tang et al. 2016). However, the filtration operations are susceptible to the development of biofilms on the membrane surface over time, undermining the efficiency of the membranes (Ogunsona et al. 2020). The intrinsic hydrophobic polymer membrane is easily fouled due to the accumulation of bacteria and pollutants on the surface and in the pores of the membrane, which becomes a constraint limiting the wide application of membranes, as it can cause a variety of negative consequences, including poorer water quality, higher energy consumption, shorter membrane lifespan and secondary pollution (Tang et al. 2016). Despite their effectiveness, filtration processes are subject to biofilm growth on the surface of membranes over time, thus limiting their function. Therefore, a strategy is required for preventing microorganism proliferation over the filtration membranes. Decorating the surfaces of the membranes with nanoparticles that have antimicrobial activity is one way that can help in this regard. Such methods allow for both filtration and disinfection of water to be done at the same time.

Nanoparticles membranes can be developed by two ways: first by synthesizing them followed by binding them to the membranes, and second by forming them in situ over the membrane surface. Blending, dip coating, grafting polymerization and interfacial polymerization are some of the approaches that can be used for incorporating nanoparticles onto polymeric membranes (Zhu et al. 2018). Graft polymerization is the most advantageous of these methods because it can covalently bond nanoparticles to filtration membranes, preventing their loss during operation while also providing the desired biocide activity. Metal oxides are the well-studied antimicrobial nanoparticles that can be incorporated into polymeric filtration membranes (Mukherjee and De 2018). Recently, Xie et al. (2020) have developed a CNT-polyacrylonitrile/polyurethane/polyaniline composite electrospun nanofibre membrane by co-electrospinning and electrochemical filtration. Upon simultaneous electrolysis with filtration, the inactivation of bacteria is substantially increased, where after 20 m of applying 3.0 V, bacteria were fully inactivated.

9.4.2.1 Mechanisms

To control the biofouling of membranes and removal of bacteria, different mechanisms have been proposed. Carbonaceous nanomaterials have antimicrobial activity primarily by physical abrasion, which stems from their morphological structures (sharp edges and tips) that are similar in size to bacteria. The production of ROS has also been documented as a disinfection mechanism of these materials. CNTs and graphene oxides have been deposited onto polymeric filtration membranes to reduce biofouling (Aslam et al. 2018). It has been proposed that the vertical alignment of GO nanosheets could cause potential mechanical damage to microorganisms through increased exposure to the sharp edges of nanosheets (Lu et al. 2018). In case of electrochemical disinfection, it is primarily based on two mechanisms. For instance, pathogens in direct contact with the CNT-polymer anode underwent a multi-electron transfer process, while pathogens that formed aqueous oxidants undergo indirect oxidation through the anode (Xie et al. 2020).

9.4.3 Direct Application

Nanoparticles having capacity to produce soluble ions in water can be applied for the disinfection of water. Metal nanoparticles are typically applied directly to the polluted water, resulting in a continuous release of metal ions into the water during the process time (Ogunsona et al. 2020). In this process, the antimicrobial property of nanoparticles has been attributed to their small size and large surface-to-volume ratio, which make them interact directly with microbial membranes instead of through solutions (Gold et al. 2018). ZnO nanoparticles have attracted a lot of attention because of their intriguing chemical and physical properties (Mostafa and Menazea 2020). TiO_2 powder is commonly regarded as one of the most important semiconductors that can also be used in direct application process. Because of the nanosize, TiO_2 NPs have many special properties such as large surface area (Menazeaa and Awwad 2020). The other antimicrobial materials related to Cu, Ag, Fe and Zn may also be used in direct application process.

9.4.3.1 Mechanisms

The dissolved metallic ions interfere with microorganisms in a number of ways. However, the actual biocidal mechanisms of action of nanoparticles and ions in antimicrobial water treatment are still unexplained. Several theories have been suggested, such as adsorption, membrane attachment, interaction with biomolecules and oxidative removal (Gold et al. 2018). For instance, in case of Ag nanoparticles the mechanism of action against bacteria has recently been published. According to that, Ag nanoparticles and ions can penetrate bacterial membranes and invade the cytosol. Upon entry, they cause oxidative damage to bio-macromolecules by direct adsorption on DNA functional group, enzymes and other essential biomolecules, as well as by the generation of ROS (Gallon et al. 2019). Considering the fact that various nanoparticles have shown similar mechanisms of action, however, Ag nanoparticles have a broad activity on a variety of microorganisms, including both bacteria and viruses; however, the other nanoparticles, on the other hand, do not have this property (Rajeshkumar et al. 2019).

9.5 LIMITATIONS AND FUTURE PERSPECTIVES

The use of nanomaterials for removal of microorganisms have various limitation which are summarized in Figure 9.4 and explained in the following section.

9.5.1 Environment

If these nanomaterials are released into the natural environment, they can destroy the microbes beneficial to or even necessary for the ecosystem. Antimicrobial nanomaterials in high concentrations in drinking water can also be harmful to human health. As a result, it's important to keep the antimicrobial nanomaterials in the environment. More studies are needed to fully comprehend the dangers they pose to the ecosystem and the flora and fauna in the vicinity of water treatment plants. Since these are nanoscale particles, the implementation of risk management techniques is challenging. As a result, more research is needed to ensure their protection while making it cost efficient and environmentally sustainable.

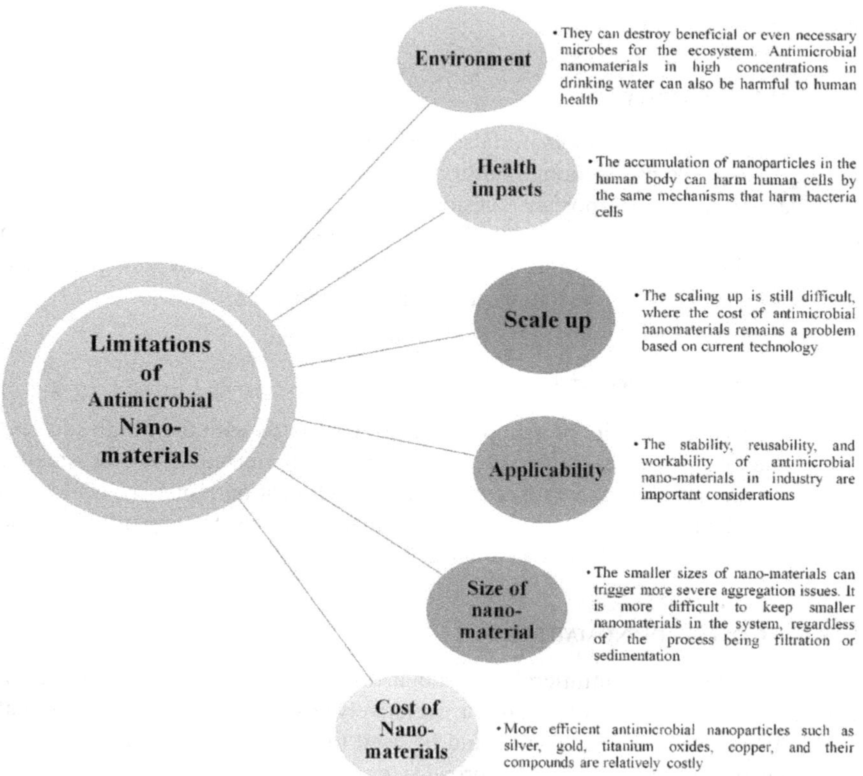

FIGURE 9.4 Impacts of antimicrobial nanomaterials.

9.5.2 Health Impacts

Antimicrobial nanomaterials are excellent defensive tools for preventing water-related epidemics, but they may have significant toxicity effects on the ecosystem and human cells. Furthermore, the accumulation of nanoparticles in the human body can harm human cells by the same mechanisms that harm bacteria cells and DNA (Hossain and colleagues, 2014). More than 400 toxicity and ecotoxicity studies on nanomaterials have been reported.

9.5.3 Scale-Up

The synthesis of nanomaterials typically involves many complex processes; thus, the scaling up is still difficult, where the cost of antimicrobial nanomaterials remains a problem based on the current technology. Water is essential for daily life, but most antimicrobial nanomaterials are still too expensive to use on a daily basis. As previously mentioned, the most commercialized ceramic filter costs about US$ 25 per unit. However, this is still too costly for people living in certain developing countries who do not have access to clean drinking water.

9.5.4 Applicability

Commercialization of nanomaterials should be cost-effective without compromising large-scale production. Stability, reusability and workability in industrial applications are also important considerations. In vitro and in vivo toxicity of nanomaterials must be measured in the short and long terms. Furthermore, due to the uncertain fate and transport of the particles in the aquifer, intentional applications for groundwater remediation face disposal challenges.

9.5.5 Size of Nanomaterials

The antimicrobial activity of nanomaterials is associated with their high surface-to-volume ratio. As a result, smaller sizes would be preferred. However, smaller sizes of nanomaterials can trigger more severe aggregation issues. It would also be more difficult to keep smaller nanomaterials in the system, regardless of the process being filtration or sedimentation.

9.5.6 Cost of Nanomaterials

Despite the fact that antimicrobial nanoparticles such as silver, gold, titanium oxides, copper and their compounds are relatively costly, it has been observed that their efficiencies are high. They could be used to stop the growth of bacteria at very low concentrations and over a broad size range, thereby reducing the process costs by minimizing material use.

9.6 CONCLUSIONS

The use of nanotechnology to target and engineer nanoparticles in water disinfection applications has resulted in the translation of basic research from the literature to functional and commercially usable materials. For wastewater treatment, the majority of nanotechnology techniques are very promising. However, incorporating nanotechnology into a large-scale wastewater treatment unit is difficult. The majority of the applications discussed here are in vitro and based on laboratory experiments. Since the application is related to water management, extra caution is needed during its implementation to ensure environmental safety and protection. In addition, they must also be made more cost-effective and superior to the current water treatment technologies. Most of the nanomaterial disinfection techniques described in this chapter are not yet cost-effective. However, focusing on the reusability of nanomaterials will help reduce the operating costs of water treatment plants.

REFERENCES

Abdel-Kareem, M. M., & Zohri, A. A. (2018). Extracellular mycosynthesis of gold nanoparticles using Trichoderma hamatum: optimization, characterization and antimicrobial activity. *Letters in Applied Microbiology*, 67(5), 465–475.

Ahamed, M., Alhadlaq, H. A., Khan, M. A., Karuppiah, P., & Al-Dhabi, N. A. (2014). Synthesis, characterization, and antimicrobial activity of copper oxide nanoparticles. *Journal of Nanomaterials*, 2014.

Ahmed, F., Santos, C. M., Vergara, R. A. M. V., Tria, M. C. R., Advincula, R., & Rodrigues, D. F. (2012). Antimicrobial applications of electroactive PVK-SWNT nanocomposites. *Environmental Science & Technology*, 46(3), 1804–1810.

Allahverdiyev, A. M., Abamor, E. S., Bagirova, M., & Rafailovich, M. (2011). Antimicrobial effects of TiO2 and Ag2O nanoparticles against drug-resistant bacteria and leishmania parasites. *Future Microbiology*, 6(8), 933–940.

Alrousan, D. M., Dunlop, P. S., McMurray, T. A., & Byrne, J. A. (2009). Photocatalytic inactivation of E. coli in surface water using immobilised nanoparticle TiO_2 films. *Water Research*, 43(1), 47–54.

Aslam, M., Ahmad, R., & Kim, J. (2018). Recent developments in biofouling control in membrane bioreactors for domestic wastewater treatment. *Separation and Purification Technology*, 206, 297–315.

Aslan, S., Loebick, C. Z., Kang, S., Elimelech, M., Pfefferle, L. D., & Van Tassel, P. R. (2010). Antimicrobial biomaterials based on carbon nanotubes dispersed in poly (lactic-co-glycolic acid). *Nanoscale*, 2(9), 1789–1794.

Azam, A., Ahmed, A. S., Oves, M., Khan, M. S., Habib, S. S., & Memic, A. (2012). Antimicrobial activity of metal oxide nanoparticles against Gram-positive and Gram-negative bacteria: a comparative study. *International Journal of Nanomedicine*, 7, 6003.

Bazaka, O., & Bazaka, K. (2014). Surface modification of biomaterials for biofilm control. *Biomaterials and Medical Device - Associated Infections*, 6, 103–132.

Benabbou, A. K., Guillard, C., Pigeot-Rémy, S., Cantau, C., Pigot, T., Lejeune, P., & Lacombe, S. (2011). Water disinfection using photosensitizers supported on silica. *Journal of Photochemistry and Photobiology A: Chemistry*, 219(1), 101–108.

Benhabiles, O., Galiano, F., Marino, T., Mahmoudi, H., Lounici, H., & Figoli, A. (2019). Preparation and characterization of TiO_2-PVDF/PMMA blend membranes using an alternative non-toxic solvent for UF/MF and photocatalytic application. *Molecules*, 24(4), 724.

Bingjun, Q., Jung, J., & Zhao, Y. (2015). Impact of acidity and metal ion on the antibacterial activity and mechanisms of β-and α-chitosan. *Applied Biochemistry and Biotechnology*, 175(6), 2972–2985.

Bitounis, D., Ali-Boucetta, H., Hong, B. H., Min, D. H., & Kostarelos, K. (2013). Prospects and challenges of graphene in biomedical applications. *Advanced Materials*, 25(16), 2258–2268.

Boman, B. M., Zschunke, M. A., & Scott, R. E. (1983). Cell cycle and cyclic AMP-dependent phosphorylation of plasma membrane proteins p14 and p24: Defects in smooth surface transformed cells. *Journal of Cellular Biochemistry*, 23(1–4), 203–209.

Bruslind, L. (2017) Bacteria: internal components. In *Microbiology*, Open Oregon Stat, Corvallis, OR.

Campodoni, E., Heggset, E. B., Rashad, A., Ramírez-Rodríguez, G. B., Mustafa, K., Syverud, K., Tampieri, A., & Sandri, M. (2019). Polymeric 3D scaffolds for tissue regeneration: Evaluation of biopolymer nanocomposite reinforced with cellulose nanofibrils. *Materials Science and Engineering: C*, 94, 867–878.

Carré, G., Hamon, E., Ennahar, S., Estner, M., Lett, M. C., Horvatovich, P., & Andre, P. (2014). TiO_2 photocatalysis damages lipids and proteins in Escherichia coli. *Applied and Environmental Microbiology*, 80(8), 2573–2581.

Chen, H., Wang, B., Gao, D., Guan, M., Zheng, L., Ouyang, H., & Feng, W. (2013). Broad-spectrum antibacterial activity of carbon nanotubes to human gut bacteria. *Small*, 9(16), 2735–2746.

Chen, Q., Ma, Z., Liu, G., Wei, H., & Xie, X. (2016). Antibacterial activity of cationic cyclen-functionalized fullerene derivatives: membrane stress. *Digest Journal of Nanomaterials and Biostructures (DJNB)*, 11, 753–761.

Cheriyamundath, S., & Vavilala, S. L. (2021). Nanotechnology-based wastewater treatment. *Water and Environment Journal*, 35(1), 123–132.

Chung, H., Son, Y., Yoon, T. K., Kim, S., & Kim, W. (2011). The effect of multi-walled carbon nanotubes on soil microbial activity. *Ecotoxicology and Environmental Safety*, 74(4), 569–575.

Crini, G., & Lichtfouse, E. (2019). Advantages and disadvantages of techniques used for wastewater treatment. *Environmental Chemistry Letters*, 17(1), 145–155.

Cui, Y., Zhao, Y., Tian, Y., Zhang, W., Lü, X., & Jiang, X. (2012). The molecular mechanism of action of bactericidal gold nanoparticles on Escherichia coli. *Biomaterials*, 33(7), 2327–2333.

Cushnie, T. T., Robertson, P. K., Officer, S., Pollard, P. M., McCullagh, C., & Robertson, J. M. (2009). Variables to be considered when assessing the photocatalytic destruction of bacterial pathogens. *Chemosphere*, 74(10), 1374–1378.

Darabitabar, F., Yavari, V., Hedayati, A., Zakeri, M., & Yousefi, H. (2020). Novel cellulose nanofiber aerogel for aquaculture wastewater treatment. *Environmental Technology & Innovation*, 18, 100786.

De France, K. J., Hoare, T., & Cranston, E. D. (2017). Review of hydrogels and aerogels containing nanocellulose. *Chemistry of Materials*, 29(11), 4609–4631.

Dhapte, V., Kadam, S., Pokharkar, V., Khanna, P. K., & Dhapte, V. (2014). Versatile SiO2 nanoparticles@ polymer composites with pragmatic properties. *International Scholarly Research Notices*, 2014.

Dizaj, S. M., Lotfipour, F., Barzegar-Jalali, M., Zarrintan, M. H., & Adibkia, K. (2014). Antimicrobial activity of the metals and metal oxide nanoparticles. *Materials Science and Engineering: C*, 44, 278–284.

Dong, H., Zeng, G., Tang, L., Fan, C., Zhang, C., He, X., & He, Y. (2015). An overview on limitations of TiO2-based particles for photocatalytic degradation of organic pollutants and the corresponding countermeasures. *Water Research*, 79, 128–146.

Dong, X., & Yang, L. (2014). Inhibitory effects of single-walled carbon nanotubes on biofilm formation from Bacillus anthracis spores. *Biofouling*, 30(10), 1165–1174.

Dostalova, S., Moulick, A., Milosavljevic, V., Guran, R., Kominkova, M., Cihalova, K., & Kizek, R. (2016). Antiviral activity of fullerene C 60 nanocrystals modified with derivatives of anionic antimicrobial peptide maximin H5. *Monatshefte für Chemie-Chemical Monthly*, 147(5), 905–918.

Dungani, R., Abdul Khalil, H. P. S., Aprilia, N. A. S., Sumardi, I., Aditiawati, P., Darwis, A., Karliati, T., Sulaeman, A., Rosamah, E., & Riza, M., (2017). Bionanomaterial from agricultural waste and its application. In *Cellulose-Reinforced Nanofibre Composites* (pp. 45–88). Woodhead Publishing.

Durán, N., Durán, M., De Jesus, M. B., Seabra, A. B., Fávaro, W. J., & Nakazato, G. (2016). Silver nanoparticles: A new view on mechanistic aspects on antimicrobial activity. *Nanomedicine: Nanotechnology, Biology and Medicine*, 12(3), 789–799.

Egger, S., Lehmann, R. P., Height, M. J., Loessner, M. J., & Schuppler, M. (2009). Antimicrobial properties of a novel silver-silica nanocomposite material. *Applied and Environmental Microbiology*, 75(9), 2973–2976.

Engler, A. C., Wiradharma, N., Ong, Z. Y., Coady, D. J., Hedrick, J. L., & Yang, Y. Y. (2012). Emerging trends in macromolecular antimicrobials to fight multi-drug-resistant infections. *Nano Today*, 7(3), 201–222.

Fang, J., Lyon, D. Y., Wiesner, M. R., Dong, J., & Alvarez, P. J. (2007). Effect of a fullerene water suspension on bacterial phospholipids and membrane phase behavior. *Environmental Science & Technology*, 41(7), 2636–2642.

Gallón, S. M. N., Alpaslan, E., Wang, M., Larese-Casanova, P., Londoño, M. E., Atehortúa, L., Pavón, J. J., & Webster, T. J. (2019). Characterization and study of the antibacterial mechanisms of silver nanoparticles prepared with microalgal exopolysaccharides. *Materials Science and Engineering: C*, 99, 685–695.

Gold, K., Slay, B., Knackstedt, M., & Gaharwar, A. K., 2018. Antimicrobial activity of metal and metal-oxide based nanoparticles. *Advanced Therapeutics*, 1(3), 1700033.

Goulhen-Chollet, F., Josset, S., Keller, N., Keller, V., & Lett, M. C. (2009). Monitoring the bactericidal effect of UV-A photocatalysis: a first approach through 1D and 2D protein electrophoresis. *Catalysis Today*, 147(3–4), 169–172.

Grinholc, M., Nakonieczna, J., Fila, G., Taraszkiewicz, A., Kawiak, A., Szewczyk, G., ... & Bielawski, K. P. (2015). Antimicrobial photodynamic therapy with fulleropyrrolidine: photoinactivation mechanism of Staphylococcus aureus, in vitro and in vivo studies. *Applied Microbiology and Biotechnology*, 99(9), 4031–4043.

Gu, H., Chen, X., Chen, F., Zhou, X., & Parsaee, Z. (2018). Ultrasound-assisted biosynthesis of CuO-NPs using brown alga Cystoseira trinodis: Characterization, photocatalytic AOP, DPPH scavenging and antibacterial investigations. *Ultrasonics Sonochemistry*, 41, 109–119.

Gurunathan, S., Han, J. W., Dayem, A. A., Eppakayala, V., & Kim, J. H. (2012). Oxidative stress-mediated antibacterial activity of graphene oxide and reduced graphene oxide in Pseudomonas aeruginosa. *International Journal of Nanomedicine*, 7, 5901.

Haghighi, F., Roudbar Mohammadi, S., Mohammadi, P., Hosseinkhani, S., & Shipour, R. (2013). Antifungal activity of TiO_2 nanoparticles and EDTA on Candida albicans biofilms. *Infection, Epidemiology and Microbiology*, 1(1), 33–38.

Hegab, H. M., ElMekawy, A., Zou, L., Mulcahy, D., Saint, C. P., & Ginic-Markovic, M. (2016). The controversial antibacterial activity of graphene-based materials. *Carbon*, 105, 362–376.

Hossain, F., Perales-Perez, O. J., Hwang, S., & Román, F., (2014). Antimicrobial nanomaterials as water disinfectant: applications, limitations and future perspectives. *Science of the Total Environment*, 466, 1047–1059.

Hosseinkhani, P., Zand, A. M., Imani, S., Rezayi, M., & Rezaei, Z. S. (2011). Determining the antibacterial effect of ZnO nanoparticle against the pathogenic bacterium, *Shigella Dysenteriae (type 1)* 279–285.

Ishaq, M., Bazaka, K., & Ostrikov, K. (2015). Pro-apoptotic NOXA is implicated in atmospheric-pressure plasma-induced melanoma cell death. *Journal of Physics D: Applied Physics*, 48(46), 464002.

Jackson, P., Jacobsen, N. R., Baun, A., Birkedal, R., Kühnel, D., Jensen, K. A., & Wallin, H. (2013). Bioaccumulation and ecotoxicity of carbon nanotubes. *Chemistry Central Journal*, 7(1), 1–21.

Jo, Y. K., Kim, B. H., & Jung, G. (2009). Antifungal activity of silver ions and nanoparticles on phytopathogenic fungi. *Plant Disease*, 93(10), 1037–1043.

Kadam, D., Momin, B., Palamthodi, S., & Lele, S. S., 2019. Physicochemical and functional properties of chitosan-based nano-composite films incorporated with biogenic silver nanoparticles. *Carbohydrate Polymers*, 211, 124–132.

Kang, S., Herzberg, M., Rodrigues, D. F., & Elimelech, M. (2008). Antibacterial effects of carbon nanotubes: size does matter! *Langmuir*, 24(13), 6409–6413.

Kang, S., Pinault, M., Pfefferle, L. D., & Elimelech, M. (2007). Single-walled carbon nanotubes exhibit strong antimicrobial activity. *Langmuir*, 23(17), 8670–8673.

Kovtun, A., Campodoni, E., Favaretto, L., Zambianchi, M., Salatino, A., Amalfitano, S., Navacchia, M.L., Casentini, B., Palermo, V., Sandri, M., & Melucci, M. (2020). Multifunctional graphene oxide/biopolymer composite aerogels for microcontaminants removal from drinking water. *Chemosphere*, 259, 127501.

Kleandrova, V., Luan, F., Speck-Planche, A., & Cordeiro, M. N. D. S. (2015). Review of structures containing fullerene-C60 for delivery of antibacterial agents. multitasking model for computational assessment of safety profiles. *Current Bioinformatics*, 10(5), 565–578.

Kumar, B., Smita, K., Cumbal, L., & Debut, A. (2014). Synthesis of silver nanoparticles using Sacha inchi (Plukenetia volubilis L.) leaf extracts. *Saudi Journal of Biological Sciences*, 21(6), 605–609.

Kundu, S., & Liang, H. (2008). Polyelectrolyte-mediated non-micellar synthesis of monodispersed 'aggregates' of gold nanoparticles using a microwave approach. *Colloids and Surfaces A: Physicochemical and Engineering Aspects*, 330(2–3), 143–150.

Li, Q., Mahendra, S., Lyon, D. Y., Brunet, L., Liga, M. V., Li, D., & Alvarez, P. J., 2008. Antimicrobial nanomaterials for water disinfection and microbial control: potential applications and implications. *Water Research*, 42(18), 4591–4602.

Lin, D., Zheng, Y., Huang, Y., Ni, L., Zhao, J., Huang, C., Chen, X., Chen, X., Wu, Z., Wu, D. and Chen, H., 2020. Investigation of the structural, physical properties, antioxidant, and antimicrobial activity of chitosan-nano-silicon aerogel composite edible films incorporated with okara powder. *Carbohydrate Polymers*, 250, 116842.

Litter, M. I., Morgada, M. E., & Bundschuh, J. (2010). Possible treatments for arsenic removal in Latin American waters for human consumption. *Environmental Pollution*, 158(5), 1105–1118.

Liu, C., Xie, X. and Cui, Y., 2012. Antimicrobial nanomaterials for water disinfection. In *Nano-Antimicrobials* (pp. 465–494). Springer, Berlin, Heidelberg.

Lok, C. N., Ho, C. M., Chen, R., He, Q. Y., Yu, W. Y., Sun, H., ... & Che, C. M. (2006). Proteomic analysis of the mode of antibacterial action of silver nanoparticles. *Journal of Proteome Research*, 5(4), 916–924.

Lokina, S., & Narayanan, V. (2013). Antimicrobial and anticancer activity of gold nanoparticles synthesized from grapes fruit extract. *Chemical Science Transactions*, 2(S1), S105–S110.

Lu, X., Feng, X., Zhang, X., Chukwu, M.N., Osuji, C.O. and Elimelech, M., 2018. Fabrication of a desalination membrane with enhanced microbial resistance through vertical alignment of graphene oxide. *Environmental Science & Technology Letters*, 5(10), 614–620.

Lv, Q., Zhang, B., Xing, X., Zhao, Y., Cai, R., Wang, W., & Gu, Q. (2018). Biosynthesis of copper nanoparticles using Shewanella loihica PV-4 with antibacterial activity: Novel approach and mechanisms investigation. *Journal of Hazardous Materials*, 347, 141–149.

Ma, H., Yin, B., Wang, S., Jiao, Y., Pan, W., Huang, S., ... & Meng, F. (2004). Synthesis of silver and gold nanoparticles by a novel electrochemical method. *ChemPhysChem*, 5(1), 68–75.

Mahadevan, S., Vijayakumar, S., & Arulmozhi, P. (2017). Green synthesis of silver nano particles from Atalantia monophylla (L) Correa leaf extract, their antimicrobial activity and sensing capability of H_2O_2. *Microbial Pathogenesis*, 113, 445–450.

Mahapatra, O., Bhagat, M., Gopalakrishnan, C., & Arunachalam, K. D. (2008). Ultrafine dispersed CuO nanoparticles and their antibacterial activity. *Journal of Experimental Nanoscience*, 3(3), 185–193.

Malato, S., Fernández-Ibáñez, P., Maldonado, M. I., Blanco, J., & Gernjak, W. (2009). Decontamination and disinfection of water by solar photocatalysis: recent overview and trends. *Catalysis Today*, 147(1), 1–59.

Manivasagan, P., Kang, K. H., Kim, D. G., & Kim, S. K., (2015). Production of polysaccharide-based bioflocculant for the synthesis of silver nanoparticles by Streptomyces sp. *International Journal of Biological Macromolecules*, 77, 159–167.

Markovic, Z., & Trajkovic, V. (2008). Biomedical potential of the reactive oxygen species generation and quenching by fullerenes (C60). *Biomaterials*, 29(26), 3561–3573.

Menazea, A. A., & Awwad, N. S., (2020). Antibacterial activity of TiO_2 doped ZnO composite synthesized via laser ablation route for antimicrobial application. *Journal of Materials Research and Technology*, 9(4), 9434–9441.

Mie, R., Samsudin, M. W., Din, L. B., Ahmad, A., Ibrahim, N., & Adnan, S. N. A. (2014). Synthesis of silver nanoparticles with antibacterial activity using the lichen Parmotrema praesorediosum. *International Journal of Nanomedicine*, 9, 121.

Moor, K. J., Osuji, C. O., & Kim, J. H. (2016). Dual-functionality fullerene and silver nanoparticle antimicrobial composites via block copolymer templates. *ACS Applied Materials & Interfaces*, 8(49), 33583–33591.

Mostafa, A. M., & Menazea, A. A. (2020). Laser-assisted for preparation ZnO/CdO thin film prepared by pulsed laser deposition for catalytic degradation. *Radiation Physics and Chemistry*, 176, 109020.

Mostafiz, F., Islam, M. M., Saha, B., Hossain, M. K., Moniruzzaman, M., & Habibullah-Al-Mamun, M. (2020). Bioaccumulation of trace metals in freshwater prawn, Macrobrachium rosenbergii from farmed and wild sources and human health risk assessment in Bangladesh. *Environmental Science and Pollution Research*, 27(14), 16426–16438.

Mukha, I., Eremenko, A., Korchak, G., & Michienkova, A. (2010). Antibacterial action and physicochemical properties of stabilized silver and gold nanostructures on the surface of disperse silica. *Journal of Water Resource and Protection*, 2010.

Mukherjee, M., & De, S. (2018). Antibacterial polymeric membranes: a short review. *Environmental Science: Water Research & Technology*, 4(8), 1078–1104.

Navarro, E., Baun, A., Behra, R., Hartmann, N. B., Filser, J., Miao, A. J., ... & Sigg, L. (2008). Environmental behavior and ecotoxicity of engineered nanoparticles to algae, plants, and fungi. *Ecotoxicology*, 17(5), 372–386.

Ogunsona, E. O., Muthuraj, R., Ojogbo, E., Valerio, O., & Mekonnen, T. H. (2020). Engineered nanomaterials for antimicrobial applications: A review. *Applied Materials Today*, 18, 100473.

Oyelami, A. O., & Semple, K. T. (2015). Impact of carbon nanomaterials on microbial activity in soil. *Soil Biology and Biochemistry*, 86, 172–180.

Pablos, C., van Grieken, R., Marugán, J., & Moreno, B. (2011). Photocatalytic inactivation of bacteria in a fixed-bed reactor: mechanistic insights by epifluorescence microscopy. *Catalysis Today*, 161(1), 133–139.

Poletti, F., Favaretto, L., Kovtun, A., Treossi, E., Corticelli, F., Gazzano, M., Palermo, V., Zanardi, C., & Melucci, M. (2020). Electrochemical sensing of glucose by chitosan modified graphene oxide. *Journal of Physics: Materials*, 3(1), 014011.

Prasad, K., Lekshmi, G. S., Ostrikov, K., Lussini, V., Blinco, J., Mohandas, M., ... & Ostrikov, K. (2017). Synergic bactericidal effects of reduced graphene oxide and silver nanoparticles against Gram-positive and Gram-negative bacteria. *Scientific Reports*, 7(1), 1–11.

Qu, X., Alvarez, P. J., & Li, Q. (2013). Applications of nanotechnology in water and wastewater treatment. *Water Research*, 47(12), 3931–3946.

Rai, M., Yadav, A., & Cioffi, N. (2012). Silver nanoparticles as nano-antimicrobials: bioactivity, benefits and bottlenecks. In *Nano-Antimicrobials* (pp. 211–224). Springer, Berlin, Heidelberg.

Rajeshkumar, S., Bharath, L. V., & Geetha, R. (2019). Broad spectrum antibacterial silver nanoparticle green synthesis: characterization, and mechanism of action. In *Green Synthesis, Characterization and Applications of Nanoparticles* (pp. 429–444). Elsevier.

Ravindranadh, M. R. K., & Mary, T. R. (2013). Development of ZnO nanoparticles for clinical applications. *Journal of Chemical, Biological and Physical Sciences (JCBPS)*, 4(1), 469.

Robertson, P. K., Robertson, J. M., & Bahnemann, D. W. (2012). Removal of microorganisms and their chemical metabolites from water using semiconductor photocatalysis. *Journal of Hazardous Materials*, 211, 161–171.

Rodrigues, D. F., & Elimelech, M. (2010). Toxic effects of single-walled carbon nanotubes in the development of E. coli biofilm. *Environmental Science & Technology*, 44(12), 4583–4589.

Sadaiyandi, K., Kennedy, A., Sagadevan, S., Chowdhury, Z. Z., Johan, M. R. B., Aziz, F. A., Rafique, R. F., & Selvi, R. T., 2018. Influence of Mg doping on ZnO nanoparticles for enhanced photocatalytic evaluation and antibacterial analysis. *Nanoscale Research Letters*, 13(1), 1–13.

Saifuddin, N., Raziah, A. Z., & Junizah, A. R. (2013). Carbon nanotubes: a review on structure and their interaction with proteins. *Journal of Chemistry*, 2013.

Sánchez-López, E., Gomes, D., Esteruelas, G., Bonilla, L., Lopez-Machado, A. L., Galindo, R., ... & Souto, E. B. (2020). Metal-based nanoparticles as antimicrobial agents: an overview. *Nanomaterials*, 10(2), 292.

Sangari, M., Umadevi, M., Mayandi, J., Anitha, K., & Pinheiro, J. P., 2015. Photocatalytic and antimicrobial activities of fluorine doped TiO_2-carbon nano cones and disc composites. *Materials Science in Semiconductor Processing*, 31, 543–550.

Sathyanarayanan, M. B., Balachandranath, R., Genji Srinivasulu, Y., Kannaiyan, S. K., & Subbiahdoss, G. (2013). The effect of gold and iron-oxide nanoparticles on biofilm-forming pathogens. *International Scholarly Research Notices*, 2013.

Sharma, S. K., Chiang, L. Y., & Hamblin, M. R. (2011). Photodynamic therapy with fullerenes in vivo: reality or a dream? *Nanomedicine*, 6(10), 1813–1825.

Singh, C., Baboota, R. K., Naik, P. K., & Singh, H., 2012. Biocompatible synthesis of silver and gold nanoparticles using leaf extract of Dalbergia sissoo.

Slavin, Y. N., Asnis, J., Häfeli, U. O., & Bach, H. (2017). Metal nanoparticles: understanding the mechanisms behind antibacterial activity. *Journal of Nanobiotechnology*, 15(1), 1–20.

Snigdha, S., Kalarikkal, N., Thomas, S., & Radhakrishnan, E. K. (2020). The need for engineering antimicrobial surfaces. In *Engineered Antimicrobial Surfaces* (pp. 1–12). Springer, Singapore.

Sondi, I., & Salopek-Sondi, B. (2004). Silver nanoparticles as antimicrobial agent: a case study on E. coli as a model for Gram-negative bacteria. *Journal of Colloid and Interface Science*, 275(1), 177–182.

Stensberg, M. C., Wei, Q., McLamore, E. S., Porterfield, D. M., Wei, A., & Sepúlveda, M. S. (2011). Toxicological studies on silver nanoparticles: challenges and opportunities in assessment, monitoring and imaging. *Nanomedicine*, 6(5), 879–898.

Sujatha, L., Govardhan, T. L., & Rangaiah, G. S., 2012. Antibacterial activity of green seaweeds on oral bacteria.

Tang, C., Bai, H., Liu, L., Zan, X., Gao, P., Sun, D. D., & Yan, W. (2016). A green approach assembled multifunctional Ag/AgBr/TNF membrane for clean water production & disinfection of bacteria through utilizing visible light. *Applied Catalysis B: Environmental*, 196, 57–67.

Teh, K. C., Tan, R. R., Aviso, K. B., Promentilla, M. A. B., & Tan, J. (2019). An integrated analytic hierarchy process and life cycle assessment model for nanocrystalline cellulose production. *Food and Bioproducts Processing*, 118, 13–31.

Tiwari, P. M., Vig, K., Dennis, V. A., & Singh, S. R. (2011). Functionalized gold nanoparticles and their biomedical applications. *Nanomaterials*, 1(1), 31–63.

Tu, Y., Lv, M., Xiu, P., Huynh, T., Zhang, M., Castelli, M., ... & Zhou, R. (2013). Destructive extraction of phospholipids from Escherichia coli membranes by graphene nanosheets. *Nature Nanotechnology*, 8(8), 594.

Usman, M. S., El Zowalaty, M. E., Shameli, K., Zainuddin, N., Salama, M., & Ibrahim, N. A. (2013). Synthesis, characterization, and antimicrobial properties of copper nanoparticles. *International Journal of Nanomedicine*, 8, 4467.

Vijayakumar, S., Krishnakumar, C., Arulmozhi, P., Mahadevan, S., & Parameswari, N. (2018). Biosynthesis, characterization and antimicrobial activities of zinc oxide nanoparticles from leaf extract of Glycosmis pentaphylla (Retz.) DC. *Microbial Pathogenesis*, 116, 44–48.

Vikneshan, M., Saravanakumar, R., Mangaiyarkarasi, R., Rajeshkumar, S., Samuel, S.R., Suganya, M., & Baskar, G., 2020. Algal biomass as a source for novel oral nano-antimicrobial agent. *Saudi Journal of Biological Sciences*, 27(12), 3753–3758.

Wang, L., Hu, C., & Shao, L. (2017). The antimicrobial activity of nanoparticles: present situation and prospects for the future. *International Journal of Nanomedicine*, 12, 1227.

Wang, Y., Tay, S. L., Wei, S., Xiong, C., Gao, W., Shakoor, R. A., & Kahraman, R. (2015). Microstructure and properties of sol-enhanced Ni-Co-TiO2 nano-composite coatings on mild steel. *Journal of Alloys and Compounds*, 649, 222–228.

Wang, Y., Xue, X., & Yang, H. (2014). Modification of the antibacterial activity of Zn/TiO_2 nano-materials through different anions doped. *Vacuum*, 101, 193–199.

Wang, Z., & Wang, G. (2004). APD: the antimicrobial peptide database. *Nucleic Acids Research*, 32(suppl_1), D590–D592.

Wu, P., Xie, R., Imlay, J. A., & Shang, J. K. (2009). Visible-light-induced photocatalytic inactivation of bacteria by composite photocatalysts of palladium oxide and nitrogen-doped titanium oxide. *Applied Catalysis B: Environmental*, 88(3–4), 576–581.

Xie, L., Shu, Y., Hu, Y., Cheng, J., & Chen, Y. (2020). SWNTs-PAN/TPU/PANI composite electrospun nanofiber membrane for point-of-use efficient electrochemical disinfection: New strategy of CNT disinfection. *Chemosphere, 251*, 126286.

Xie, Y., He, Y., Irwin, P. L., Jin, T., & Shi, X. (2011). Antibacterial activity and mechanism of action of zinc oxide nanoparticles against Campylobacter jejuni. *Applied and Environmental Microbiology*, 77(7), 2325–2331.

Yamamoto, O., Ohira, T., Alvarez, K., & Fukuda, M. (2010). Antibacterial characteristics of CaCO3–MgO composites. *Materials Science and Engineering: B*, 173(1–3), 208–212.

Yang, C., Mamouni, J., Tang, Y., & Yang, L. (2010). Antimicrobial activity of single-walled carbon nanotubes: length effect. *Langmuir, 26*(20), 16013–16019.

Yuan, P., Ding, X., Yang, Y. Y., & Xu, Q. H. (2018). Metal nanoparticles for diagnosis and therapy of bacterial infection. *Advanced Healthcare Materials*, 7(13), 1701392.

Yun, H., Kim, J. D., Choi, H. C., & Lee, C. W. (2013). Antibacterial activity of CNT-Ag and GO-Ag nanocomposites against gram-negative and gram-positive bacteria. *Bulletin of the Korean Chemical Society*, 34(11), 261–3264.

Zawadzka, K., Kądzioła, K., Felczak, A., Wrońska, N., Piwoński, I., Kisielewska, A., & Lisowska, K. (2014). Surface area or diameter–which factor really determines the antibacterial activity of silver nanoparticles grown on TiO_2 coatings? *New Journal of Chemistry*, 38(7), 3275–3281.

Zhang, L., Ding, Y., Povey, M., & York, D. (2008). ZnO nanofluids–A potential antibacterial agent. *Progress in Natural Science*, 18(8), 939–944.

Zhang, Y., Zhang, M., Jiang, H., Shi, J., Li, F., Xia, Y., Zhang, G., & Li, H. (2017). Bio-inspired layered chitosan/graphene oxide nanocomposite hydrogels with high strength and pH-driven shape memory effect. *Carbohydrate Polymers*, 177, 116–125.

Zhao, W., Chen, I. W., & Huang, F., (2019). Toward large-scale water treatment using nanomaterials. *Nano Today*, 27, 11–27.

Zhou, H., Yang, D., Ivleva, N. P., Mircescu, N. E., Niessner, R., & Haisch, C. (2014). SERS detection of bacteria in water by in situ coating with Ag nanoparticles. *Analytical Chemistry*, 86(3), 1525–1533.

Zhou, R., & Gao, H. (2014). Cytotoxicity of graphene: recent advances and future perspective. *Wiley Interdisciplinary Reviews: Nanomedicine and Nanobiotechnology*, 6(5), 452–474.

Zhou, R., Zhou, R., Zhang, X., Li, J., Wang, X., Chen, Q., ... & Ostrikov, K. K. (2016). Synergistic effect of atmospheric-pressure plasma and TiO_2 photocatalysis on inactivation of Escherichia coli cells in aqueous media. *Scientific Reports*, 6(1), 1–10.

Zhu, J., Hou, J., Zhang, Y., Tian, M., He, T., Liu, J., & Chen, V. (2018). Polymeric antimicrobial membranes enabled by nanomaterials for water treatment. *Journal of Membrane Science*, 550, 173–197.

Zhu, J., Wang, J., Hou, J., Zhang, Y., Liu, J., & Van der Bruggen, B. (2017). Graphene-based antimicrobial polymeric membranes: a review. *Journal of Materials Chemistry A*, 5(15), 6776–6793.

10 Soil Remediation
Application of Nanoparticles

E Lokesh Goud, Prasann Kumar, and Bhupendra Koul
Lovely Professional University

CONTENTS

10.1 INTRODUCTION

Ensuring a healthy environment for the living organism by restoring the deteriorated quality of water, air and soil is a global challenge of the present time (Susanto 2020; Kumar and Tortajada 2020; Brevik et al. 2020). In order to create a pollutant-free environment, conventional approaches and pollution treatment techniques have found a limited effectiveness with the generation of novel contaminants (e.g., pharmaceuticals, microplastics) challenging the existing technologies (Nandhini et al. 2019; Fanourakis et al. 2020). The strategic intervention of nanotechnology can mitigate the problem of environmental pollution. The present researches and studies have proven that the smallest particles can be a solution to the largest problem (Mueller and Nowack 2010).

Environmental problems, such as heavy metal contamination, biomagnification of toxic elements, polluted water resources and contaminated air have emerged due to technological and industrial advancements (Zhou et al. 2020; David et al. 2012). Soil entitled to be a 'universal sink' in all aspects is more prone to pollution (Ashraf et al. 2014). The uncontrolled release of such materials may lead to unfertile and unproductive soil, leading to a decline in the agricultural production (Mishra et al. 2016). Thus, soil remediation must be done with sustainable and economically efficient techniques.

DOI: 10.1201/9781003129042-10

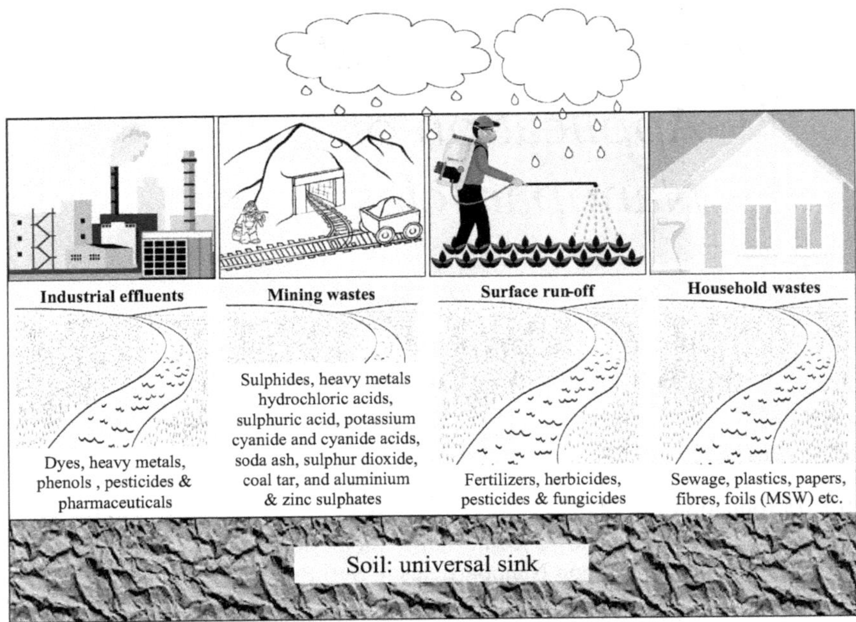

FIGURE 10.1 Contamination of soil by various pollutants from different sources.

The particles and structures in the nanoscale range are studied under nanotechnology. Nanoparticles are those having dimensions in the range of 1–100 nm (Lin et al. 2014). The smaller size of the nanoparticles offers them large specific surface area due to which they have wider range of applications in food technology, medicine, industries and environmental remediation (Khan et al. 2019). Nanotechnology can facilitate in removing organic pollutants and inorganic substances from groundwater as well as soil (Araujo et al. 2015).

Some industrial processes can yield nanoparticles as their by-product accidentally, or they have natural occurrence with geogenic, atmospheric or biogenic origin (Mueller and Nowack 2010) (Figure 10.1). Besides that, engineered nanoparticles (ENPs) are also used in these sectors. Nowadays, biotechnology has also been found compatible with the nanotechnology to serve the purpose of soil remediation (Gong et al. 2018). Despite having a number of uses, nanotechnology can also create environmental problems if not used with proper care (Qian et al. 2020). This chapter deals with the application of nanoparticles (nanotechnology) in soil remediation.

10.2 TYPES OF NANOPARTICLES USED IN SOIL REMEDIATION PROCESS

Contaminants present in the soil vary widely in their chemical characteristics (Hillel 2008), and nanomaterials to be used for the removal of a particular contaminant molecule vary from one contaminant to other as well as with the reaction mechanism of each nanoparticle (Amin et al. 2014). Remediation process by nanotechnology has to deal with ubiquitous chemicals characterized for their persistency, such as heavy metals (mercury, arsenic, lead, chromium and cadmium), chlorinated solvents, polyfluoroalkyl substances, alkylated polycyclic aromatic hydrocarbons (PAHs) or substituted PAH with nitrogen, oxygen and sulphur (Gupta et al. 2016; Nandhini et al. 2019). These complex molecules demand different reagents, reaction times and reaction conditions to get immobilized (Sun et al. 2020). These dissimilar requirements and simultaneous occurrence of complex contaminants limit the efficiency of single technology, making it a time-consuming and tedious process. But target-specific nanoparticles can be designed (with multiple reactions combined) for a broad-spectrum action towards contaminants (Sarkar et al. 2019; Sun et al. 2020).

The conventional soil remediation strategies are shown in Figure 10.2. Soil remediation can be attained through *ex situ* (Kim et al. 2019) as well as *in situ* (Cai et al. 2019) methodologies with the help of nanoparticles. Through pressurized spray or injection or by gravity, nanoparticles are allowed into the contaminated soil as aqueous slurry or colloidal solution (Wang et al. 2019; Galdames et al. 2020). A treatment zone is created after the injection of nanoparticles and remains as a suspension. Nanoparticles adsorb and immobilize the heavy metals from the soil (Rabbani et al. 2016). The mechanism adopted by the nanomaterials for the remediation has classified them into reactive and adsorptive materials (Mueller and Nowack 2010). Adsorbents remove the contaminants from the site by retaining them on their surface or internal structure, and this property is due to their extremely small size and large specific surface area. Reactive nanomaterials offer chemical reactions such as redox, dissolution, precipitation, acid--base, photocatalysis and ion exchange, for the contaminants to get excluded from the site (Zhang et al. 2019; Lu and Astruc 2020).

The nanomaterials should necessarily be selective; otherwise, their finite reactive capacity may exhaust by reacting with other metal ions and naturally occurring organic molecules (Fan et al. 2016). Selectivity of the remediation techniques will reduce the number of agents deployed for the removal of contaminants from the targeted site as this limits the off-target reaction (Zhang et al. 2016; Guerra et al. 2018). Manipulation of the surface chemistry aspects such as charge, acidity or basicity and hydrophilicity or hydrophobicity as well as the physical aspects such as crystal facets and pore structure enables the selective action of ENPs (Sadegh et al. 2017). Enzymes, aptamers, carbohydrates, peptides or conductive materials are capable of identifying specific molecules which can be functionalized on the surface of the ENPs with the help of ligands (Farzin et al. 2017; Mehndiratta et al. 2013). For example, ENPs became selective for Hg(II) among many other divalent metal cations when the surface of ENPs are functionalized with magnetite particles using amine groups (Zhang et al. 2019).

Also, ENPs are capable of conducting indirect degradation of contaminants by stimulating the native microbes performing the biodegradation of the contaminant by delivering nutrition (Cecchin et al. 2016; Zhang et al. 2019). This has been observed from the action of calcium peroxide, which is a nanoparticle having the

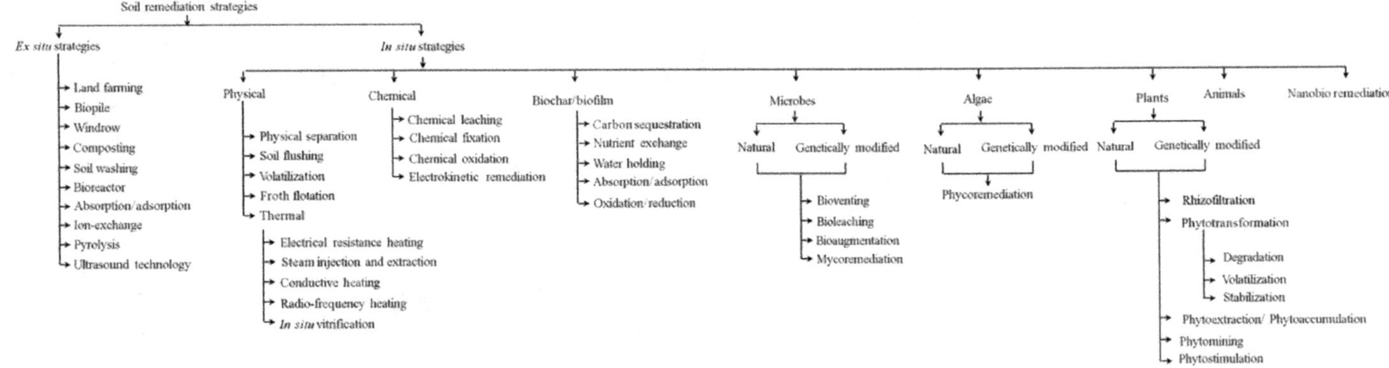

FIGURE 10.2 Soil remediation strategies.

capability to release oxygen. Heterotrophs present in the groundwater are limited by the dissolved oxygen in water to perform biodegradation of the pollutants. Calcium peroxide nanoparticles exhibited Fenton's reaction to perform *in situ* chemical oxidation (ISCO) by generating reactive oxygen species (Khodaveisi et al. 2011). Emerging trends in nanotechnology include some important nanoparticles in soil remediation process, such as metal oxide nanoparticles (Fe_3O_4, ZnO and TiO_2), carbon nanomaterials (graphene, carbon nanotubes, C_{70} and fullerenes), nanocomposites and nanoscale zero-valent iron (nZVI).

10.3 APPLICATIONS OF CARBON NANOPARTICLES IN SOIL REMEDIATION

The release of carbon nanoparticles (CNPs) into the living environment can be from natural or Anthropogenic-activities. Combustion of fossil fuel, forest fires, transportation of fossil coal (petrogenic CNPs), application of biochars (pyrogenic CNPs), filtration using activated carbon and industrial release of engineered CNPs (eCNPs) are the sources of environmental exposure of CNPs (Sigmund et al. 2018; Patil and Lekhak 2020). Fullerenes, carbon nanotubes and graphene are commercially exploited eCNPs (Figure 10.3). Carbon nanomaterials (CNMs) are graphite-based materials with a range of size varying in between 1 and 100 nm, which encompasses single-walled carbon nanotubes (SWCNTs), multi-walled carbon nanotubes (MWCNTs) and fullerene soot. An internal cavity with the single roll of graphene sheet forms the structure of SWCNTs, while an internal cavity with inter-wall spaces formed by several rolls of graphite sheets is the structure of MWCNTs. Fullerenes have only one external surface as it has a spherical configuration formed by a graphite ball (Yang and Xing 2007).

Organic contaminants are accumulated over CNPs due their high adsorption affinities by which the bioavailability of most contaminants are significantly impacted (Shrestha et al. 2015; Ren et al. 2018). Morphology, size and surface functionality are the physicochemical properties of the CNPs, in which significant changes have been observed due to their chemical, physical and biological

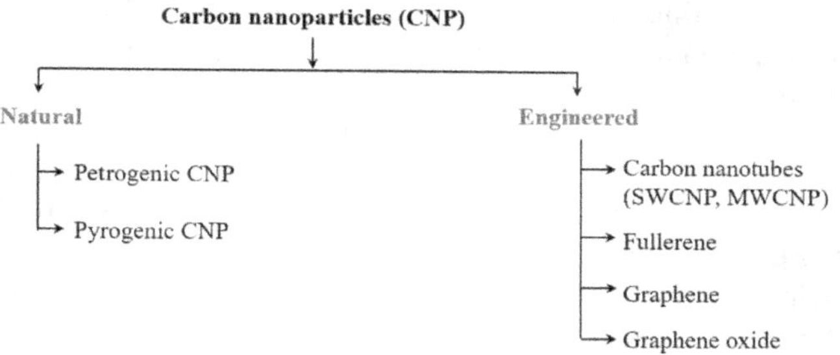

FIGURE 10.3 Types of carbon nanoparticles.

transformation after release into the environment. eCNPs have reported an approximate value for the global production as 20 Gg in 2016, whereas until 2023, a 20% increase has been predicted per year (Sigmund et al. 2018). More than 99% of naturally occurring CNPs are pyrogenic or petrogenic, while eCNPs constitute <1% in the environment (Patil and Lekhak 2020). The aspects on which both type of the CNPs share their similarities are high electron donor–acceptor capacity and aromaticity, while significant difference exists for porosity, surface functionality, structural irregularities and inorganic phases. Petrogenic CNPs have a lower affinity for adsorption of organic pollutants as compared to eCNPs because they have small aromatic sheets and more negative charge for the surface groups (Sigmund et al. 2018).

The CNPs take up the organic molecules by π–π interaction and van der Waals forces, by which they immobilize such particles within the soil matrices (Hou et al. 2013). In contaminated water and soil, these can be used based on their adsorption capacities and surface hydrophobicity (Matos et al. 2017; Baby et al., 2019). Carbon nanotubes (CNTs) show more affinity to adsorb the organic molecules due to large surface area, pore structure and wide range of functional groups attached to the surface, which help increase their efficiency even more and assist them in overcoming their limitations due to hydrophobicity and insolubility (Adewoye et al. 2021). The unique properties of the CNTs enabled their wide application since their discovery by Iijima in 1991. Pesticides and pharmaceuticals are some ionizable organic compounds that are adsorbed and sequestered by CNTs with the help of cation–π-assisted π–π interactions and charge-assisted hydrogen bonds (Kah et al. 2017). The wide acceptability of CNTs among other adsorbents is due to their large surface area with unique chemical structure and high thermal stability. Their excellent binding affinity enables them to remove PAHs. PAHs include pyrene, phenanthrene and naphthalene and belong to priority pollutants as they can induce mutagenic, toxic, carcinogenic and teratogenic impacts on ecosystem (Patel et al. 2020). It has been found that soils having organic carbon greater than 2% significantly reduced the bioaccessibility and extractability of the PAHs due to the sequestration ability of organic carbon. Black carbon also limits the bioaccessibility and ensures restricted desorption of PAHs (Semple et al. 2013). The type and concentration of the CNMs applied to the PAHs-contaminated soil have significant impact on the extraction and mineralization of PAHs (Towell et al. 2011).

For the remediation of polluted water resources, uses of CNMs are widely studied as compared to that for contaminated soil. Heavy metals are an emerging concern of the ecological health. Even though variety of techniques such as membrane filtration, electrochemical technologies, ion exchange and chemical precipitation are efficiently operated for the removal of heavy metals, adsorption technique holds more efficiency due to its operational and design related flexibility (Fu and Wang 2011). Multiple use of the same adsorbents is possible by their regeneration or reversible use through desorption process. Thus, effluent after treatment can be expected to be of high quality. A study conducted by Kabbashi et al. (2009) by using CNTs in aqueous solution observed 96% removal of Pb(II). As adsorption is an endothermic process, increased solution temperature increased the adsorption percentage as found in an experiment by Abdel Salam (2012), in which Zn^{2+}, Cd^{2+}, Cu^{2+} and Pb^{2+} were removed by MWCNTs from the water successfully and the capacity for the adsorption increased in the order $Cd^{2+} < Zn^{2+} < Pb^{2+} < Cu^{2+}$. It was also assessed that the adsorption of heavy metals by MWCNTs is an entropy-driven process. Thus, the scope of CNPs in remediating heavy metal-contaminated soil is wide. The mobility of each heavy metal differs between species and soil properties (Violante et al. 2010). The efficiency of CNTs in immobilizing the heavy metals depends on those factors also. Acidic pH can increase the mobility of heavy metals. Immobilization of heavy metals are observed more in fine-grained soils than in soils with coarse particles (Correia et al. 2020). Compared to nickel and zinc, copper and lead have high affinity for organic matter. The least adsorption capacity of zinc is attributed to low electronegativity value, which offers it high mobility and makes it the weakest competitor for the sites of adsorption on CNPs (Gomes et al. 2001).

The pretreatment of nanoparticles is essential before their release into the soil as there are chances of reduced efficiency of the particles due to the formation of agglomerates attributed to van der Waals force generated and surface attraction, which leaves low sites for adsorption (Darlington et al. 2009). Chemical methods such as covalent modification or non-covalent modification or mechanical methods such as ultrasonic energy or the use of surfactants can reduce agglomeration (Yu et al. 2020). The dispersal of CNTs is immediately possible with ultrasonic sound. But the duration of the exposure and quantity of energy used can impact the normal properties of CNTs as excess duration or energy may break the CNTs (Nguyen et al. 2012). Surfactant application is a non-covalent chemical method. They get adsorbed on the surface of CNTs and make alterations in the outer structure and get dispersed through electrostatic and steric interactions. The combinations of these methods are acceptable to ensure a dispersion efficiency in the long term (Matos et al. 2017).

Functional groups present on the CNTs and surface area determine the adsorptive properties. Pyrene gets adsorbed on CNTs when surface area increases (Peng et al. 2014), and oxygen as a functional group on CNTs enhances the adsorption of chemicals, which may also be due to π–π electron donor–acceptor interactions (Kurwadkar et al. 2019). For non-polar chemicals such as phenanthrene, hydrophobic effect (50%–85%) serves as the major adsorptive control mechanism over MWCNTs, while for polar molecules such as 9,10-phenanthrenequinone π–π electron donor–acceptor interactions enabled through functional groups and hydrogen bonds regulate the adsorptive mechanism of 9-phenanthrol (Peng et al. 2016).

There are chances of risk of strong capacity of adsorption, as changes may occur in bioavailability, uptake, bioaccumulation and toxicity of hydrophobic organic compounds. Phenanthrene, fluorene and naphthalene are some hydrophobic PAHs whose sorption mechanism was not influenced by the MWCNTs, as reported in a study conducted by Li et al. (2013). Phenolic compounds can also be adsorbed on CNTs, but their affinity is strongly related to the number of –OH functional groups (hydroxyl), the number of layers, size and shape. The π–π interaction is responsible for the stronger adsorption as CNTs can attract π-receptor and π-donor to the surface at same time (Perez and Martin 2012).

10.4 APPLICATIONS OF METAL OXIDE NANOPARTICLES IN SOIL REMEDIATION

High adsorption capacity at fast kinetic rates is the promising feature of using nanomaterials based on certain metals and their oxides. Inorganic and organic contaminants such as heavy metals, halogenated organic pollutants and dyes can be removed from soil and water using nanoscale particles of TiO_2, CdO, MgO, ZnO, MnO_2 and Fe_3O_4 (Sadegh et al. 2017). Nanoparticles of silver are eminent in imparting antimicrobial mechanism, i.e. antiviral, antifungal and antibacterial. They bind to glycoproteins of virus and thus prevent the host cell from binding with virus (Guerra et al. 2018). Zinc oxide (ZnO) and titanium dioxide (TiO_2) have semiconducting properties (Rehman et al. 2009). Inorganic and organic pollutants are transformed by the oxidative reaction of TiO_2. Other applications of titanium dioxide nanoparticle are filters used for air purification, disinfection of tiles and self-cleaning glass. Hydroxyl radical, a highly reactive oxidant that is produced by these metal oxide nanoparticles is responsible for their disinfectant property (Haider et al. 2017). Electrodes made of TiO_2 determine COD of water, i.e. chemical oxygen demand (Ge et al. 2016). The interaction of these metal oxides with the contaminants is induced by ultraviolet rays by both reduction and oxidation reactions. Electron–hole pairs develop due to the excitation by UV rays, and pollutants are oxidized with the oxidizing potential of these electrons (Pawar et al. 2018) (Figure 10.4). The combination of both metal oxide and UV radiation imparts a bactericidal action that can kill pathogens. The use of titanium dioxide is relatively inexpensive as it can be used in repetition without any deterioration in its catalytic property due to its resistance against corrosion and the property to remain chemically and biologically inert. The effectiveness of TiO_2 is limited to transparent water (Bhawana and Fulekar 2012).

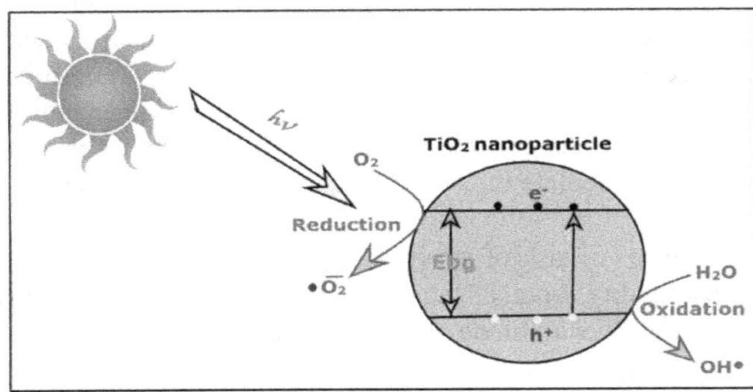

FIGURE 10.4 Photocatalysis reaction of TiO_2 nanoparticles.

In order to overcome this limitation, several studies have been performed involving doping of TiO_2 with another transition metal ion. This is in a view of increasing its performance in terms of photocatalysis. A comparative study conducted (Chang Chien et al. 2011) in red soil, quartz sand and alluvial soil regarding the degradation of the pollutant, pyrene, with the assistance of TiO_2 and irradiation inferred that the degradation of the pyrene by the nanoparticles of TiO_2 by photocatalysis is more promising under the sunlight and the percentage of photocatalytic degradation in these three soils varies as alluvial soil (23.4%)>red soil (31.8%)>quartz sand (78.3%). Iron oxide (Fe_3O_4) nanoparticles are also efficient in removing contaminants from soil environment. Adsorption capacity of Fe_3O_4 is strong to immobilize cadmium and arsenic. At pH 6.0 and temperature 10°C–20°C, iron oxide nanoparticles have a maximum adsorption capacity of 37.03 mg/g (Qian et al. 2020). Phenanthrene, acenaphthene and fluorene are carcinogenic PAHs. A nanocomposite was synthesized, namely titanium dioxide-based zinc hexacyanoferrate, and applied in water and soil to observe the photocatalytic degradation of PAHs. PAHs were reduced by 82%–86% in soil and 93%–96% in water when exposed with sunlight that induced the activity of the nanocomposite. Interaction between PAHs and the organic content of soil lowers down the diffusion, and it causes slower degradation rate in soil than in water (Rachna et al. 2019).

As a consequence of World War II, arsenic-contaminated sites were developed due to the leakage from arsenical weapons used at that time. Diphenylarsinic acid (DPAA) has been derived from diphenylcyanoarsine and diphenylchloroarsine through oxidation and hydrolysis reactions (Li et al. 2016). Japan, Europe and China were found victims of the DPAA-contaminated soil, which has toxicity at neural and genetic level (Tanaka et al. 2013). Organic pollutants are efficiently removed through photocatalytic oxidation, which degrades organic form to inorganic. In TiO_2 solution, phenyl arsenic acid, dimethyl arsenic acid and monomethyl arsenic acid degrades into arsenates through photocatalytic degradation. Organic arsenics or inorganic arsenics can also be removed by the adsorption mechanism by nanocrystalline TiO_2 (Ashraf et al. 2019). The adsorbed contaminants on soil can be treated by TiO_2 in two ways. After washing the contaminated soil with nonpolar solvents, washings are treated with TiO_2 under photocatalytic oxidation. This method encompasses multiple treatments. Another method is the direct addition of TiO_2 for the contaminant degradation in the presence of UV rays. Limitation of this method is that light has a poor penetration into the depth of the soil. The efficiency of photocatalytic degradation of DPAA was 90% in soil treated with slurry of TiO_2 and UV rays whose end product was arsenate (Akakuru et al. 2020). The removal efficiency is a function of soil properties and external factors. The most important factor that impacts the degradation is soil-to-water ratio because it enhances the penetration of light and dilutes the soil slurry. Soil pH, soil organic matter, EC and total P have an inverse relation to the removal efficiency of DPAA by TiO_2. From an experiment, it was observed that in order to treat 4 g soil having 20 mg/kg DPAA the optimum conditions required were soil-to-water ratio (1:10), TiO_2 (5%), irradiation time (3 h) and light intensity (40 mW/cm^2), which resulted in 82.7% removal efficiency (Wang et al. 2016).

10.5 APPLICATIONS OF ZERO-VALENT IRON NANOPARTICLES IN SOIL REMEDIATION

For the *in situ* methods of remediation, zero-valent iron nanoparticles have a wide acceptance over others such as alumina-supported noble metals and non-ionic amphiphilic polyurethane as they have improved chemistry and deployment strategies (Tratnyek and Johnson 2006). This is a cost-effective technology due to the abundance of iron in earth and will not harm the environment as they remain non-toxic in the elemental form (Rabbani et al. 2016). The very small size of the particles permits their entry into the contaminated site, and the pollutant remains closely attached to the iron nanoparticles due to the large surface area with a lot of reactive sites (Zhang 2003). Direct injection of iron nanoparticles into sediments, contaminated soils and aquifers is possible as they suspend in their colloidal solution (Comba et al. 2010). The availability of heavy metals in soil declines by the conversion of active toxic metal ions into less toxic species through the adsorption and reduction reaction of iron nanoparticles.

There are two procedures for synthesis of nZVI: bottom-up approach and top-down process (Pasinszki and Krebsz 2020). Top-down processes through which nanoscale materials are formed from large-sized materials through chemical or mechanical techniques comprise milling, machining, noble gas sputtering and pulsed laser ablation. On the contrary, bottom-up approach is concerned with growing nanostructures in an atom-by-atom manner or molecule-by-molecule manner through positional assembling, self-assembling and chemical synthesis. The structure of nZVI consists of an iron core (Fe^0) with a metallic or zero-valent iron and an oxide shell of Fe^{2+} and Fe^{3+} of mixed valency as it is oxidized easily (Mukherjee et al. 2016). Zero-valent iron nanoparticles are manufactured unlike naturally occurring Fe^{2+} and Fe^{3+}. The remediation property of nZVI is due to its electron-donating property. The core of the nanoparticles acts as the electron source and thus reducing power is provided to carry out the reaction, while the site of all chemisorption reactions is provided by the shell (Monga et al. 2020). The chemical properties of these nanoparticles rely on the core–shell structure. The properties of the oxide layer are similar to those of a semiconductor (Mukherjee et al. 2016), and facile transfer of charges is possible due to its thin layer and defective sites. Under the neutral pH, core is protected from rapid oxidation by the insoluble property of the oxide shell. nZVI should be handled as a slurry due to its high reactivity.

The remediation techniques by nZVI in soil or in groundwater have to be performed in the following ways: (1) injection of nZVI to form a reactive barrier comprised of Fe particles. (2) Surfactants, polysaccharides, polyelectrolytes or cellulose is coated on the surface to modify the nZVI surface before being injected into the site to create a plume of reactive Fe that is capable of destroying any kind of organic contaminant (Latif 2006; Mueller and Nowack 2010). As a reactive barrier, nZVI was found more effective in the degradation of halogenated solvents through reduction reaction. They include trihalomethanes, brominated benzenes, chlorinated benzenes, chlorinated methanes, polychlorinated hydrocarbons and chlorinated ethenes. nZVI particles are efficient in removing pollution-causing dyes (e.g., decolourization of the azo dye named Acid Black 24) and pesticides (Rahman et al. 2014). The remediation of atrazine-contaminated soil is also possible with nZVI, of which the primary mechanism is the reductive dechlorination by which atrazine is destroyed, and the rate of removal by nZVI particles is seven times higher as compared to microscale particles of the same species (Satapanajaru et al. 2008). In a laboratory experiment, soil contaminated with pyrene is successfully remediated by the synthesized nZVI particles as compared to the microscale ZVI particles as they can induce strong reducing conditions (Chang and Kang 2009). At room temperature, it was reported that PAHs adsorbed on soil particles were efficiently removed by the nZVI particles, while the destruction of polychlorinated biphenyls which were held by the soil matrix very strongly was observed to be 38%. The behaviour exhibited by nZVI and associated compounds is similar to environmental colloids rather than a true nanosized particle even under the laboratory condition as they have the tendency to form clusters with size in the range of microns (Tratnyek and Johnson 2006). Polymers or surfactants are used for functionalizing the nZVI to avoid the aggregation. Combination of susfactants with the carbon platelets or nZVI embedded in oil droplets (emulsions) enable their delivery in contaminated sites (Mackenzie and Georgi 2019).

The factors determining the reactivity of any nanoparticles are surface area, reactive sites and their intrinsic reactivity on the surface, and for nZVI, the interaction of these factors gives three different operational results: (1) contaminants such as polychlorinated biphenyls will get degraded with nZVI even they will not react with larger particles of similar materials; (2) contaminants such as chlorinated ethylenes will be rapidly degraded by nZVI as they have already reacted with larger particles at useful rates; and (3) contaminants such as carbon tetrachloride that yielded some undesirable by-products by degradation with larger particles will give more favourable products when degraded by nZVI (Tratnyek and Johnson 2006). Among these reactions, the second one is more frequently observed. The surface area of the nZVI is the reason for the higher rate of CCl_4 degradation rather than the intensity of reactive sites. Chances for the inefficiency of the applied nZVI are higher due to the reaction of off-target particles such as H_2O and dissolved O_2, which indirectly increase the cost of remediation (Ahmed et al. 2017). So, it is preferable to have short duration for the *in situ* remediation to avoid off-target exposures.

There are four species of heavy metals in soil that determine their toxicity, such as oxidizable fraction, reducible fraction, acid-soluble fraction and residual fraction (Xue et al. 2018). Highly toxic, mobile and bioavailable form is the acid-soluble fraction. Potential toxicity depending upon the redox potential is the characteristic feature of reducible and oxidizable fractions. More stable forms are those metals remaining in residual fraction. nZVI has the role of converting unstable and toxic forms of metal ions into stable forms. Heavy metals removal depends on the standard redox potential of the metal to be removed, which determines the mechanism of reaction between metal ions and zero-valent iron

(Mu et al. 2016). Adsorption of zinc and cadmium on the iron shell is because they have a more negative or equal value of standard redox potential as that of nZVI. But, in the case of copper, arsenic and chromium which have the standard redox potential with more positive values as compared to the nZVI will undergo reduction reaction and precipitation in soil. Another mechanism is observed for nickel and lead that have only a slightly positive standard redox potential compared to that of nZVI, which enables both adsorption and reduction reactions of those metals by nZVI. These nanoparticles are more prone to oxidation with air due to their reactivity and surface area. This is effectively controlled by attachment with some noble metals such as platinum or palladium, by which it becomes bimetallic iron. In a comparative study made by Nasiri et al. (2013) between uncoated nZVI and carboxymethyl cellulose-coated nZVI, to determine their colloidal stability and cadmium removal efficiency, it was observed that as compared to uncoated particles, coated nanoparticles had more colloidal stabilization and direct relation had been established between the concentration of nZVI, i.e. adsorbent, and the removal of cadmium from soil.

In Kanpur, India, accumulation of chromium has been a major risk due to the illegal disposal of waste from tanning industry. Singh et al. (2012) conducted an experiment in this site with nZVI and observed the efficiency of the nanoparticles in soil remediation. They reported that 0.1 g/L of nZVI completely reduced the chromium within a duration of 120 min. The existence of chromium in soil is in two forms: they are the immobile form Cr(III) and the mobile form Cr(VI). If a stabilized suspension of nZVI particles is added directly to the soil, as an *in situ* approach, it will reduce the mobility of Cr(VI) by its precipitation as an insoluble hydroxide compound. The use of zero-valent iron nanoparticles stabilized by carboxymethyl cellulose in chromium-contaminated soil reduced the toxicity caused by Cr(VI) by enhancing its immobilization, reducing the bioaccumulation and bioavailability within the plants and also lowering the leachability (Wang et al. 2014). With the assistance of nZVI, the following reactions occur:

$$Cr^{6+} \rightarrow Cr^{3+}[\text{reduction}]$$

$$3Fe^{0} + Cr_2O_7^{2-} + 7H_2O \rightarrow 3Fe^{2+} + 2Cr(OH)_3 + 8OH^-$$

Otherwise, in the iron oxide shell of the nZVI, chromium gets immobilized as Cr^{3+}–Fe^{3+} hydroxide which will be in an alloy-like form.

From various studies, Xue et al. (2018) have summarized the interaction between nZVI particles and the metals into five groups. They are as follows:

i. Reduction: Ag, As, Cr, Cu, Ni, Pb, U, Hg, Pd, Pt, Se and Co.
ii. Precipitation: Co, Pb, Cu, Cd and Zn.
iii. Oxidation/reoxidation: U, As, Pb and Se.
iv. Adsorption: Cd, Cr, Pb, As, Ni, Zn, Ba, Se, U and Co.
v. Co-precipitation: Ni, As, Se and Cr.

Apart from metals, nZVI have the capability to degrade highly toxic organic contaminant groups such as chlorinated organic compounds (COCs), which include many organochlorine pesticides, trichloroethylene, polychlorinated biphenyls and tetrachloroethylene. Lindane, an organochlorine pesticide that is proved to be a persistent organic pollutant, can effectively be degraded by nZVI progressed through dihaloelimination and reductive degradation (Elliott et al. 2009). Sodium persulphate is used in hazardous waste management as an oxidant. Sulphate radicals are produced from persulphate by its activation through transition metals, heat, light or ultrasound, and they are able to destroy organic pollutants. nZVI can assist the activation of persulphate as a catalyst and thus help the oxidation of PAHs that create harmful effects on environment (Chen et al. 2015). Organic acids are natural chelators that can be used in soil washing to remove the heavy metals. Oxalic acid, tartaric acid and citric acid were used to remediate heavy metal-contaminated sites, and their removal efficiency were found to be higher with higher concentration of nanoparticles of ZVI (Cao et al. 2018). DDT, an organochlorine insecticide, poses serious risks due to its persistence, and the removal of the same from the contaminated site is under study even long after its ban. Bioremediation, permeable reactive barriers, photocatalytic treatments, incineration and excavation are some already existing methods. nZVI can also be used to degrade the DDT within the site, and more efficient degradation was observed in aqueous solution as compared to soil (El-Temsah et al. 2016).

The removal of organic pollutants can be done through advanced oxidation processes (AOPs) as an alternative to conventional methods (Garrido-Cardenas et al. 2019). Homogenous Fenton oxidation is a prominent method among other AOPs to remove petroleum effluents from the industries as the mechanism involves the release of hydroxyl radical (OH•) that can convert non-biodegradable pollutants to biodegradable particles. Catalysts such as zero-valent iron, magnetite, goethite and lepidocrocite are being used in Fenton oxidation, but apart from all these catalysts, nZVI particles are preferable because they do not require any assistance such as irradiation and ultrasounds (Nidheesh 2015). nZVI can be used as a catalyst in heterogeneous Fenton-like oxidation systems in which two electrons are lost generating H_2O_2. This heterogeneous Fenton-like oxidation completely removed high molecular weight PAHs and partially removed low molecular weight PAHs, when nZVI is used as a catalyst (Haneef et al. 2020). The removal rate of polychlorinated biphenyls from contaminated water was found to be 90%, while in soil, it was 38% with the assistance of nZVI (Xue et al. 2018).

10.6 CONCLUSIONS

Nanotechnology has a wide scope in environmental remediation. The use of nanoparticles can be a cost-effective approach to the removal of contaminants as compared to the existing technologies and conventional approaches. Consideration of external factors and soil properties is essential in achieving successful soil remediation through nanoparticles. Even though nanoparticles are smaller particles in the size range of more than 1 nm and less than 100 nm, they have large specific surface area, which attributes to their contaminant removal efficiency from the site. Adsorption of the pollutants on the surface of nanoparticles will immobilize or stabilize the adsorbate and converts them to less toxic substances. Reduction and oxidation reactions are carried out by nZVI with heavy metal ions and other organic contaminants for their removal. Other reactions such as Fenton-like oxidation, precipitation and co-precipitation are also efficient mechanisms. Many of the studies inferred that the removal efficiency of contaminants are higher for aqueous solution as compared to soil. Future research is necessary to explore the optimum conditions of nanoparticles to act in each type of soils and for all types of contaminants. Repetitive use of some nanoparticles is possible by desorption and other mechanisms, which determines the cost efficiency of the treatment.

REFERENCES

Abdel Salem, M. 2012. Removal of heavy metal ions from aqueous solutions with multi-walled carbon nanotubes: kinetic and thermodynamic studies. *Int. J. Environ. Sci. Technol.* doi: 10.1007/s13762-012-0127-6.

Adewoye, T.L., Ogunleye, O.O., Abdulkareem, A.S., Salawudeen, T.O., and Tijani, J.O. 2021. Optimization of the adsorption of total organic carbon from produced water using functionalized multi-walled carbon nanotubes. *Heliyon* 7: e05866.

Ahmed, M.A., Bishay, S.T., Ahmed, F.M., and El-dek, S.I. 2017. Effective Pb^{2+} removal from water using nanozerovalent iron stored 10 months. *Appl. Nanosci.* 7: 407–416.

Akakuru, O.U., Iqbal, Z.M., and Wu, A. 2020. TiO_2 nanoparticles: properties and applications. In *TiO_2 Nanoparticles: Applications in Nanobiotechnology and Nanomedicine*. Wu, A. and Ren, W. (Eds.), pp. 1–66. doi: 10.1002/9783527825431.ch1.

Amin, M.T., Alazba, A.A., and Manzoor, U. 2014. A review of removal of pollutants from water/wastewater using different types of nanomaterials. *Adv. Mater. Sci. Eng.* doi: 10.1155/2014/825910.

Araujo, R., Castro, A.C.M., and Fiuza, A. 2015. The use of nanoparticles in soil and water remediation processes. *Mater. Today: Proc.* 2: 315–320.

Ashraf, M.A., Maah, M.J., and Yusoff, I. 2014. Soil contamination, risk assessment and remediation. In *Environmental Risk Assessment of Soil Contamination*. InTech, Croatia, pp. 2–47. doi: 10.5772/57287.

Ashraf, S., Siddiqa, A., Shahida, S., and Qaisar, S. 2019. Titanium-based nanocomposite materials for arsenic removal from water: a review. *Heliyon* 5: e01577.

Baby, R., Saifullah, B. and Hussein, M.Z. 2019. Carbon nanomaterials for the treatment of heavy metal-contaminated water and environmental remediation. *Nanoscale Res Lett* 14:341. doi:10.1186/s11671-019-3167-8

Bhawana, P., and Fulekar, M.H. 2012. Nanotechnology: remediation technologies to clean up the environmental pollutants. *Res. J. Chem. Sci.* 2(2): 90–96.

Brevik, E.C., Slaughter, L., Singh, B.R., Steffan, J.J., Collier, D., Barnhart, P., and Pereira, P. 2020. Soil and human health: current status and future needs. *Air Soil Water Res.* 13: 1–23.

Cai, C., Zhao, M., Yu, Z., Rong, H., and Zhang, C. 2019. Utilization of nanomaterials for in-situ remediation of heavy metal(loid) contaminated sediments: a review. *Sci. Total Environ.* 662: 205–217.

Cao, Y., Zhang, S., Zhong, Q., Wang, G., Xu, X., Li, T., Wang, L., Jia, Y., and Li, Y. 2018. Feasibility of nanoscale zero-valent iron to enhance the removal efficiencies of heavy metals from polluted soils by organic acids. *Ecotoxicol. Environ. Safety* 162: 464–473.

Cecchin, I., Reddy, K.R., Thome, A., Tessaro, E.F., and Schnaid, F. 2016. Nanobioremediation: integration of nanoparticles and bioremediation for sustainable remediation of chlorinated organic contaminants in soils. *Int. Biodeterior. Biodegrad.* 1–10.

Chang Chien, S.W., Chang, C.H., Chen, S.H., Wang, M.C., Rao, M.M., and Veni, S.S. 2011. Effect of sunlight irradiation on photocatalytic pyrene degradation in contaminated soils by micro-nano size TiO_2. *Sci. Total Environ.* 409: 4101–4108.

Chang, M.C., and Kang, H.Y. 2009. Remediation of pyrene-contaminated soil by synthesized nanoscale zero-valent iron particles. *J. Environ. Sci. Health Part A* 44(6): 576–582.

Chen, C.F., Binh, N.T., Chen, C., and Dong, C. 2015. Removal of polycyclic aromatic hydrocarbons from sediments using sodium persulfate activated by temperature and nanoscale zero-valent iron. *J. Air Waste Manage. Assoc.* 65(4): 375–383, doi: 10.1080/10962247.2014.996266.

Comba, S., Molfetta, A.D., and Sethi, R. 2010. A comparison between field applications of nano-, micro-, and millimetric zero-valent iron for the remediation of contaminated aquifers. *Water Air Soil Pollut.* doi: 10.1007/s11270-010-0502-1.

Correia, A.A.S., Matos, M.P.S.R., Gomes, A.R. and Rasteiro, M.G. 2020. Immobilization of heavy metals in contaminated soils—performance assessment in conditions similar to a real scenario. *Appl. Sci.* 10: 7950.

Darlington, T.K., Neigh, A.M., Spencer, M.T., Guyen, O.T.N., and Oldenburg, S.J. 2009. Nanoparticle characteristics affecting environmental fate and transport through soil. *Environ. Toxicol. Chem.* 28(6). doi: 10.1897/08-341.1.

David, I.G., Matache, M, L., Tudorache, A., Chisamera, G., Rozylowicz, L., and Radu, G.L. 2012. Food chain biomagnification of heavy metals in samples from the lower prut floodplain natural park. *Environ. Eng. Manage. J.* 11(1): 69–73.

Elliott, D.W., Lien, H.L., and Zhang, W.X. 2009. Degradation of lindane by zero-valent iron nanoparticles. *J. Environ. Eng.* 135: 317–324.

El-Temsah, Y.S., Sevcu, A., Bobcikova, K., Cernik, M., and Joner, E.J. 2016. DDT degradation efficiency and ecotoxicological effects of two types of nano-sized zero-valent iron (nZVI) in water and soil. *Chemosphere* 144: 2221–2228.

Fan, D., O'Carroll, D., Elliott, D.W., Tratnyek, P.G., Johnson, R.L., and Garcia, A.N. 2016. Selectivity of nano zerovalent iron in *In Situ* chemical reduction: challenges and improvements. *Remed. J.* 26(4).

Fanourakis, S.K., Pena-Bahamonde, J., Bandara, P.C., and Rodrigues, D.F. 2020. Nano-based adsorbent and photocatalyst use for pharmaceutical contaminant removal during indirect potable water reuse. *Npj Clean Water* 3: 1–15.

Farzin, L., Shamsipur, M., and Sheibani, S. 2017. A review: aptamer-based analytical strategies using the nanomaterials for environmental and human monitoring of toxic heavy metals. *Talanta* 174: 619–627.

Fu, F. and Wang, Q. 2011. Removal of heavy metal ions from wastewaters: a review. *J. Environ. Manage.* 92: 407–418.

Galdames, A., Ruiz-Rubio, L., Orueta, M., Sanchez-Arzalluz, M., and Vilas-Vilela, J.L. 2020. Zero- valent iron nanoparticles for soil and groundwater remediation. *Int. J. Environ. Res. Public Health* 17: 5817.

Garrido-Cardenas, J.A., Esteban-Garcia, B., Aguera, A., Sanchez-Perez, J.A., and Manzano-Agugliaro, F. 2019. Wastewater treatment by advanced oxidation process and their worldwide research trends. *Int. J. Environ. Res. Public Health* 17: 170.

Ge, Y., Zhai, Y., Niu, D., Wang, Y., Fernandez, C., Ramakrishnappa, T., Hu, X., and Wang, L. 2016. Electrochemical determination of chemical oxygen demand using Ti/TiO_2 electrode. *Int. J. Electrochem. Sci.* 11: 9812–9821.

Gomes, P.C., Fontes, M.P.F., da Dilva, A.G., Mendonca, E.S., and Netto, A.R. 2001. Selectivity sequence and competitive adsorption of heavy metals by Brazilian soils. *Soil Sci. Soc. Am. J.* 65(4).

Gong, X., Huang, D., Liu, Y., Peng, Z., Zeng, G., Xu, P., Cheng, M., Wang, R., and Wan, J. 2018. Remediation of contaminated soils by biotechnology with nanomaterials: bio-behavior, applications, and perspectives. *Crit. Rev. Biotechnol.* 38(3): 455–468.

Guerra, F.D., Attia, M.F., Whitehead, D.C., and Alexis, F. 2018. Nanotechnology for environmental remediation: materials and applications. *Molecules* 23: 1760. doi: 10.3390/molecules23071760.

Gupta, H., Kumar, R., Park, H., and Jeon, B. 2016. Photocatalytic efficiency of iron oxide nanoparticles for the degradation of priority pollutant anthracene. *Geosyst. Eng.*

Haider, A.J., Al-Anbari, R.H., Kadhim, G.R., and Salame, C.T. 2017. Exploring potential environmental applications of TiO_2 nanoparticles. *Energy Procedia* 119: 332–345.

Haneef, T., Mustafa, M.R.U., Rasool, K., Ho, Y.C., and Kutty, S.R.M. 2020. Removal of polycyclic aromatic hydrocarbons in a heterogeneous Fenton like oxidation system using nanoscale zero-valent iron as a catalyst. *Water* 12(9): 2430. doi: 10.3390/w12092430.

Hillel, D. 2008. Soil pollution and remediation. *Soil Environ.* 211–222.

Hou, L., Zhu, D., Wang, X., Wang, L., Zhang, C., and Chen, W. 2013. Adsorption of phenanthrene, 2-naphthol, and 1-naphthylamine to colloidal oxidized multiwalled carbon nanotubes: effects of humic acid and surfactant modification. *Environ. Toxicol. Chem.* 32(3): 493–500.

Iijima, S. 1991. Helical microtubules of graphitic carbon. *Nature* 354: 56–58.

Kabbashi, N.A., Atieh, M.A., Al-Mamun, A., Mirghami, M.E.S., Alam, M.D.Z., and Yahya, N. 2009. Kinetic adsorption of application of carbon nanotubes for Pb(II) removal from aqueous solution. *J. Environ. Sci.* 21: 539–544.

Kah, M., Sigmund, G., Xiao, F., and Hofmann, T. 2017. Sorption of ionizable and ionic organic compounds to biochar, 2 activated carbon and other carbonaceous materials. *Water Res.* 124(48): 673–692.

Khan, I., Saeed, K., and Khan, I. 2019. Nanoparticles: properties, applications and toxicities. *Arabian J. Chem.* 12: 908–931.

Khodaveisi, J., Banejad, H., Afkhami, A., Olyaie, E., Lashgari, S., and Dashti, R. 2011. Synthesis of calcium peroxide nanoparticles as an innovative reagent for in situ chemical oxidation. *J. Hazard. Mater.* 192: 1437–1440.

Kim, N., Kwon, K., Park, J., Kim, J., and Choi, J. 2019. Ex situ soil washing of highly contaminated silt loam soil using core – crosslinked amphiphilic polymer nanoparticles. *Chemosphere* 224: 212–219.

Kumar, M.D., and Tortajada, C. 2020. Health impacts of water pollution and contamination. In *Assessing Wastewater Management in India*. Springer, London, pp. 23–30.

Kurwadkar, S., Hoang, T.V., Malwade, K., Kanel, S.R., Harper, W.F., and Struckhoff, G. 2019. Application of carbon nanotubes for removal of emerging contaminants of concern in engineered water and wastewater treatment systems. *Nanotechnol. Environ. Eng.* 4(12).

Latif, B. 2006. Nanotechnology for site remediation: fate and transport of nanoparticles in soil land water systems. Prepared for US Environmental Protection Agency, Washington, DC. http://www.clu-in.org/download/ studentpapers/B_Latif_Nanotechology.pdf.

Li, C., Srivastava, R.K., and Athar, M. 2016. Biological and environmental hazards associated with exposure to chemical warfare agents: arsenicals. *Ann. N Y Acad. Sci.* 1378(1): 143–157.

Li, S., Anderson, T.A., Green, M.J., Maul, J.D., and Canas-Carrell, J.E. 2013. Polyaromatic hydrocarbons (PAHs) sorption behavior unaffected by the presence of multi-walled carbon nanotubes (MWNTs) in a natural soil system. *Environ. Sci. Process. Impacts.* 15(6): 1095–1292.

Lin, P.C., Lin, S., Wang, P.C., and Sridhar, R. 2014. Techniques for physicochemical characterization of nanomaterials. *Biotechnol. Adv.* 32(4): 711–726.

Lu, F., and Astruc, D. 2020. Nanocatalysts and other nanomaterials for water remediation from organic pollutants. *Coord. Chem. Rev.* 408: 213180.

Mackenzie, K., and Georgi, A. 2019. NZVI synthesis and characterization. In Phenrat, T. and Lowry, G.V. (Eds.), *Nanoscale Zerovalent Iron Particles for Environmental Restoration*. Springer.

Matos, M.P.S.R., Correia, A.A.S., and Rasteiro, M.G. 2017. Application of carbon nanotubes to immobilize heavy metals in contaminated soils. *J. Nanopart. Res.* 19: 126. doi: 10.1007/s11051-017-3830-x.

Mehndiratta, P., Jain, A., Srivastava, S., and Gupta, N. 2013. Environmental pollution and nanotechnology. *Environ. Pollut.* 2(2).

Mishra, R.K., Mohammad, N., and Roychoudhury, N. 2016. Soil pollution: causes, effects and control. *Van Sangyan* 3(1).

Monga, Y., Kumar, P., Sharma, R.K., Fiip, J., Varma, R.S., Zboril, R., and Gawande, M.B. 2020. Sustainable synthesis of nanoscale zerovalent iron particles for environmental remediation. *ChemSusChem* doi: 10.1002/cssc.202000290.

Mu, Y., Jia, F., Ai, Z., and Zhang, L. 2016. Iron oxide shell mediated environmental remediation properties of nano zero-valent iron. *Environ. Sci.: Nano* doi: 10.1039/C6EN00398B.

Mueller, N.C., and Nowack, B. 2010. Nanoparticles for remediation: solving big problems with little particles. *Elements* 6: 395–400.

Mukherjee, R., Kumar, R., Sinha, A., Lama, Y., and Saha, A.K. 2016. A review on synthesis, characterization, and applications of nano zero valent iron (nZVI) for environmental remediation. *Crit. Rev. Environ. Sci. Technol.* 46(5): 443–466, doi: 10.1080/10643389.2015.1103832.

Nandhini, N.T., Rajeshkumar, S., and Mythili, S. 2019. The possible mechanism of eco – friendly synthesized nanoparticles on hazardous dyes degradation. *Biocatal. Agricult. Biotechnol.* 19: 101138.

Nasiri, J., Gholami, A., and Panahpour, E. 2013. Removal of cadmium from soil resources using stabilized zero-valent iron nanoparticles. *J. Civ. Eng. Urban.* 3(6): 338–341.

Nguyen, V.S., Rouxel, D., Hadji, R., Vincent, B., and Fort, Y. 2010. Effect of ultrasonication and dispersion stability on the cluster size of alumina nanoscale particles in aqueous solution. *Ultrason. Sonochem.* 18: 382–388.

Nidheesh, P.V. 2015. Heterogenous Fenton catalysts for the abatement of organic pollutants from aqueous solution: a review. *RSC Adv.* 5(51): 40552–40577.

Pasinszki, T., and Krebsz, M. 2020. Synthesis and application of zero-valent iron nanoparticles in water treatment, environmental remediation, catalysis, and their biological effects. *Nanmaterials* 10: 917.

Patel, A.B., Shaikh, S., Jain, K.R., Desai, C., and Madamwar, D. 2020. Polycyclic aromatic hydrocarbons: sources, toxicity, and remediation approaches. *Front. Microbiol.* 11: 562813.

Patil, S.S., and Lekhak, U.M. 2020. Toxic effects of engineered carbon nanoparticles on environment. *Carbon Nanomater. Agri-Food Environ. Appl.* 237–260. doi: 10.1016/B978-0-12-819786-8.00012-8.

Pawar, M., Sendogdular, S.T. and Gouma, P. 2018. A brief overview of TiO_2 photocatalyst for organic dye remediation: case study of reaction mechanisms involved in Ce-TiO_2 photocatalysts system. *J. Nanomater.* doi: 10.1155/2018/5953609.

Peng, H., Zhang, D., Li, H., Wang, C., and Pan, B. 2014. Organic contaminants and carbon nanoparticles: sorption mechanisms and impact parameters. *J. Zhejiang Univ. Sci. A* 15(8): 606–617.

Peng, H., Zhang, D., Pan, B., and Peng, J. 2016. Contribution of hydrophobic effect to the sorption of phenanthrene, 9-phenanthrol and 9, 10-phenanthrenequinone on carbon nanotubes. *Chemosphere* 1–9.

Perez, E.M., and Martin, N. 2012. π–π interactions in carbon nanostructures. *Journal Name* 1–3.

Qian, Y., Qin, C., Chen, M., and Lin, S. 2020. Nanotechnology in soil remediation – applications vs. implications. *Ecotoxicol. Environ. Safety* 201: 110815.

Rabbani, M, Ahmed, I., and Park, S.J. 2016. Application of nanotechnology to remediate contaminated soils. Hasegawa, H., Rahman, I.M.M., and Rahman, M.A. (Eds.), *Environmental Remediation Technologies for Metal-Contaminated Soils*, Springer, 219–229.

Rachna, Rani, M., Shanker, U. 2019. Degradation of tricyclicpolyaromatic hydrocarbons in water, soil and river sediment with a novel TiO2 based heterogeneous nanocomposite. *J. Environ. Manag.* 248: 109340. doi:10.1016/j.jenvman.2019.109340

Rahman, N., Abedin, Z., and Hossain, M.A. 2014. Rapid degradation of azo dyes using nano – scale zero valent Iron. *Am. J. Environ. Sci.* 10(2): 157–163.

Rani, M., and Shanker, U. 2019. Degradation of tricyclic polyaromatic hydrocarbons in water, soil and river sediment with a novel TiO_2 based heterogeneous nanocomposite. *J. Environ. Manage.* 248: 109340.

Rehman, S., Ullah, R., Butt, A.M., and Gohar, N.D. 2009. Strategies of making TiO_2 and ZnO visible light active. *J. Hazard. Mater.* 170: 560–569.

Ren, X., et al., 2018. Effect of exogenous carbonaceous materials on the bioavailability of organic pollutants and their ecological risks. *Soil Biol. Biochem.* 116: 70–81.

Sadegh, H., et al., 2017. The role of nanomaterials as effective adsorbents and their applications in wastewater treatment. *J. Nanostruct. Chem.* 7: 1–14.

Sarkar, A., Sengupta, S., and Sen, S. 2019. Nanoparticles for soil remediation. In Gothandam K., Ranjan S., Dasgupta N., Lichtfouse E. (Eds.), *Nanoscience and Biotechnology for Environmental Applications. Environmental Chemistry for a Sustainable World.* Vol. 22, pp. 249–262, Springer, Cham. doi: 10.1007/978-3-319-97922-9-9.

Satapanajaru, T., Anurakpongsatorn, P., Pengthamkeerati, P., and Boparai, H. 2008. Remediation of Atrazine – contaminated soil and water by nano zerovalent iron. *Water Air Soil Pollut.* 192: 349–359.

Semple, K.T., Riding, M.J., McAllister, L.E., Sopena-Vazquez, F., and Bending, G.D. 2013. Impact of black carbon on the bioaccessibility of organic contaminants in soil. *J. Hazard. Mater.* 261: 808–816.

Shrestha, B., Anderson, T.A., Acosta-Martinez, V., Payton, P., and Canas-Carrell, J. 2015. The influence of multiwalled carbon nanotubes on polycyclic aromatic hydrocarbon (PAH) bioavailability and toxicity to soil microbial communities in alfalfa rhizosphere. *Ecotoxicol. Environ. Safety* 116: 143–149.

Sigmund, G., Jiang, C., Hofmann, T., and Chen, W. 2018. Environmental transformation of natural and engineered carbon nanoparticles and implications for the fate of organic contaminants. *Environ. Sci. Nano* 5: 2500–2518.

Singh, R., Misra, V., and Singh, R.P. 2012. Removal of Cr(VI) by nanoscale zero-valent iron (nZVI) from soil contaminated with tannery wastes. *Bull. Environ. Contam. Toxicol.* 88: 210–214.

Sun, B., Li, Q., Zheng, M., Su, G., Lin, S., Wu, M., Li, C., Wang, Q., Tao, Y., Dai, L., Qin, Y., and Meng, B. 2020. Recent advances in the removal of persistent organic pollutants (POPs) using multifunctional materials: a review. *Environ. Pollut.* 265 doi: 10.1016/j.envpol.2020.114908.

Susanto, D.A. 2020. Air pollution and human health. *Med. J. Indones.* 29(1): 8–10.

Tanaka, M., Takahashi, Y., and Yamaguchi, N. 2013. A study on adsorption mechanism of organoarsenic compounds on ferrihydrite by XAFS. *J. Phys: Conf. Ser.* 430: 012100.

Towell, M.G., Browne, L.A., Paton, G.I., and Semple, K.T. 2011. Impact of carbon nanomaterials on the behaviour of ^{14}C-phenanthrene and ^{14}C-benzo-[a] pyrene in soil. *Environ. Pollut.* 159: 706–715.

Tratnyek, P.G. and Johnson, R.L. 2006. Nanotechnologies for environmental cleanup. *Nanotoday* 1(2): 44–48.

Violante, A., Cozzolino, V., Perelomov, L., Caporale, A.G., and Pigna, M. 2010. Mobility and Bioavailability of heavy metals and metalloids in soil environments. *J. Soil Sci. Plant Nutr.* 10(3): 268–292.

Wang, A., Teng, Y., Hu, X., Wu, L., Huang, Y., Luo, Y., and Christie, P. 2016. Diphenylarsinic acid contaminated soil remediation by titanium dioxide (P25) photocatalysis: degradation pathway, optimization of operating parameters and effects of soil properties. *Sci. Total Environ.* 541: 348–355.

Wang, Y., Fang, Z., Kang, Y., and Tsang, E.P. 2014. Immobilization and phytotoxicity of chromium in contaminated soil remediated by CMC-stabilized nZVI. *J. Hazard. Mater.* 275: 230–237.

Wang, Y., O'Connor, D., Shen, Z., Lo, I.M.C., Tsang, D.C.W., Pehkonen, S., Pu, S., and Hou, D. 2019. Green synthesis of nanoparticles for the remediation of contaminated waters and soils: constituents, synthesizing methods, and influencing factors. *J. Clean. Prod.* 226: 540–549.

Xue, W., Huang, D., Zeng, G., Wan, J., Cheng, M., Zhang, C., Hu, C., and Li, J. 2018. Performance and toxicity assessment of nanoscale zero valent iron particles in the remediation of contaminated soil: a review. *Chemosphere* 210: 1145–1156.

Yang, K. and Xing, B. 2007. Desorption of polycyclic aromatic hydrocarbons from carbon nanomaterials in water. *Environ. Pollut.* 145: 529–537.

Yu, W., Sisi, L., Haiyan, Y., and Jie, L. 2020. Progress in the functional modification of grapheme/grapheme oxide: a review. *RSC Adv.* 10: 15328.

Zhang, F., Chen X., Wu, F., and Ji, Y. 2016. High adsorption capability and selectivity of ZnO nanoparticles for dye removal. *Colloids Surf. A: Physicochem. Eng. Aspects* 509: 474–483.

Zhang, T., et al., 2019. In situ remediation of subsurface contamination: opportunities and challenges for nanotechnology and advanced materials. *Environ. Sci.: Nano.* 6(5).

Zhang, W. 2003. Nanoscale iron particles for environmental remediation: an overview. *J. Nanopart. Res.* 5: 323–332.

Zhang, W., Zhang, D., and Liang, Y. 2019. Nanotechnology in remediation of water contaminated by poly- and perfluoroalkyl substances: a review. *Environ. Pollut.* 247: 266–276.

Zhao, X., Liu, W., Cai, Z., Han, B., Qian, T., and Zhao, D. 2016. An overview of preparation and applications of stabilized zero-valent iron 2 nanoparticles for soil and groundwater remediation. *Water Res.* 100: 245–266.

Zhou, Q., Yang, N., Li, Y., Ren, B., Ding, X., Bian, H., and Yao, X. 2020. Total concentrations and sources of heavy metal pollution in global river and lake water bodies from 1972 to 2017. *Global Ecol. Conserv.* 22: e00925.

11 Nanomaterials for Air Purification

Advances and Challenges

Daniel Pramudita and Veinardi Suendo
Institut Teknologi Bandung

Muhammad Mufti Azis
Universitas Gadjah Mada

Antonius Indarto
Institut Teknologi Bandung

CONTENTS

DOI: 10.1201/9781003129042-11

11.1 INTRODUCTION

Air pollution has been one of the world's major problems in the past decades. Air pollutants are mainly generated from human activities such as use of fossil fuel; production of gases, vapours and sewage from industries; and chemical warfare agents. Industrial revolution and mass production processes along with a rapid growth in population and the number of vehicles have been seen as the main reasons for this. Air pollution is especially worse in regions where low-tech manufacturing is still dominant [1]. Air pollution is also responsible for faster global warming, as it enhances the greenhouse gas effect. This indirectly results in many natural disasters and threatens the life of living beings. The World Health Organization (WHO) reported around 7 million annual casualties due to air contamination in 2012–2014 [2]. Ironically, long-existing activities that are crucial for human existence such as agriculture and livestock production have also been shown to contribute significantly to the generation of pollutants in respect of global warming.

Air pollution is defined as the change in the atmospheric composition due to the release of chemical, biological or physical pollutants sourced from human activity or industrial processes to the atmosphere [2]. Common examples are the use of fossil fuels for vehicles, forest burning, generation of industrial and household wastes, and leakage of industrial chemicals. Some pollutant substances are considered toxic, while some are not. Among toxic pollutants are H_2S, CO, NOx, HONO, SOx, volatile organic compounds (VOCs), halogens and chlorofluorocarbons (CFCs) [3]. They can cause significant harm to human health directly. H_2S is a poisonous gas produced from sewages, farms, sewers and ports. It may cause skin burn, eye inflammation and respiratory diseases and can be deadly at concentrations above 100 ppm [4]. SO_2 is a toxic gas released from burning. It is commonly found in volcanic areas and in smokes from sulphur-containing fuels. It has the same toxicity level as H_2S. CO is more dangerous as it is, unlike H_2S and SO_2, odourless and can be fatal at concentrations higher than 35 ppm.

VOCs are organic chemicals emitted as gases from certain solids or liquids due to their high vapour pressure at room temperature. Major groups of VOCs are aromatic compounds, oxygenated volatile organic compounds (OVOCs), alkanes/alkenes, nitrogen-containing organic compounds and sulphur-containing compounds [5]. Halogenated aromatic compounds are released mainly from agricultural areas, as they are found in agricultural chemicals such as herbicides, insecticides and fungicides [6]. Some of them, such as dioxins and chloroform, are known for being toxic. OVOCs are mostly more polar and reactive than the other groups. Common OVOCs in the atmosphere are aldehydes, ketones, alcohols and ethers. Aldehydes are even more toxic than CO as they can be extremely dangerous even at ppb levels [7]. Formaldehyde, the most well known of aldehydes, is known for its carcinogenic effect. OVOCs mainly result from oxidation of aliphatic hydrocarbons with hydroxyl radicals in the atmosphere [5].

Particulate matters (PMs) are all non-gaseous objects suspended in air. They include solid particles such as dust, pollen, soot, smoke and liquid droplets. They are mixtures of organic and inorganic compounds, with various sizes, forms, and compositions. Dangerous heavy metals such as arsenic, chromium, lead, cadmium, mercury and zinc are found in PMs released from various industrial activities [3]. PM particles are highly polar in air [8]. This is due to the existence of metal cations [9] and functional groups such as C–N, $-NO_3$, $-SO_3H$ and C–O on the surface [10]. They are usually classified according to the size of the particulate, namely coarse particles with diameter of 10 μm or less (PM10), fine particles of 2.5 μm or less (PM2.5) and ultrafine dust smaller than 100 nm (PM0.1). Categorization into PM1.0 can also be found. PM particles are dangerous to human health because they are inhalable and can easily enter respiratory organs. They are often responsible for diseases such as asthma, influenza, chronic lung disease and lung cancer. The increased mortality rate in urban areas is associated with the exposure to PMs in polluted air. In year 2015, 4.2 million deaths and 103.1 million disabilities due to exposure to PMs, especially PM2.5, were reported [1].

Non-toxic gases such as hydrocarbons and other greenhouse gases (CO_2, N_2O and fluorinated gases) can also be considered pollutants [11] when their concentrations in the atmosphere exceed some certain thresholds, above which they are deemed to negatively affect the environment. Greenhouse gases, in particular, have been regarded as the main culprit of global warming having the capability to remain in the atmosphere for years, which leads to long-term negative effects on the environment [11]. Generation of CO_2 can be found in many essential human activities and thus cannot be fully suppressed. Carbon capture and storage (CCS) is one of the hottest topics of research in environmental studies and is considered one of the most urgent technologies. Separation and capture of CO_2 can be done via chemical absorption, physical adsorption, membrane separation and cryogenic distillation. The captured CO_2 can be conversed by solar-assisted catalytic conversion or by directly reducing CO_2 using H_2O and solar energy [12]. The process can also be done biologically with the help of microalgae [13]. Another important greenhouse gas is CH_4, which is released mainly (>80%) from agricultural areas. Its global warming potential is actually far higher than that of CO_2, with 1 kg of CH_4 being equal to 25 kg of CO_2 in 100 years. However, since CH_4 is not as abundant as CO_2, its overall contribution to global warming is only one-third that of CO_2.

There have been many commercialized air purification technologies for dealing with the pollutants. The main techniques used include membrane separation, incineration and oxidation. Currently available technologies are still considered inefficient and ineffective. Moreover, they are still relatively expensive and thus not yet feasible for large-scale use, not to mention the possibility of generating secondary pollutants [5]. In response to this problem, some new technologies based on adsorption, plasma [14] and photocatalysis have been introduced. They are relatively more economically feasible, easier to use and more environmentally friendly. By adsorption, hazardous gases are captured based on the selectivity of the adsorbent material. The gases are trapped on the pore surface of the adsorbent via van der Waals' forces (physical adsorption) and/or chemical bonds with the surface species/functional groups (chemical adsorption) [5]. In case of plasma, the organic pollutant is exposed to the electric or high energetic condition [15,16]. Photocatalytic remediation happens when semiconducting materials are exposed to a light and oxidize toxic organic pollutants into non-toxic materials. The oxidation process is triggered when an electron is transferred (excitation) from the valence band to the conduction band of the photocatalytic material. The effectivity and efficiency still need to be improved. As an example, for photocatalysis, a certain sufficient level of light is required, and this could become an issue as the optimum time for the process is limited. Adsorption efficiency and capability decrease over time with the amount of adsorbed species and available free surface. Moreover, the issues related to secondary pollutants still persist. Release of toxins during disruption of VOCs using TiO_2-doped paints has been reported. Materials with sufficiently solid performance and design feasibility are highly desirable [5].

Nanotechnology has emerged in the last century as one of the most promising technologies for many aspects in life. The term refers to any technology that uses materials on nanoscale that offer enhanced or unique properties not present in their bulk size. The use of nanomaterials for environmental remediation is very promising, as it shows considerably better cost–performance and little risks compared to previously existing materials [14]. This is due to the enhanced properties of the material. Nanomaterials have been used as catalysts for environmental problems and show some advantages of being more effective and more efficient in energy and material consumption. Other applications of nanotechnology in this field include removal of pollutants via adsorption and absorption, chemical reaction and filtration. The use of nanomaterials in general improves the *in situ* performance of these processes [15].

In this chapter, use of nanomaterials in various technologies as a strategy for dealing with air pollution problems is discussed. The development and applications of common nanomaterials used for this purpose as well as the technologies in which the materials are used are discussed. A brief introduction to nanotechnology is also given to help novice readers. A special section on the potential of nanotechnology for the prevention of COVID-19, which became a global pandemic in 2020, is also presented.

11.2 NANOTECHNOLOGY AND NANOMATERIALS

11.2.1 Definition, Classification and Production of Nanomaterials

The term nanotechnology refers to the design, production and application of materials in a form of structures or devices on atomic, molecular or supramolecular scale. It has been regarded as a breakthrough technology from the 20th century and is currently a leading field of research and innovation, thanks to its potential and commercialization. It has found applications in many industrial sectors such as energy, food, health, information and communication. It improves the quality of processes, materials and systems by offering products that are more precise, more efficient, cleaner and more effective.

A material is usually coined as a 'nanomaterial' when at least one of its dimension is in nanoscale, i.e. in the range of 1–100 nm. Figure 11.1 compares the size range of nanomaterials and other small objects. According to the number of dimensions that are not in that range, it can further be classified as 0-D (all dimensions in nanoscale), 1-D, 2-D and 3-D nanomaterials, as depicted in Figure 11.2. Especially for the last one, it refers to a bulk of smaller nanomaterials, and not a single object, of which the three dimensions are not in nanoscale. The material can also be classified based on the nature of the composing material: organic, inorganic, biological and also hybrid (combination of the previous) [16].

Nanomaterials can be produced by using top-down or bottom-up approaches. The first refers to the production of nanoscale materials or devices from bulk materials mechanically or chemically. Copies of devices that are already available at macroscale can be made at nanoscale by using this approach. A famous example is computer chips, which now has reached down to below 10 nm. In the top-down production approach, repetitive downsizing of a device is performed until the desired size (nanoscale, in this case) is reached [18]. This is made

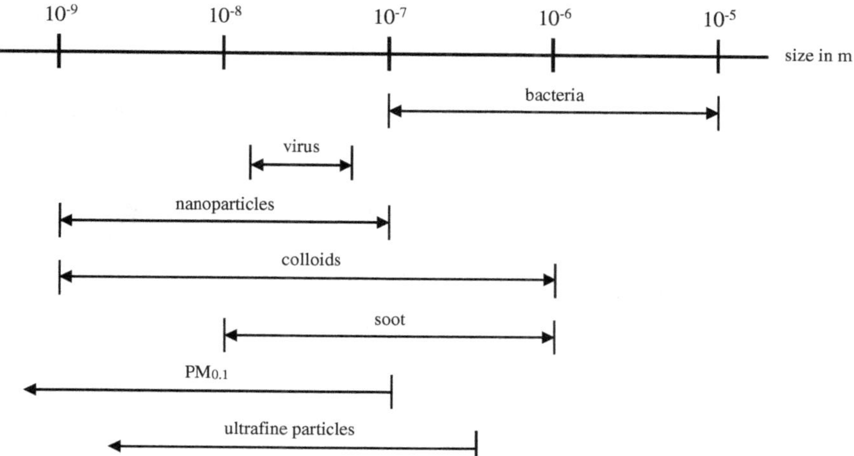

FIGURE 11.1 A comparison of the size range of nanoparticles and other particles. (See Ref. [17].)

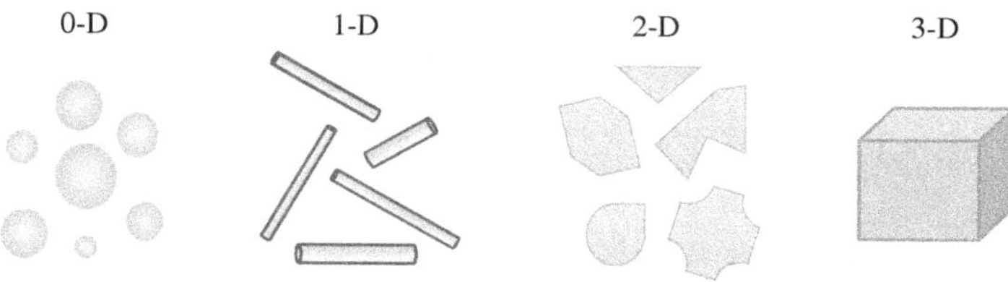

FIGURE 11.2 The classification of nanomaterials based on the number of dimensions being at nanoscale. (See Ref. [17].)

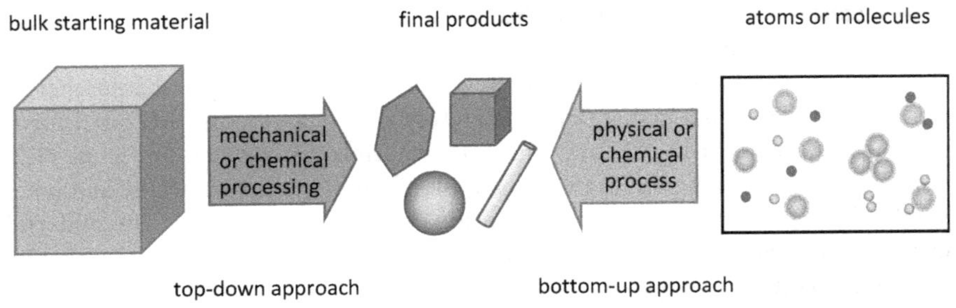

FIGURE 11.3 Two routes of nanotechnology product manufacturing. (See Ref. [17].)

possible by automation of the used manufacturing devices. The process becomes more difficult when the size reaches nanoscale, at which extreme accuracy is required. There is thus a limit of how small the product can be made. These, in addition to the excessive amount of time and resource required, are the main challenges for this approach. The bottom-up approach first came to answer to those problems. In this approach, a device is made from atoms or molecules assembled via chemical reactions and engineered to result in the desired size and structure [16]. An illustration of top-bottom and bottom-up approaches is given in Figure 11.3. Other than physical, mechanical and chemical processes, nanomaterials can also be produced by using biotechnology [18].

The main characteristics of nanomaterials are their large surface-to-volume ratio, as well as the quantum effect resulting from it. At nanoscale, the fraction of surface molecules or atoms is big enough and intermolecular forces at the surface become more dominant. The surface atoms are more reactive and less stable than the ones in the bulk, and they play a bigger role in determining the overall properties of the material. This results in enhanced or specialized properties that do not exist at larger size-scale, which are as follows [17]:

1. Mechanical properties
2. Repellent or bonding
3. Reactivity
4. Electrical conductivity
5. Magnetic property
6. Thermal properties
7. Optical properties
8. Antimicrobial
9. Delivery ability
10. Absorption ability.

The novel or improved properties of the materials are in fact used as an important criterion for nanomaterials in industry, aside from only the size.

When nanomaterials are embedded with a bulk material to improve its properties, a new material called nanocomposite is produced. This can be achieved by using methods such as dispersion (in a solution) and polymerization, and by

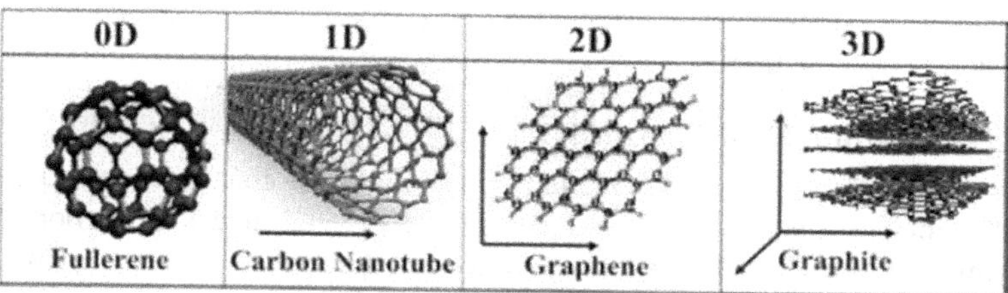

FIGURE 11.4 Some examples of carbon nanostructures classified by the dimensions. (See Ref. [2].)

insertion via melting. The following are some properties that are commonly improved by the addition of nanomaterials [19].

1. Mechanical strength
2. Thermal endurance
3. Flame retardancy
4. Barrier properties
5. Abrasion resistance
6. Reduced shrinkage and residual loss
7. Altered electrical, electronic and optical properties.

Which properties are improved depends on the combination of the main and added materials. Even though new characteristics might be added by the embedding process, in most cases the bulk material already has some certain important properties. Moreover, it is usually nanomaterials that are used to improve a certain bulk material and not the other way around. The main reason is its economic feasibility. Four main classes of nanocomposites are (1) thermoplastics, (2) thermosets, (3) elastomers and (4) natural and biodegradable polymers [20]. One of the oldest and most popular nanocomposites is carbon-based nanocomposites, which encompass a range of structures (Figure 11.4).

Another important advantage of nanoparticles is the possibility of customization. They can be designed to have specific properties for particular applications, through chemical bonding with other molecules. This process is called functionalization, which can be achieved by attaching other molecules or atoms to the nanoparticles. Bonding with the secondary species is also often required when a nanocomposite is made, as it facilitates attachment between nanomaterials and the primary material [21]. The bond can be either a covalent bond or a van der Waals one. Each offers different characteristics, and thus, the selection depends on the purpose, although the former is more common.

11.2.2 Common Nanomaterials Used for Air Purification

In this section, some nanomaterials that are commonly used for dealing with air pollutants are introduced. They are carbon nanotubes (CNTs), zeolites, titanium dioxide (TiO_2), zinc oxide (ZnO) and metal–organic frameworks (MOFs).

Their applications in various air purification technologies are discussed in more detail in the next section.

11.2.2.1 Carbon Nanotubes (CNTs)

Carbon nanotubes are tubes with nanoscale diameter that are created from a rolled graphene sheet, which is a surface of hexagonal arrays of carbon atoms. They are thus considered as 1-D nanomaterials. The tube can be composed of a single rolled sheet, giving it the name single-walled carbon nanotubes (SWCNTs), or multiple sheets (multi-walled carbon nanotubes, MWCNTs). CNTs are known for their enhanced properties such as electrical and thermal conductivities, strength, hardness and adsorption capacity [6]. CNTs have been used as adsorbents, quantum nanowires, electron field emitters, support materials for catalysts, storage materials for hydrogen, etc. [3]. Their highly graphitic structure makes them have adsorption capacity that is far higher than even activated carbons [22]. Adsorption selectivity and stability of CNTs can be increased by combining surface functional groups (OH, COOH and C=O) [23]. Pore size and geometry govern the interaction between the carbon atoms and the functional groups, as well as between the surface groups and the adsorbed molecules, which influence the maximum adsorption capacity [23,24]. In practical applications, CNTs can exist as bundles of individual tubes. The adsorption properties of a CNT bundle are influenced by the adsorption sites. Figure 11.5 shows four possible absorption sites of a bundle of SWCNTs, which are as follows:

1. Inner surface of individual tubes, which are hardly accessible if the open ends are blocked.
2. Interstitial channels between individual tubes.

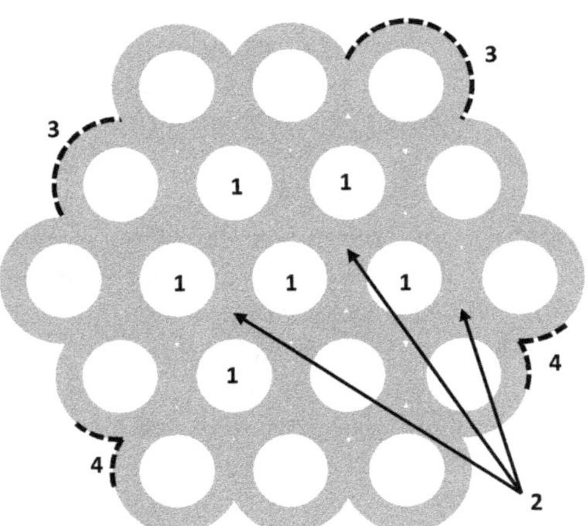

FIGURE 11.5 Possible absorption sites of a bundle of SWCNTs: (1) internal sites; (2) interstitial channels; (3) external surface; and (4) external grooves sites. (See Ref. [14]. Used with permission.)

3. Outside surface of the outermost nanotubes of the bundles.
4. External grooves sites between two adjacent outer tubes.

11.2.2.2 Nanozeolites

Zeolites are hydrated aluminosilicate minerals that are characterized by microporous structures made of polyhedral interlinks of alumina (AlO_4) and silica (SiO_4). They are commonly found in nature, can be made synthetically and have been used in various industrial processes for a very long time. Their application as a fertilizer is quite common and the performance is very good, thanks to their high ion exchange capacity [25]. Their availability also makes them very competitive from price point of view. Zeolites are relatively lighter than other materials, thanks to the presence of channels and pores [26]. The large surface area due to the structure makes them a very efficient adsorbent. This unique structure also allows modifications to improve the activity and selectivity by introducing various functional groups [27]. Natural zeolites can also be modified with other nanomaterials. The capacity, flexibility, price and availability make zeolites more preferred over other materials. Moreover, they are also heat resistant and chemically stable, which broaden the range of applicable process conditions [28]. Applications of zeolites for air purification include air separation, sensors and air quality monitoring, and control of effluents and exhaust gases [28–30].

11.2.2.3 Titanium Dioxide (TiO_2)

Titanium dioxide (TiO_2), also called titania, is a natural oxide of titanium. There are three crystalline forms of TiO_2: rutile, anatase and brookite. The first two are more common than the last and have tetragonal crystal structure. Rutile is more stable than anatase and is the equilibrium polymorph of anatase. Brookite, on the other hand, has an orthorhombic structure. When heated above 750°C, the crystal structure changes to that of rutile. The differences in crystal structure are depicted in Figure 11.6. The methods used for producing synthetic TiO_2 nanoparticles (mostly rutile and anatase) are sol–gel process, chloride process and sulphate process. TiO_2 nanoparticles are sometimes coated with other metal oxides depending on the purpose [31].

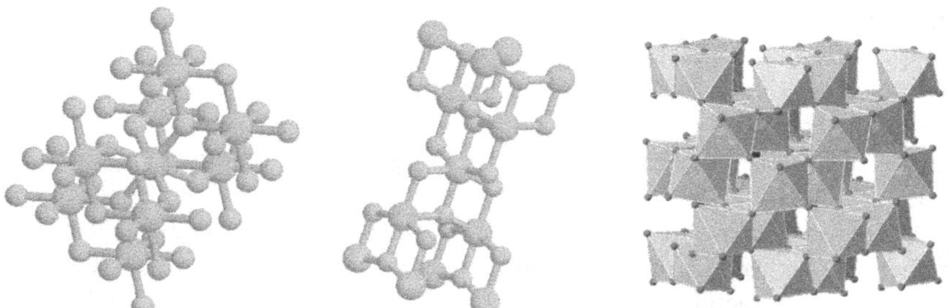

FIGURE 11.6 Crystal structures of rutile (left), anatase (middle) and brookite (right). (See Ref. [32].)

TiO_2 is known for its wide range of applications, in particular as a colouring material. It is found in paints, papers, inks, food products, cosmetics and pharmaceuticals. In fact, it is the most used pigment material in the world, thanks to its high refractive index and brightness. It is also an effective UV blocker. TiO_2 nanoparticles are often one of the key ingredients in sunscreen. The nanoparticles have also gained reputation for their potential as a photocatalyst for decomposing contaminants, thanks to their high UV-absorbing capabilities. Further applications of TiO_2 in photocatalysis will be given later in this chapter.

11.2.2.4 Zinc Oxide

Another metal oxide that is commonly found in air remediation is zinc oxide (ZnO). It is an n-type semiconductor from group II–IV. As nanomaterials, it can take forms of nanowires, nanorods, nanobelts, nanoparticles, etc. While also found in the nature like TiO_2, mostly it is produced synthetically. It can be produced by various techniques including electron beam evaporation, vapour-phase transport, pulsed laser deposition, chemical vapour deposition, electrochemical deposition, hydrothermal synthesis and chemical bath deposition [33]. ZnO is characterized by its wide direct band gap, which makes it used as diodes and varistors. The band gap can be further tuned by combining it with MgO or CdO as alloys. Other applications cover personal care products, plastics, ceramics, glass, cement, rubber, lubricants, paints, pigments, food, batteries, cigarette filters and fire retardants [34]. Similar to TiO_2, ZnO nanoparticles are also used in photocatalysis. Despite being less stable compared to TiO_2 [35], they are relatively cheaper and still competitive as a photocatalyst, thanks to their high photosensitivity and photocatalytic activity [36]. They also have a high chemical sensitivity that makes them a good sensing material for chemicals [37] and biomolecules [38]. Their uses in biomedical field as a sensing material include cell labelling, tumour targeting and diagnostics [39].

11.2.2.5 Metal–Organic Frameworks

Metal–organic frameworks (MOFs) are a new class of organic–inorganic coordination polymers. They are highly porous crystalline materials and semiconductors with tuneable pore size [8] and outstanding photocatalytic activity and thermal stability [5]. MOFs consist of organic ligand parts and metal secondary units (metal oxide clusters or metal ions) [40], as shown in Figure 11.7. They are mostly produced by hydrothermal or solvothermal synthesis, where they are crystallized from a hot solution. Production without use of solvent is also possible due to remarkable mechanochemical reactivity between the organic ligands and metal salts [41].

MOFs are effective for the removal of gaseous pollutants because of their large surface area as well as the porous and diverse structure [5]. The structure offers enormous open metal sites and functionalities for various types of interaction, including acid–base interaction, electrostatic interaction, π-complex formation, π–π interaction, hydrogen bonding and coordination bonds [5]. The open transitional metal sites of MOFs generate highly active radicals for oxidizing hazardous gas and serve as the adsorption site [5]. The more the number of open metal sites is, the

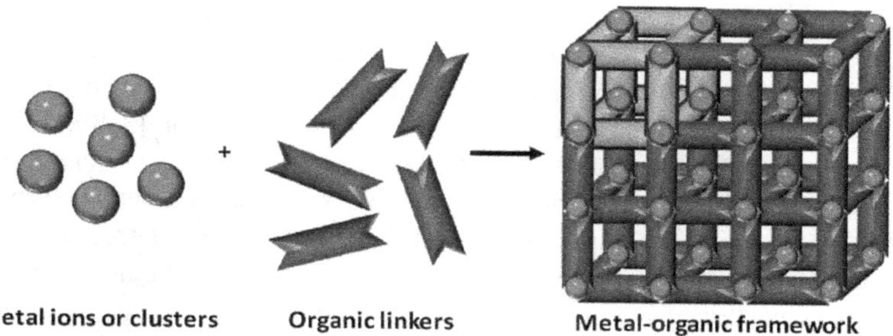

FIGURE 11.7 An illustration of how metal ions or clusters and organic linkers form on MOF. (See Ref. [42].)

better they are in binding toxic gases. A large number of sites for adsorption are also available in quantum-sized nodes, of which each is formed by the binding between a transitional metal ion and oxygen from an organic linker, and is electrondeficient [5]. The high adsorption capacity also means that MOFs can be further utilized as a sensor for specific substances depending on the selectivity [42]. However, the adsorption capacity of MOFs is reduced in humid environment, as molecular water will form irreversible coordination bonds with the open metal sites. To deal with this, some adjustment could be made to the hydrophobicity. Other possibilities are to create a crowding effect by increasing the coordination number between metal cations and organic linkers, and to simply increase the number of binding sites [5,43–45]. Another problem associated with MOFs is the possibility of decomposition when the interaction between toxic substances and the open metal sites is too strong [5]. Moreover, fabrication of MOFs is still expensive.

The properties of MOFs can be modified and adjusted to the applications. Post-modification of MOFs can be done by grafting functional groups on open metal sites and/or on organic linkers or by incorporating ionic elements and other functional materials into the MOFs; thus, the electron density can be tuned [5]. Grafting functional groups on the open metal sites of MOFs also improves the stability as it prevents the hydrolysis of metal clusters. Other than for adsorption and sensing, MOFs are also used for catalysis, separation, drug delivery and gas storage [42]. MOFs have similarities to zeolites as structured composites and in how they are produced. They can coexist in a combined form called zeolitic imidazolate frameworks (ZIFs). They are a class of MOFs with isomorphic topology with zeolites, as transition metal ions are also connected in tetrahedral network by imidazolate linkers [46]. ZIFs also show a promising potential in CO_2 capture and storage [47,48].

11.3 TECHNOLOGIES OF AIR PURIFICATION USING NANOMATERIALS

Nanomaterials are very promising as additions to or even substitutes for commonly used materials. Small size, high surface-to-volume ratio, sensitivity, selectivity and other enhanced properties of nanomaterials allow better detection, capture and treatment of pollutants. They can be designed specifically to effectively react with specific pollutants and turn them into non-toxic substance.

There are three major applications of nanotechnology in the fields of environment, namely pollution prevention, detection and remediation [3,49]. Prevention of pollution in the sense that pollutants are prevented from being released to the air is out of scope of this chapter, which focuses on the remediation of polluted air. Pollutant detection, on the other hand, is considered indispensable for effective remediation. Air purification using nanotechnology can be performed in the following ways [2]:

1. Filtration/separation by nanofilters
2. Adsorption by nanoadsorptive materials (nanoadsorbents)
3. Degradation by nanocatalysts.

11.3.1 Membrane Air Filtration

Conventional air filters are used to physically separate particulates from an airflow passing through. The filters are porous membranes made of stacked and woven fibres of micron-sized diameter. For the filters to be able to remove particulates, a thick filter or layers of membrane are usually required. This, however, suffers from significant pressure drop and is often still insufficient as a large fraction of particulates can still escape [50]. Membranes with high adhesive interactions with the particulates but at the same time thin and transparent are thus necessary. Nano-air filters are made of nanofibrous membrane with fibre diameters between 10 and 1000 nm. Compared to normal air filters, they have higher specific surface area, porosity and mechanical strength, as well as better air flowability. They are far more effective in filtration, thanks to the superior number of active sites, large surface area and better selectivity due to their easy functionalization. An example is filtration of PM2.5 using a nanofibrous membrane with high transparency and low airflow resistance [50]. PM2.5 particulates are trapped on the fibre surfaces due to high adhesive force. The fraction and size of smaller particles (e.g. PM0.1) that pass through the fibre depend on the fibre density and surface properties. As the filtration continues, incoming particulates may also attach on the already captured particulates, forming larger aggregates around the fibres [10]. Over time, the available surface for capture increases, and at the same time, the bond with the fibre becomes stronger, which makes the filtration performance better. However, the air passability slowly decreases and filter handling is required.

Nanofibrous air filters for PM2.5 removal have been made from various polymers such as proteins, silk, nylon 6, polyacrylonitrile (PAN), polyvinylpyrrolidone (PVP), polysulphides, polyvinyl alcohol (PVA), polyiodide, polyurethane (PU), polylactic acid (PLA), poly(m-phenylene isophthalamide), polycarbonates, poly(methyl methacrylate) (PMMA) and polyvinylidene fluoride (PVDF). The use of composite materials such as PVDF/PTFE (polytetrafluoroethylene),

PVC/PU, PAN/fluorinated PU, PAN/silica, nylon 6/PAN, PVA/PAN, PAN/ionic liquid, polysulphone/TiO_2, PAN/polysulphone, PLA/TiO_2 and PAN/MOF have also been reported [50].

Electrospun polymeric nanofibrous membranes (EPNMs) are membranes that are fabricated by electrospinning technique. They have unique nanofibrous architectures and can be incorporated with various multifunctional materials [51]. Nanofibrous air filters equipped with highly transparent PAN EPNMs have shown an efficiency of >95% in PM2.5 removal [10]. Carbon nanofibre membrane made of electrospun PAN with high porosity and plenty nitrogen-containing functional groups has been used for the removal of formaldehyde [52].

Potential use of MOFs as materials for nanofibrous filters has also been reported. PMs mostly have similar composition with MOFs, which are composed of inorganic and organic matters. The charged nature of MOFs serves as an advantage for capturing polar PMs via dipole–dipole or electrostatic interactions [5]. MOF filters made of nanocrystals of four unique MOF structures have been designed and shown very efficient for removing PM2.5 and PM10. [8]. MOF air filters have also been produced commercially from plastic mesh, metal mesh, non-woven fabric and glass cloth by using a roll-to-roll fabrication method [53]. By using this method, window screens are rapidly coated with transparent nanofibrous filter films. The coating can be easily removed by gentle wiping [54]. While quite a lot have been developed for PM10 and PM2.5, very few can be found for smaller particulates such as PM1 and PM0.1. Development of MOF membranes with a hierarchical pore structure can be a good strategy for dealing with these fine particulates [5].

The concept of intelligent nanomaterials, which are characterized by automated capability to react to some certain changes in the environment, can be applied to air filters. Gu et al. used triboelectrification effect, which is the exchange of electrostatic charges between two materials in contact, to design self-powered air filter in automobiles [55]. A triboelectric nanogenerator was used to form a space electric field for PM2.5 removal. The same technology was also used with polyiodide electrospun nanofibrous filters for PM0.1 [56]. Nanofibrous air filters with antibacterial capabilities have also been made using Ag [57], ZnO [58] and TiO_2 [59] nanoparticles. The antibacterial properties are attributed to the highly reactive radicals generated under UV irradiation [50]. The mechanism is called photocatalysis, which will be discussed in detail in the next section.

One of the biggest challenges in the membrane air filtration is the low mechanical strength of the fibrous layers. Mostly, they still require non-woven substrates and exist as composite filters. The optimal filtration capability thus cannot be reached [60]. The classic problem associated with commercial-scale production and affordability is also there. In regard to intelligent nanotechnology, there are still some discrepancies between the current development and the practical applications [50]. Performance tests were carried out mostly in laboratory, and the external triggers were in fact controlled instead of being natural. Studies on the durability and long-term efficacy are also still scarce.

11.3.2 Photocatalysis

Photocatalysis happens when semiconducting materials are exposed to light irradiation of specific energy levels, which should be equal to the band gap energy level. The semiconducting materials act as a catalyst for the reduction reaction of pollutants. The process starts when a polluting substance is adsorbed on the surface of a photocatalyst. Photon energy absorbed by the catalyst leads to generation of electron–hole pair – excitation of an electron from the valence band to the conduction band of the catalyst, leaving a positive hole in the valence band. The electron migrates to the catalyst surface and reacts to produce radical ions, which will oxidize the pollutants [61]. The radical ions are usually generated from the reaction between excited electrons and water molecules. For the case of microorganism, the oxidation by radical ions damages the cell walls/membranes and the DNAs [62].

Catalysts in general can be classified into homogeneous catalysts and heterogeneous catalysts. The former are catalysts that exist in the same phase of the reactants, for example a soluble catalyst mixed with reactants in a solution. They have the merits of high activity and selectivity. The disadvantage is that the product purification and catalyst recovery can be quite difficult. Heterogeneous catalysts do not have this problem. They are also relatively more stable than the homogeneous one. However, the catalytic performance is inferior, since they have lower activity and selectivity compared to the homogeneous one, leading to ineffective and slower reactions [63]. The use of nanomaterials as photocatalysts has the advantage of increased reaction efficiency due to the large active surface [64]. They possess the advantages of both homogeneous and heterogeneous catalysts and do not suffer from the aforementioned demerits. Moreover, the properties are in general superior, which makes them more resource efficient.

The most well-known photocatalyst is TiO_2, which is relatively inert and resistant to photocorrosion [65]. It takes the energy from UV light to oxidize pollutants on its surface into less toxic substances. The process of TiO_2 photocatalysis for pollutant removal is illustrated in Figure 11.8. The photocatalytic ability of TiO_2 makes it used as a coating material for self-cleaning surfaces [66]. It has long been used in paints as colouring material as well as due to its photocatalytic ability for removing VOCs, although recently release of some certain toxins from the paint due to the catalytic reaction has been reported [67]. Photocatalytic efficiency of TiO_2 can be improved by using CNTs, which have more positive conduction bands than TiO_2 [61].

MOFs have been used for self-cleaning surfaces and show good photocatalytic ability and acceptable thermal stability [5]. In MOFs, electrons migrate from the highest occupied molecular orbital (HOMO) to the lowest unoccupied molecular orbital (LUMO), analogous to electron excitation from the valence band to the conduction band in other photocatalysts. The charges then migrate to the surface of metal oxide cluster, on which adsorbed pollutants are oxidized. The charge transfer pathways can be ligand-to-metal, ligand-to-ligand, metal-to-ligand or metal-to-metal-to-ligand depending on the composition of the MOFs [68]. This is, however, not fully understood yet, and further investigations are required. The photocatalytic performance of MOFs for decomposing toxic gases, however, is

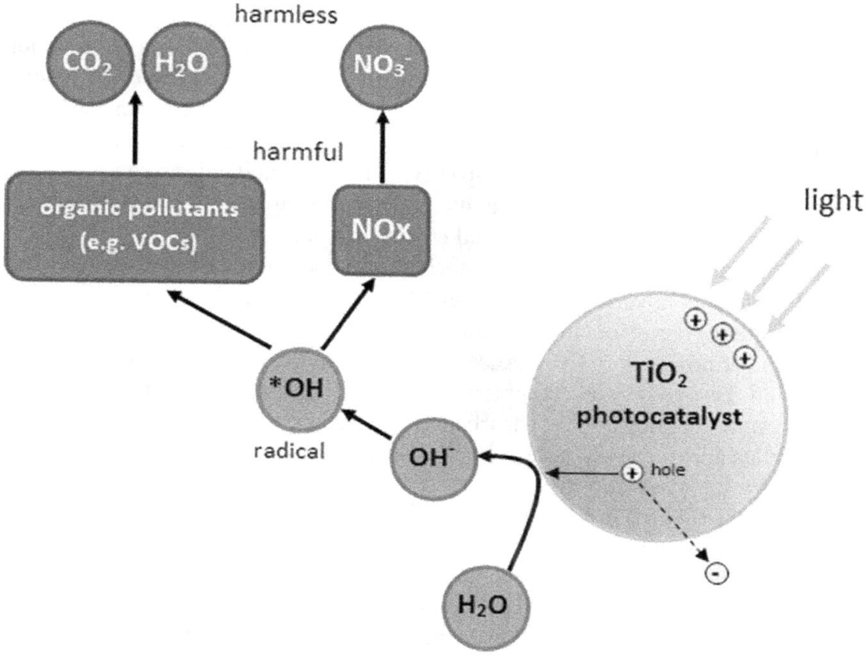

FIGURE 11.8 Mechanism of pollutant reduction by using TiO_2 photocatalyst. (See Ref. [17].)

still inconsistent. Further improvement can be done by coupling them with other semiconducting materials or electronic materials [69]. MOFs still have the problem of stability, and it is still unclear how the stability may be influenced by the active species generated.

Several nanomaterial-based photocatalysts have also been used for dealing with toxic gases. Bismuth oxybromide (BiOBr) nanoplate microspheres have been used as a catalyst under visible light for reducing NO concentration to below 400 ppb, which is the acceptable level [70]. The removal of carbon monoxide can be done at room temperature by using nanogold-based catalysts [63]. ZnO nanostructures combined with bone char have been used as a photocatalyst for removing formaldehydes [71]. Highly effective photocatalysis effect has been attributed to strong adsorption of formaldehyde on bone char. The removal of formaldehyde using photocatalysts can, however, be risky, as ozone might be released as a secondary pollutant [72]. Other applications of nanocatalysts for the reduction of air pollution have been overviewed in a report by the Ministry of Environment and Food of Denmark [73].

11.3.3 Adsorption

Adsorption is a phenomenon of adhesion of atoms, ions or molecules on a surface. The adsorbed substances are called adsorbates, whereas the material on which the adsorbates are adsorbed is called adsorbent. When the adsorbent is a nanomaterial or contains nanomaterials, the term nanoadsorbent is often used. Adsorption can occur through physical or chemical interactions between the surface species of the adsorbent and the adsorbates. Physical interactions occur via van der Waals and electrostatic forces. Pore volume, cavity size and surface area of an adsorbent influence the physical adsorption capacity. Chemical interactions can be due to hydrogen interaction and acid–base interaction. The capacity depends on surface functionality, atomic coordination and electron density.

Adsorption is one of the major mechanisms for separation and decomposition of pollutants in various environmental systems. Gas separation is a common and important process in chemical industries. An example of gas separation process in the atmosphere is CCS. Adsorption-regeneration is more developed than absorption, cryogenic and membrane separation as a CO_2 capture technology [74,75]. Among effective CO_2 adsorbent materials are zeolites, activated carbon, CNTs, silica adsorbent and molecular baskets based on nanoporous silica [3]. The CO_2 adsorption efficiency has been found to increase when modified with ethylene diamine (EDA), polyethylenimine (PEI) and (3-aminopropyl)triethoxysilane (APTES) [3]. The rate of CO_2 adsorption on modified CNTs is positively influenced by relative humidity, but decreases as the temperature is increased.

Adsorption of toxic gas is also an important mechanism in air purification. Some prominent examples of nanoadsorbents for toxic gas removal include zeolites, CNTs, gold nanoparticles, metal oxide nanoparticles and MOFs. Zeolites, for example, can be used to remove NOx at low temperatures [3]. Nanoadsorbents can also exist as composite materials. For example, Sinha and Kazuki developed a composite material made of highly porous manganese oxide and gold nanoparticles [76]. It has been shown very effective for the removal VOCs, nitrogen, SOx, acetaldehyde, toluene and hexane at room temperature.

CNTs are probably the most popular nanoadsorbent out there. They have extremely high adsorption selectivity, affinity and capacity due to the large surface area created by small average diameter and a large number of pores [2]. They also have high resistance to oxidation, which allows regeneration at high temperatures [3]. The interaction between CNTs and organic compounds can occur via hydrogen bonding, π–π and hydrophobic interactions, electrostatic forces and van der Waals forces [77]. The unique structure of CNTs allows modification with additional functional groups, which results in further enhanced properties [23,78].

CNTs have been used for removal of toxic and non-toxic gases from the atmosphere. CNTs have been shown three times more effective than activated carbon in adsorbing dioxins [79]. The efficiency of activated carbon in removing dioxin is in fact already far superior compared to other common commercial adsorbents such as clay, γ-Al_2O3 and zeolites [80]. The strong interaction is due to the presence of two benzene rings in dioxine, which interact strongly with CNT surface. This enhancement effect is also attributed to interaction with the possibility of overlapping event in the tubes. Long and Yang used CNTs as an adsorbent for removing NO [81], which is adsorbed in its oxidized form, NO_2. Adsorption of isopropyl alcohol vapour using SWCNTs has been studied by Hsu and Lu [82]. Oxidation of SWCNTs by HCl, HNO_3 and NaClO solution results in reduction in pore size and subsequent increase in the surface areas of micropores and active functional groups, which leads to improved physicochemical properties.

MOFs are promising for the selective adsorption of gaseous pollutants, thanks to the large number of open metal sites and possibilities of functionalization and framework modification. MOFs doped with semiconductors/POMs have been considered ideal for toxic gas reduction [5], as the pore size and chemical properties can be modified at molecular level [83]. Pore sizes determine the size and shape of adsorbate molecule that may diffuse into the pores. MOFs can be used for inorganic pollutants, N- or S-containing organic compounds, aromatic compounds and halogenated aromatic compounds [84]. The adsorptions can be physical or chemical. Physical adsorptions occur via van der Waals' (e.g. in the framework of MOF) or electrostatic (e.g. with unsaturated metal cations in open metal sites) interactions. Van der Waals' interaction is the binding mechanism in adsorption of short-chain hydrocarbons in alkenes–alkanes separation. Acid–base (chemical) interactions with N- or S-containing organic compounds can occur in hard Lewis acid sites or soft Lewis acid sites, whereas for aromatic pollutants, adsorption occurs through π–π and cation–π interactions [5].

There have been studies on the application of MOFs for adsorption of alcohols and ketones [85]. MOFs showed much higher adsorption capacities compared to Y-zeolite, ACFC (activated carbon fibre cloth) and PCHs (porous clay heterostructures), which are already available adsorbents in the market [86]. Two MOFs, namely MIL-88(B) and amino-functionalized MIL-88(B), have been used in CO removal [87]. CO is adsorbed via interaction with the open metal sites of iron cations in the MOFs. Open metal sites in MOFs also allow efficient adsorption for NH_3, CO, NOx, formaldehyde and other aldehydes [5], as well as adsorption-separation of alkene from alkanes [88], which occurs through the cation–π interaction. In adsorption of halogenated aromatic compounds, it is the π–π interaction that serves as the binding mechanism. The same also plays a role, along with cation–π interaction, in adsorption of VOCs [5,6].

Excellent interaction with S atoms makes some MOFs potential for H_2S adsorption. H_2S could react with metal cations such as Cu, Zn or Fe in the open metal sites to form metal sulphide. Liu et al. tested 11 MOFs of different combinations and properties for adsorbing H_2S [89]. Among them, Mg-MOF-74 showed the most stable adsorption capability, which is due to the presence of strong O–Mg–O bond. MOFs applications for adsorption of SO_2, another S-containing gas, are, however, still limited as the interactions with metal clusters and organic linkers are relative weak. Moreover, there is a possibility of H_2SO_3 and H_2SO_4 formation. They can bind strongly with the metal sites, thus reduce the adsorption capacity and negatively influence the structure stability [90].

As in the case of filtration, MOFs still have a problem with their stability. As mentioned earlier, the effectivity is usually limited in humid environment, with some exceptions like in the removal of warfare agents. The framework of MOFs might collapse when exposed to light halogens at high concentration due to corrosive acids generated [91]. The adsorption capacity and stability of MOFs can be increased by strengthening the coordination bond between the organic linker and the metal cluster, by further optimizing the pore and cavity structures to make new binding sites available, and by adding new functionality via combination with other materials [5]. A common functionalization is by using materials with amino groups, which forms hydrogen bonds with adsorbates. For example, improved formaldehyde adsorption in highly humid conditions has been achieved with a diamine-appended MOF [92]. The increase in adsorption capacity depends on the increase of amino groups to some certain extent. Reversible interactions between formaldehyde and amino groups are the reason for the increased recyclability and hydrophobicity. High Cl_2 adsorption efficiency has also been reported with the use of amino-functionalized IRMOF-3, which is composed of Cu and 2-aminoterephthalate [93].

11.3.4 Nanosensors

Nanomaterials-based sensors are promising due to quick and accurate sensing ability, better selectivity and cost and energy efficiency [2]. They can be used for detecting toxic compounds at ppm and ppb levels in various air, water and soil [94]. Nanotechnology-based sensors can take forms of thick films, multilayer architectures and nanofabricated devices [95]. Nanoparticles improve sensor selectivity as they can be coated with various chemical and biological ligands. The conductivity and sensitivity can be enhanced by using a combination of different metals [96]. Some examples of nanotechnology application for pollution detection have been tabulated by Bhawana and Fulekar [97] and are given in Table 11.1. An intensive review on detection and treatment of toxic gases has also been done by Lin et al. [98].

Metal oxides such as SnO_2, ZnO and WO_3 have intensively been studied as sensor materials for NOx gases [98]. They have the advantages of being highly

TABLE 11.1
Pollution Detection and Sensing – Nanostructural Materials [97]

Nanostructure Material	Function
Silver nanoparticle array membranes	Water quality monitoring
Carbon nanotubes (CNTs)	Electrochemical sensors
CNTs with enzymes	Establish a fast electron transfer from the active site of the enzyme through the CNT to an electrode, in many cases enhancing the electrochemical activity of the biomolecules
CNT sensors	Developed for glucose, ethanol, sulphide and sequence-specific DNA analysis
Magnetic nanoparticles coated with antibodies	Useful for the rapid detection of bacteria in complex matrices

sensitive, non-toxic, relatively cheap and thermally and chemically stable. Nanosensors of metal sulphide and metal oxide have also been used for the detection of SO_2. Some materials such as Polyaniline/WO_3 nanocomposite [99], metal-doped MoS_2 nanoflowers [100], SnO_2 nanostructure and Ni-doped SnO_2 [101] can even operate at room temperature. SnO_2, TiO_2 and WO_3 are the most common nano-metal oxides used for detecting formaldehydes [98]. They work based on response to UV-illumination, cataluminescence and resistance. They are also used in hybrid forms or doped with other metals. Most of them, however, require high temperatures. Only very few can be operated at room temperature, for example TiO_2 [102, 103] and indium tin oxide [104]. ZnO has been used as a sensing material for detection of NH_3, NO_2 and CO, due to its high chemical sensitivity and large surface area [37,105]. In detection of NH_3, the reaction between NH_3 and negative ions on ZnO surface generates N_2. The surface ions were formed from surface electrons that react with oxygen in the air [105].

CNTs, especially SWCNTs, have been used as sensors for NH_3, NO_2 and O_3 [106], with a quicker response and higher sensitivity compared to those of the conventional probes. Exposure to the gases changes the electrical resistance of CNTs. Unlike conventional sensors that require high temperatures, SWCNTs are still effective event at room temperature [3]. Modification with specific functional groups can be done to widen the range of chemical and biological species that can be detected. However, the flexibility will be affected by the type of the functional group [3].

Rajagopalan et al. [107] developed a nanosensor for the detection of heavy metal ions and radioactive elements at any concentration. Portable biosensors with enhanced sensitivity have been used to detect human exposure to chemicals [108]. Wang et al. [109] produced a thin label of nanoparticles that can improve the detection of biomarker signals. It is based on the electrochemical immunoassay method, in which biomarkers are attracted by specific antibodies in the label.

A cantilever sensor is a device made of a series of silicon cantilevers coated with nanomaterials [110]. Each is usually 10–500 μm long, but less than a few microns thick. The detectable molecules depend on the type of nanomaterials, which provide probe molecules. When target molecules are captured by the probe molecules, surface pressure changes and, as a result, the cantilever bends, as shown in Figure 11.9 [3]. The degree of bending is measured by a laser beam, allowing quantitative measurement of the pollutants. Cantilever sensors have been applied for the detection of VOCs, heavy metals, pesticides and pathogenic bacteria.

Some challenges remain in the development and application of nanosensors [98]. Although the selectivity and sensing speed are superior, stability and reproducibility are still in question. Power consumption and durability can be improved by modifying the device sensor structure. Selectivity also still needs to be improved, which can be done by designing specific sensor arrays and using hybrid materials, as well as developing a database of fingerprints of various gases. More theoretical and experimental investigations are still required.

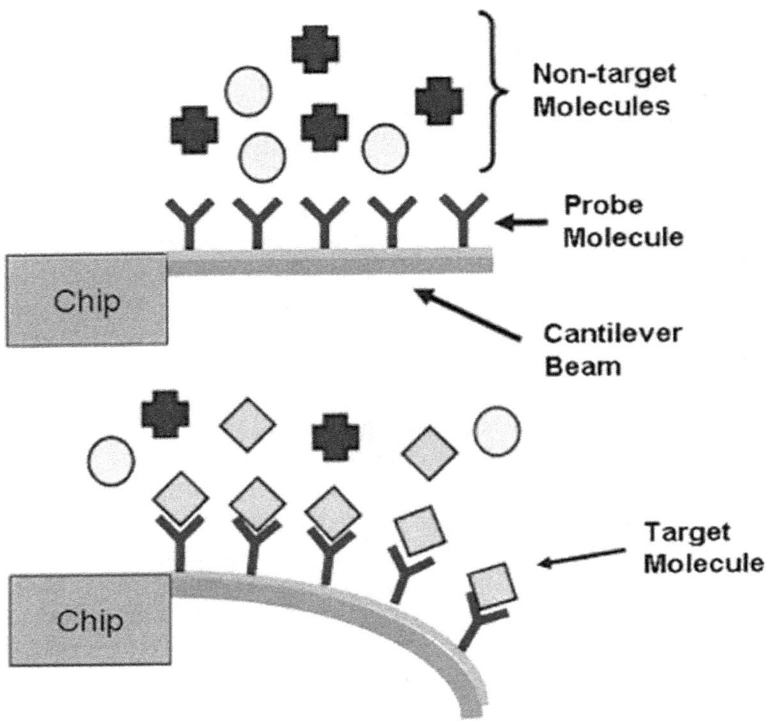

FIGURE 11.9 The mechanism of cantilever-based biosensor in molecular sensing: *above*: before interaction with the target molecules; *below*: after the target molecules have been captured (From Ref. [3]. Courtesy of www.nmji.in. Used with permission.)

11.4 NANOTECHNOLOGY FOR PREVENTION OF COVID-19

Some diseases can be transmitted in the air through small solid and liquid particulates. The pathogen can be viruses, bacteria or fungi in the particulates. The infection usually occurs via respiratory system, although other mucous membranes can also be the route. Among the major airborne diseases are influenza, tuberculosis, measles, diphtheria, chickenpox and SARS diseases. In this section, we take a look at nanotechnology-based air purification as a potential measure for dealing with disease transmission in the air. A special attention is given to COVID-19, a SARS disease and the most recent global pandemic that has taken millions of lives and caused huge economic loses.

11.4.1 SARS-CoV-2 and COVID-19

In 2019, some cases of infection of a new corona virus was found in Wuhan, China, and since then the disease has spread over the world and in 2020 become a global pandemic regarded as the greatest in the last few decades. The disease, well known as COVID-19 (coronavirus disease 2019), is caused by a new type of coronavirus, which is named severe acute respiratory syndrome coronavirus 2 (SARS-CoV-2). The disease is transmitted mainly via oral, nasal and eye cavities. Common symptoms of COVID-19 include fever, dry cough, fatigue, muscle aches and pain, loss of smell and appetite and shortness and difficulty of breathing. The disease has spread in over 220 countries, and by the end of 2020, around 80 million infection cases and 1.7 million deaths have been reported [111]. People already with chronic or heavy diseases are more at risk of death.

A SARS-CoV-2 virus can have spherical to pleomorphic shape, and its size ranges from 65 to 125 nm. The RNA is tightly coiled and coated by the nucleocapsid protein [111,112]. The size and structure make the virus in fact a functional core–shell nanoparticle [113]. The virus is highly stable (remains bioactive) at room temperature. Its surface and aerosol stability is relatively similar to that of SARS-CoV-1 [114], which remains active for more than 3 h. On some surface, it can even remain active until 9 h. Viruses are in general easier to inactivate than bacteria [115]. Virus inactivation could be done by heating at 65°C or higher, or by exposing to extreme pH conditions [116,117]. Since the sensitivity and stability are quite similar, it can be hypothesized that the treatment effective for SARS-CoV-1 would also be effective for SARS-CoV-2. Inactivation by daily means can be performed by using common disinfectants such as bleach, alcohols, povidone-iodine, chloroxylenol, chlorhexidine and benzalkonium chloride [118].

The virus is spread in the form of droplets resulting from respiratory and oral cavities. The droplets are in micron size, and the lifetime depends on the size. Large droplets (>150 μm) need more time to evaporate, but at the same time fall quickly onto nearby surfaces. Smaller droplets evaporate more quickly, but fall more slowly as their settling velocity rapidly decreases. Their dusty nature makes them more airborne and could potentially be an important means of disease transmission, as indicated by several epidemiological studies [119]. Thus, although they are not fully airborne, their aerosol nature and relatively long lifetime on surfaces have caused a great concern [120]. An old question of how effective measures for air purification from bacteria and viruses can be made resurfaces and is now becoming more urgent.

Applications of nanotechnology cover for dealing with diseases cover wide technological areas. Of importance related to COVID-19 are antiviral agents, vaccines, diagnosis and devices for infection control and treatment [113]. Nanotechnology allows the development of self-cleaning and self-disinfecting surfaces that automatically deal with contamination. Some nanomaterials such as metal-based NPs and graphene are known for their antimicrobial properties and can be used as a coating material for surface treatment. The inactivation of microorganisms can be performed photothermally or via photocatalysis-induced ROS generation [117]. Obviously, chemical sensing capability of nanomaterials and nanodevices means that nanotechnology can also be used for diagnosis.

Figure 11.10 shows a distribution of patents related to SARS-CoV viruses until mid-2020. It was compiled from ESPACENET by using the keyword ‘nano AND coronavirus’ [113,121], which was further refined by using additional keywords

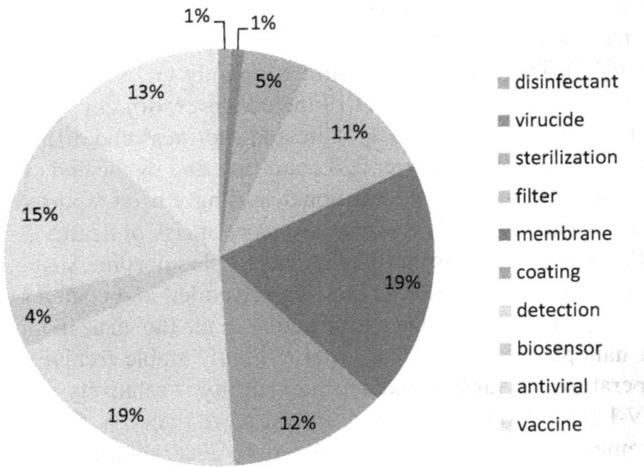

FIGURE 11.10 Distribution of patents related to SARS-CoV viruses available in ESPACENET database. The data were obtained by using the keyword 'nano AND coronavirus', which was further refined by using additional keywords indicated in the chart. (See Refs. [113,121].)

indicated in the chart. Although there is no single product that dominates the distribution, it must be remembered that same patents can be found with different keywords that are in fact closely related, such as virucide-antiviral-disinfectant-sterilization, detection-biosensor and filter-membrane. Although there is no category that dominates the distribution, as expected the use in detection is among the top, which is in accordance with the increasing popularity of nanosensors for diagnostics. It is important to note that the data are mostly contributed by SARS-CoV-1. However, given the similarities between SARS-CoV-1 and SARS-CoV-2, it can be expected that the distribution does not differ much except for vaccines, which distribution has started in 2021.

11.4.2 Masks and Filters

The most common and easy measure for COVID-19 is the use of face mask. The material used in masks and other mechanical ventilation systems is composed of myriad interwoven fibres which allow air to pass through, but capture small objects with size larger than the pores. The capture process occurs by three mechanisms, namely impaction, interception and diffusion [117]. The penetration depth of droplets and micro-objects into the filter depends on the size. The filtration capability of a mask can be enhanced by using a combination of cloth materials with different filtration mechanisms. Konda et al. [122] showed that combining cotton (mechanical filtration) with silk, chiffon and flannel (electrostatic filtration) can greatly increase the filtration efficiency for aerosols smaller than 0.3 μm.

Nanotechnology can further increase the protection capability of masks and other protective gears. Nanoparticles are effective for dealing with microorganisms because they can accumulate in the cells without getting expelled by efflux pumps [117]. Potential improvements due to nanotechnology-based air filters and protection devices include the following:

1. Enhanced particle capture and retention properties (e.g. size range, penetration depth)
2. Reduced particle redistribution due to exhaled humid air
3. Fast inactivation of the captured microorganisms
4. Better material durability and reusability
5. Thinner mask layer for a better user experience (e.g. easiness to breath, voice projection).

Addition of coating materials with antiviral properties allows the masks to not only prevent and capture aerosol, but also kill the virus. This technique has been shown successful even for daily clothes. A famous example is a nanocoating technology from Japan, which prevents suits from getting wet. The technology has been commercialized in the form of fashion products and sprays. Similar strategies should be done for antiviral and antibacterial purpose, as illustrated in Figure 11.11. Nanoparticles adsorb on a fibre surface and act as an inhibitor for the microorganisms. In situ inactivation of pathogen is an advantage as the risk of infection during filter handling can be minimized [113]. Ag and Cu nanoparticles are known for their antimicrobial properties and are widely used in air conditioner filters for the treatment of bioaerosols [2]. Joe et al. used Ag nanoparticles as coating material to give an antiviral property to an air filter media [123]. They also developed a model that can theoretically predict the antiviral ability, optimum coating density and lifetime of an air filter coated with various antiviral materials. TiO_2 nanoparticles, another popular antiviral agent, have also been

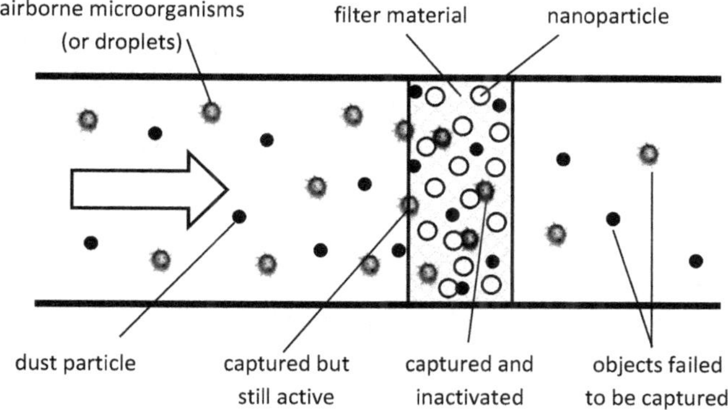

FIGURE 11.11 Antiviral air filtration using nanocoated filter material. (Adapted from [123], with permission.)

used to imbue high-efficiency particle air (HEPA) filters with the capability to destroy microorganisms [124]. Another example is ultrafine glass fibre air filters (ultra-low penetration air [ULPA] filter), which are very effective for particles of around 100 nm [124,125], which are comparable to the size of virus particles. Other nanomaterials potential for this application include Ag, Ag-hybrid or Zn-MOF nanoparticles [113].

As wearing mask in public spaces has been made compulsory during the pandemic in many countries, comfortability for long-time use becomes an important issue. Complaints during daily use include less comfortable breathing, worse voice projection and heat build-up. This gives an opportunity for face masks designed to have better airflow and sound propagation, as well as heating/cooling properties. A nanofibrous face mask with thermal management properties has been made of PE/nylon composite [126]. It can generate a radiative cooling effect, thanks to its high IR transparency and heat dissipation. Oppositely, warming (and possibly antiviral) effect is also obtained when a thin layer of Ag coating is applied on the nano-PE substrate, which increases the IR reflectance [126].

11.4.3 Surface Treatments

Nanotechnology can provide antipathogenic properties to surfaces, and it could be a crucial measure for dealing with COVID-19. The use of nanomaterials for surface treatment has been shown effective for SARS-CoV-1, and the same thing is expected for SARS-CoV-2 [117]. This has in fact been realized in Milan, a city that was once the epicentre of COVID-19 in Europe, as TiO_2/Ag mixture is used to disinfect the streets [127]. Ag has long been known for its antimicrobial properties [128], while TiO_2 is well known as a photocatalyst. For photocatalytic inactivation of SARS-CoV-2, the nanoparticles need to be fixated on surfaces, either by simple physical entrapment in matrices, or with the help of other chemicals in coating [113]. The quality of attachment is one of the key factors that influence how long the self-sterilizing effect would last. For the case in Milan, it is expected to last from 6 months to 2 years.

Nanomaterials can be used to promote surface oxidation by releasing toxic ions and therefore preventing viral dissemination by inhibiting binding/penetration of viral particles. The virus's membrane is destroyed either by heat generated from photothermal-based reactions, or by reactive oxygen species (ROS) and other toxic ions from surface oxidation. The mechanisms of virus inactivation are the following [129]:

1. Surface oxidation
2. Nanoparticle degradation (by oxidation on NPs' shell and surface)
3. Release of toxic ions/free radicals
4. Inhibition of interaction with viral glycoproteins
5. Prevention of binding and penetration
6. Oxidation of the viruses by ROS and other toxic (radical) ions
7. Dissolution of virus's membrane.

The antimicrobial property of Ag has been used since ancient times for medical applications [128]. In some traditional treatments in Asia, silver rings are used as a treatment medium, as they are believed to emanate positive energy and able to cure diseases. As science and technology reigned over myths and traditions, the number of possibilities of Ag utilization as an antiviral and antibacterial agent has rapidly increased. Recently, Ag nanomaterials have been used as biocide in paints (as silver zeolites) [67] and food trays [130]. A combination of activated carbon powder and Ag and CuO nanoparticles was shown very efficient for the adsorption of bacteria and virus particles [131]. Inactivation of viruses by Ag nanoparticles happens through three mechanisms [50]:

1. Ag nanoparticles dissolve and release some toxic Ag(I) forms (including Ag^+ ions), which may interact with surface proteins of viruses or accumulate in host cells. The ions inhibit virus replication by interacting with thiol-containing enzymes required for the process.
2. The presence of Ag nanoparticles on the surface of viruses interferes their docking on host cells.
3. ROS released from the nanoparticle surface damage the virus's membrane.

Another precious metal, Au, has also been used as an antimicrobial agent. It has found some applications in the medical field. Both Ag and Au are used for treatment of cancer in a process called plasmonic photothermal treatment [132]. In forms of nanoparticles and nanorods, they have the ability to generate heat under illumination at certain wavelengths that correspond to the plasmon resonance condition. The required wavelength can be tuned by modifying the size and shape of nanoparticles, which makes the application under solar irradiation possible [133]. Similarly, plasmon-enhanced shockwave generated by Au nanorods under femtosecond pulsed laser irradiation could alter virus membrane and/or surface groups, thus weakening the bind between virus and host cell [134]. One limitation is that a close contact between the nanoparticles and the target cells is required. For surface treatment, this will not be an issue. However, the obvious problem of cost still lingers. A strategy is to use Ag as a secondary material in a catalyst. For example, Ag/Al_2O_3 (5% Ag) catalysts have been shown to be able to inactivate the virus within 5 min [135].

Cu has been shown quite effective for both SARS-CoV-1 and SARS-CoV-2 [136]. Cu-containing alloys can be a replacement of stainless steel for metal surfaces in buildings. Experiments by van Doremalen et al. [114] showed that the coronaviruses could last for only less than 4 h on Cu surfaces, whereas they persist for almost a day on cardboard and 2–3 days on plastic and stainless steel. Coronavirus can be inactivated in less than 20 min by using Al_2O_3-based catalyst containing Cu (10%) [135]. The main mechanisms of virus inactivation are the breakdown of protein and lipid by Cu ions and the shielding of proteins by ROS [114]. Release of Cu ions from Cu and CuO nanoparticles to live cells in contact has been observed by Shi et al. [137]. The virus genes get fragmented as the virus is in contact with Cu-containing surfaces, due to the toxicity of the Cu ions released and the ROS generated from Fenton-like or Haber–Weiss reaction between Cu and exogenous hydrogen or molecular oxygen [138].

TiO_2 nanoparticles are often used as an antipathogen. Bare TiO_2 has been shown as an effective photocatalyst for oxidizing various gram-positive and gram-negative bacteria [115] and viruses. The performance of photocatalytic titanium apatite filter (PTAF) for the treatment of coronavirus has been tested by Han et al. [139]. The result was still not satisfying, as 6 h of exposure under UV light was required for an effective inactivation. This could be due to the high recombination rate of electron–hole pairs and low solar light activity of TiO_2. To improve the efficiency, other metal compounds can be combined with TiO_2 in the catalyst. Byrne et al. showed that the photocatalytic capability under visible light has been shown to improve when TiO_2 is doped with S and N [62].

Other metal oxides that have been used for surface treatment are graphene oxides. An example of application is inhibition of herpes simplex virus 1 (HSV-1) attachment by using functionalized graphene oxide in the form of sheets and sulphated derivatives [140]. The first inactivation mechanism is electrostatic binding between virus particles (positive charge) and the catalyst surface (negative charge). The second inactivation mechanism is shielding of protein by radicals, just like in the case of other metal oxides. Sharp edges in graphene oxide sheets could also give an additional antiviral effect [140].

11.5 CONCLUSIONS

In this chapter, the potential of nanotechnology for air purification has been discussed. Nanomaterials are effective for dealing with pollutants, thanks to the large surface area, selectivity, sensitivity, enhanced adsorptive and catalytic capabilities and modification possibilities. Specific features of some potential nanomaterials as well as various applications of these materials in air filters, catalysts, adsorbents and sensors have been discussed. Nanotechnology also holds promises for dealing with diseases transmitted by spreading of microorganisms in the atmosphere, such as COVID-19. Some nanomaterials based on metals and metal oxides possess antiviral properties that make them effective as an antiviral agent. The inactivation mechanisms include attachment inhibition, release of toxic ions, tearing down of membrane and protein enveloping. As with other applications of nanotechnology in environmental remediation, the challenges include improving the stability of nanomaterials and devices in various conditions, scaling up for commercial production, decreasing production costs and preventing the release of nanomaterials as secondary pollutants.

REFERENCES

[1] Cohen, A. J.; Brauer, M.; Burnett, R.; Anderson, H. R.; Frostad, J.; Estep, K.; Balakrishnan, K.; Brunekreef, B.; Dandona, L.; Dandona, R.; et al. Estimates and 25-Year Trends of the Global Burden of Disease Attributable to Ambient Air Pollution: An Analysis of Data from the Global Burden of Diseases Study 2015. *Lancet* 2017, 389(10082), 1907–1918. https://doi.org/10.1016/S0140–6736(17)30505-6.

[2] Mohamed, E. F. Nanotechnology: Future of Environmental Air Pollution Control. *Environ. Manag. Sustain. Dev.* 2017, 6(2), 429. https://doi.org/10.5296/emsd.v6i2.12047.

[3] Yunus, I. S.; Harwin Kurniawan, A.; Adityawarman, D.; Indarto, A. Nanotechnologies in Water and Air Pollution Treatment. *Environ. Technol. Rev.* 2012, 1(1), 136–148. https://doi.org/10.1080/21622515.2012.733966.

[4] Barea, E.; Montoro, C.; Navarro, J. A. R. Toxic Gas Removal-Metal-Organic Frameworks for the Capture and Degradation of Toxic Gases and Vapours. *Chem. Soc. Rev.* 2014, 43, 5419–5430. https://doi.org/10.1039/c3cs60475f.

[5] Wen, M.; Li, G.; Liu, H.; Chen, J.; An, T.; Yamashita, H. Metal-Organic Framework-Based Nanomaterials for Adsorption and Photocatalytic Degradation of Gaseous Pollutants: Recent Progress and Challenges. *Environ. Sci. Nano* 2019, 6(4), 1006–1025. https://doi.org/10.1039/c8en01167b.

[6] Wang, L.; Li, Y. A.; Yang, F.; Liu, Q. K.; Ma, J. P.; Dong, Y. B. Cd(II)-MOF: Adsorption, Separation, and Guest-Dependent Luminescence for Monohalobenzenes. *Inorg. Chem.* 2014, 53(17), 9087–9094. https://doi.org/10.1021/ic501100p.

[7] Woellner, M.; Hausdorf, S.; Klein, N.; Mueller, P.; Smith, M. W.; Kaskel, S. Adsorption and Detection of Hazardous Trace Gases by Metal–Organic Frameworks. *Adv. Mater.* 2018, 30(37), 1704679. https://doi.org/10.1002/adma.201704679.

[8] Zhang, Y.; Yuan, S.; Feng, X.; Li, H.; Zhou, J.; Wang, B. Preparation of Nanofibrous Metal-Organic Framework Filters for Efficient Air Pollution Control. *J. Am. Chem. Soc.* 2016, 138(18), 5785–5788. https://doi.org/10.1021/jacs.6b02553.

[9] Gawhane, R. D.; Rao, P. S. P.; Budhavant, K. B.; Waghmare, V.; Meshram, D. C.; Safai, P. D. Seasonal Variation of Chemical Composition and Source Apportionment of PM2.5 in Pune, India. *Environ. Sci. Pollut. Res.* 2017, 24(26), 21065–21072. https://doi.org/10.1007/s11356-017-9761-3.

[10] Liu, C.; Hsu, P. C.; Lee, H. W.; Ye, M.; Zheng, G.; Liu, N.; Li, W.; Cui, Y. Transparent Air Filter for High-Efficiency PM 2.5 Capture. *Nat. Commun.* 2015, 6, 6205. https://doi.org/10.1038/ncomms7205.

[11] IPCC. Summary for Policymakers, In: Climate Change 2014, Mitigation of Climate Change. Contribution of Working Group III to the Fifth Assessment Report of the Intergovernmental Panel on Climate Change; 2014.

[12] Herron, J. A.; Kim, J.; Upadhye, A. A.; Huber, G. W.; Maravelias, C. T. A General Framework for the Assessment of Solar Fuel Technologies. *Energy Environ. Sci.* 2015, 8, 126–157. https://doi.org/10.1039/c4ee01958j.

[13] Singh, J.; Dhar, D. W. Overview of Carbon Capture Technology: Microalgal Biorefinery Concept and State-of-the-Art. *Front. Mar. Sci.* 2019, 6, 29. https://doi.org/10.3389/fmars.2019.00029.

[14] Pramudita, D.; Iskandar, I.; Indarto, A. Nano-Enhanced Materials for Reclamation of Mine Spoils. In *Bio-Geotechnologies for Mine Site Rehabilitation*. Prasad, M. N. V., Favas, P. J., de C., Maiti, S. K. (Eds.), Elsevier Inc., 2018, pp. 201–214. https://doi.org/10.1016/B978-0-12-812986-9.00012-9.

[15] Guerra, F. D.; Attia, M. F.; Whitehead, D. C.; Alexis, F. Nanotechnology for Environmental Remediation: Materials and Applications. *Molecules* 2018, 23(7), 1760. https://doi.org/10.3390/molecules23071760.

[16] Kumar, N.; Kumbhat, S. *Essentials in Nanoscience and Nanotechnology*. John Wiley and Sons Inc., New Jersey, 2016. https://doi.org/10.1002/9781119096122.

[17] Abdillah, F.; Fitriana Pramudita, D.; Indrto, A.; Handojo, L. A. Nanotechnology - an Emerging Technology for Bioremediation of Environmental Pollutants. In *Recent Advances in Environmental Management*, Bharagava, R. N. (Ed.), CRC Press, 2018, pp. 109–143. https://doi.org/10.1201/9781351011259.

[18] Ashby, M.; Ferreira, P.; Schodek, D. *Nanomaterials, Nanotechnologies and Design*. Elsevier Butterworth-Heinemann, Oxford, 2009. https://doi.org/10.1016/B978-0-7506-8149-0.X0001-3.

[19] Collister, J. Commercialization of Polymer Nanocomposites. In *Polymer Nanocomposites*. Krishnamoorti, R., Vaia, R. A. (Eds.), American Chemical Society, Washington DC, 2001, pp. 7–14.

[20] Bhattacharya, S. N.; Kamal, M. R.; Gupta, R. K. *Polymeric Nanocomposites: Theory and Practice*. Carl Hanser Verlag GmbH & Co. KG, Munich, 2006. https://doi.org/10.3139/9783446418523.

[21] Boysen, E.; Boysen, N. *Nanotechnology for Dummies*. 2nd ed., Wiley Publishing, Indiana, 2011.

[22] Ren, X.; Li, J.; Tan, X.; Wang, X. Comparative Study of Graphene Oxide, Activated Carbon and Carbon Nanotubes as Adsorbents for Copper Decontamination. *Dalt. Trans.* 2013, 42, 5266–5274. https://doi.org/10.1039/c3dt32969k.

[23] Gupta, V. K.; Saleh, T. A. Sorption of Pollutants by Porous Carbon, Carbon Nanotubes and Fullerene- An Overview. *Environ. Sci. Pollut. Res.* 2013, 20, 2828–2843. https://doi.org/10.1007/s11356-013-1524-1.

[24] Zhao, Y. L.; Stoddart, J. F. Noncovalent Functionalization of Single-Walled Carbon Nanotubes. *Acc. Chem. Res.* 2009, 42(8), 1161–1171. https://doi.org/10.1021/ar900056z.

[25] Ming, D. W.; Allen, E. R. Use of Natural Zeolites in Agronomy, Horticulture, and Environmental Soil Remediation. *Rev. Miner. Geochem.* 2001, 45(1), 619–654. https://doi.org/10.2138/rmg.2001.45.18.

[26] Liu, R.; Lal, R. Nanoenhanced Materials for Reclamation of Mine Lands and Other Degraded Soils: A Review. *J. Nanotechnol.* 2012, 2012, 1–17. https://doi.org/10.1155/2012/461468.

[27] Inglezakis, V. J.; Grigoropoulou, H. Effects of Operating Conditions on the Removal of Heavy Metals by Zeolite in Fixed Bed Reactors. *J. Hazard. Mater.* 2004, 112(1–2), 37–43. https://doi.org/10.1016/j.jhazmat.2004.02.052.

[28] Kosinov, N.; Gascon, J.; Kapteijn, F.; Hensen, E. J. M. Recent Developments in Zeolite Membranes for Gas Separation. *J. Memb. Sci.* 2016, 499, 65–79. https://doi.org/10.1016/j.memsci.2015.10.049.

[29] Byrappa, K.; Yoshimura, M. Hydrothermal Synthesis and Growth of Zeolites. In *Handbook of Hydrothermal Technology*. 2013. https://doi.org/10.1016/b978-0-12-375090-7.00006-2.

[30] De Vos, R. M.; Maier, W. F.; Verweij, H. Hydrophobic Silica Membranes for Gas Separation. *J. Memb. Sci.* 1999, 158(1–2), 277–288. https://doi.org/10.1016/S0376-7388(99)00035-6.

[31] International Organization for Standardization. ISO/TS 11937:2012 Nanotechnologies - Nanoscale Titanium Dioxide in Powder Form - Characteristics and Measurement; Geneva, 2012.

[32] CDC-NIOSH, U. Current Intelligence Bulletin 63: Occupational Exposure to Titanium Dioxide. 2011–160, 2011.

[33] Poornajar, M.; Marashi, P.; Haghshenas Fatmehsari, D.; Kolahdouz Esfahani, M. Synthesis of ZnO Nanorods via Chemical Bath Deposition Method: The Effects of Physicochemical Factors. *Ceram. Int.* 2016, 42, 173–184. https://doi.org/10.1016/j.ceramint.2015.08.073.

[34] Naqvi, S.; Kumar, V.; Gopinath, P. Nanomaterial Toxicity: A Challenge to End Users. In *Applications of Nanomaterials*, 2018, pp. 315–343. https://doi.org/10.1016/b978-0-08-101971-9.00012-0.

[35] Boyjoo, Y.; Sun, H.; Liu, J.; Pareek, V. K.; Wang, S. A Review on Photocatalysis for Air Treatment: From Catalyst Development to Reactor Design. *Chem. Eng. J.* 2017, 310(2), 537–559. https://doi.org/10.1016/j.cej.2016.06.090.

[36] Cappelletti, G.; Pifferi, V.; Mostoni, S.; Falciola, L.; Di Bari, C.; Spadavecchia, F.; Meroni, D.; Davoli, E.; Ardizzone, S. Hazardous O-Toluidine Mineralization by Photocatalytic Bismuth Doped ZnO Slurries. *Chem. Commun.* 2015, 51, 10459–10462. https://doi.org/10.1039/c5cc02620b.

[37] Park, S. High-Response and Selective Hydrogen Sensing Properties of Porous ZnO Nanotubes. *Curr. Appl. Phys.* 2016, 16(10), 1263–1269. https://doi.org/10.1016/j.cap.2016.07.005.

[38] Voon, C. H.; Sam, S. T. Physical Surface Modification on the Biosensing Surface. In *Nanobiosensors for Biomolecular Targeting.* 2018, pp. 23–50. https://doi.org/10.1016/B978-0-12-813900-4.00002-6.

[39] Burns, A.; Self, W. T. Antioxidant Inorganic Nanoparticles and Their Potential Applications in Biomedicine. In *Smart Nanoparticles for Biomedicine.* 2018, pp. 159–169. https://doi.org/10.1016/B978-0-12-814156-4.00011-2.

[40] Wang, J. L.; Wang, C.; Lin, W. Metal-Organic Frameworks for Light Harvesting and Photocatalysis. *ACS Catal.* 2012, 2, 2630–2640. https://doi.org/10.1021/cs3005874.

[41] Pichon, A.; James, S. L. An Array-Based Study of Reactivity under Solvent-Free Mechanochemical Conditions - Insights and Trends. *CrystEngComm* 2008, 10, 1839–1847. https://doi.org/10.1039/b810857a.

[42] Carrasco, S. Metal-Organic Frameworks for the Development of Biosensors: A Current Overview. *Biosensors* 2018, 8(4), 92. https://doi.org/10.3390/bios8040092.

[43] Petit, C.; Bandosz, T. J. Enhanced Adsorption of Ammonia on Metal-Organic Framework/Graphite Oxide Composites: Analysis of Surface Interactions. *Adv. Funct. Mater.* 2010, 20(1), 111–118. https://doi.org/10.1002/adfm.200900880.

[44] Petit, C.; Mendoza, B.; Bandosz, T. J. Reactive Adsorption of Ammonia on Cu-Based MOF/Graphene Composites. *Langmuir* 2010, 26, 15302–15309. https://doi.org/10.1021/la1021092.

[45] Petit, C.; Bandosz, T. J. Exploring the Coordination Chemistry of MOF-Graphite Oxide Composites and Their Applications as Adsorbents. *Dalt. Trans.* 2012, 41, 4027–4035. https://doi.org/10.1039/c2dt12017h.

[46] Wikipedia.org. Zeolitic imidazolate framework. https://en.wikipedia.org/wiki/Zeolitic_imidazolate_framework (Last edited on 24 Sept. 2021).

[47] Banerjee, R.; Phan, A.; Wang, B.; Knobler, C.; Furukawa, H.; O'Keeffe, M.; Yaghi, O. M. High-Throughput Synthesis of Zeolitic Imidazolate Frameworks and Application to CO_2 Capture. *Science* 2008, 319(5865), 939–943. https://doi.org/10.1126/science.1152516.

[48] Wang, B.; Côté, A. P.; Furukawa, H.; O'Keeffe, M.; Yaghi, O. M. Colossal Cages in Zeolitic Imidazolate Frameworks as Selective Carbon Dioxide Reservoirs. *Nature* 2008, 453, 207–211. https://doi.org/10.1038/nature06900.

[49] Yadav, K. K.; Singh, J. K.; Gupta, N.; Kumar, V. A Review of Nanobioremediation Technologies for Environmental Cleanup: A Novel Biological Approach. *J. Mater. Environ. Sci.* 2017, 8(2), 740–757.

[50] Chang, J.; Zhang, L.; Wang, P. Intelligent Environmental Nanomaterials. *Environ. Sci. Nano* 2018, 5(4), 811–836. https://doi.org/10.1039/c7en00760d.

[51] Liao, Y.; Loh, C. H.; Tian, M.; Wang, R.; Fane, A. G. Progress in Electrospun Polymeric Nanofibrous Membranes for Water Treatment: Fabrication, Modification and Applications. *Prog. Polym. Sci.* 2018, 77, 69–94. https://doi.org/10.1016/j.progpolymsci.2017.10.003.

[52] Lee, K. J.; Shiratori, N.; Lee, G. H.; Miyawaki, J.; Mochida, I.; Yoon, S. H.; Jang, J. Activated Carbon Nanofiber Produced from Electrospun Polyacrylonitrile Nanofiber as a Highly Efficient Formaldehyde Adsorbent. *Carbon* 2010, 48 (15), 4248–4255. https://doi.org/10.1016/j.carbon.2010.07.034.

[53] Chen, Y.; Zhang, S.; Cao, S.; Li, S.; Chen, F.; Yuan, S.; Xu, C.; Zhou, J.; Feng, X.; Ma, X.; et al., Roll-to-Roll Production of Metal-Organic Framework Coatings for Particulate Matter Removal. *Adv. Mater.* 2017, 29(15), 1606221. https://doi.org/10.1002/adma.201606221.

[54] Khalid, B.; Bai, X.; Wei, H.; Huang, Y.; Wu, H.; Cui, Y. Direct Blow-Spinning of Nanofibers on a Window Screen for Highly Efficient PM2.5 Removal. *Nano Lett.* 2017, 17(2), 1140–1148. https://doi.org/10.1021/acs.nanolett.6b04771.

[55] Han, C. B.; Jiang, T.; Zhang, C.; Li, X.; Zhang, C.; Cao, X.; Wang, Z. L. Removal of Particulate Matter Emissions from a Vehicle Using a Self-Powered Triboelectric Filter. *ACS Nano* 2015, 9(12), 12552–12561. https://doi.org/10.1021/acsnano.5b06327.

[56] Gu, G. Q.; Han, C. B.; Lu, C. X.; He, C.; Jiang, T.; Gao, Z. L.; Li, C. J.; Wang, Z. L. Triboelectric Nanogenerator Enhanced Nanofiber Air Filters for Efficient Particulate Matter Removal. *ACS Nano* 2017, 11(6), 6211–6217. https://doi.org/10.1021/acsnano.7b02321.

[57] Wang, C.; Wu, S.; Jian, M.; Xie, J.; Xu, L.; Yang, X.; Zheng, Q.; Zhang, Y. Silk Nanofibers as High Efficient and Lightweight Air Filter. *Nano Res.* 2016, 9, 2590–2597. https://doi.org/10.1007/s12274-016-1145-3.

[58] Zhong, Z.; Xu, Z.; Sheng, T.; Yao, J.; Xing, W.; Wang, Y. Unusual Air Filters with Ultrahigh Efficiency and Antibacterial Functionality Enabled by ZnO Nanorods. *ACS Appl. Mater. Interfaces* 2015, 7(38), 21538–21544. https://doi.org/10.1021/acsami.5b06810.

[59] Wang, Z.; Pan, Z.; Wang, J.; Zhao, R. A Novel Hierarchical Structured Poly(Lactic Acid)/Titania Fibrous Membrane with Excellent Antibacterial Activity and Air Filtration Performance. *J. Nanomater.* 2016, 2016, 6272983. https://doi.org/10.1155/2016/6272983.

[60] Wang, N.; Mao, X.; Zhang, S.; Yu, J.; Ding, B. Electrospun Nanofibers for Air Filtration. In *Electrospun Nanofibers for Energy and Environmental Applications*, Ding, B., Yu, J. (Eds.), Springer Verlag, Berlin, Heidelberg, 2014, pp. 299–323.

[61] Low, J.; Cheng, B.; Yu, J. Surface Modification and Enhanced Photocatalytic CO_2 Reduction Performance of TiO_2: A Review. *Appl. Surf. Sci.* 2017, 392(15), 658–686. https://doi.org/10.1016/j.apsusc.2016.09.093.

[62] Byrne, J. A.; Dunlop, P. S. M.; Hamilton, J. W. J.; Fernández-Ibáñez, P.; Polo-López, I.; Sharma, P. K.; Vennard, A. S. M. A Review of Heterogeneous Photocatalysis for Water and Surface Disinfection. *Molecules* 2015, 20(4), 5574–5615. https://doi.org/10.3390/molecules20045574.

[63] Singh, S. B.; Tandon, P. K. Catalysis: A Brief Review on Nano-Catalyst. *J. Energy Chem. Eng.* 2014, 2(3), 106–115.

[64] Özkar, S. Enhancement of Catalytic Activity by Increasing Surface Area in Heterogeneous Catalysis. *Appl. Surf. Sci.* 2009, 256(5), 1272–1277. https://doi.org/10.1016/j.apsusc.2009.10.036.

[65] Maeda, K.; Domen, K. New Non-Oxide Photocatalysts Designed for Overall Water Splitting under Visible Light. *J. Phys. Chem. C* 2007, 111(22), 7851–7861. https://doi.org/10.1021/jp070911w.

[66] Shen, W.; Zhang, C.; Li, Q.; Zhang, W.; Cao, L.; Ye, J. Preparation of Titanium Dioxide Nano Particle Modified Photocatalytic Self-Cleaning Concrete. *J. Clean. Prod.* 2015, 87, 762–765. https://doi.org/10.1016/j.jclepro.2014.10.014.

[67] Kaiser, J. P.; Zuin, S.; Wick, P. Is Nanotechnology Revolutionizing the Paint and Lacquer Industry? A Critical Opinion. *Sci. Total Environ.* 2013, 442, 282–289. https://doi.org/10.1016/j.scitotenv.2012.10.009.
[68] Wen, M.; Mori, K.; Kuwahara, Y.; An, T.; Yamashita, H. Design of Single-Site Photocatalysts by Using Metal–Organic Frameworks as a Matrix. *Chem. - Asian J.* 2018, 13(14), 1767–1779. https://doi.org/10.1002/asia.201800444.
[69] Zhang, Y.; Li, Q.; Liu, C.; Shan, X.; Chen, X.; Dai, W.; Fu, X. The Promoted Effect of a Metal-Organic Frameworks (ZIF-8) on Au/TiO_2 for CO Oxidation at Room Temperature Both in Dark and under Visible Light Irradiation. *Appl. Catal., B Environ.* 2018, 224, 283–294. https://doi.org/10.1016/j.apcatb.2017.10.027.
[70] Ai, Z.; Ho, W.; Lee, S.; Zhang, L. Efficient Photocatalytic Removal of NO in Indoor Air with Hierarchical Bismuth Oxybromide Nanoplate Microspheres under Visible Light. *Environ. Sci. Technol.* 2009, 43, 4143–4150. https://doi.org/10.1021/es9004366.
[71] Rezaee, A.; Rangkooy, H.; Khavanin, A.; Jafari, A. J. High Photocatalytic Decomposition of the Air Pollutant Formaldehyde Using Nano-ZnO on Bone Char. *Environ. Chem. Lett.* 2014, 12(2), 353–357. https://doi.org/10.1007/s10311-014-0453-7.
[72] Miyawaki, J.; Lee, G. H.; Yeh, J.; Shiratori, N.; Shimohara, T.; Mochida, I.; Yoon, S. H. Development of Carbon-Supported Hybrid Catalyst for Clean Removal of Formaldehyde Indoors. *Catal. Today* 2012, *185*, 278–283. https://doi.org/10.1016/j.cattod.2011.09.036.
[73] Christensen, F. M.; Brinch, A.; Kjølholt, J.; Mines, P. D.; Schumacher, N.; Jørgensen, T. H.; Hummelshøj, R. M. *Nano-Enabled Environmental Products and Technologies - Opportunities and Drawbacks.* Copenhagen, 2015.
[74] Aaron, D.; Tsouris, C. Separation of CO_2 from Flue Gas: A Review. *Sep. Sci. Technol.* 2005, 40, 321–348. https://doi.org/10.1081/SS-200042244.
[75] White, C. M.; Strazisar, B. R.; Granite, E. J.; Hoffman, J. S.; Pennline, H. W. Separation and Capture of Co_2 from Large Stationary Sources and Sequestration in Geological Formations—Coalbeds and Deep Saline Aquifers. *J. Air Waste Manag. Assoc.* 2003, *53*, 645–715. https://doi.org/10.1080/10473289.2003.10466206.
[76] Sinha, A. K.; Suzuki, K. Novel Mesoporous Chromium Oxide for VOCs Elimination. *Appl. Catal. B Environ.* 2007, 70(1–4), 417–422. https://doi.org/10.1016/j.apcatb.2005.10.035.
[77] Ren, X.; Chen, C.; Nagatsu, M.; Wang, X. Carbon Nanotubes as Adsorbents in Environmental Pollution Management: A Review. *Chem. Eng. J.* 2011, 170, 395–410. https://doi.org/10.1016/j.cej.2010.08.045.
[78] Wang, S.; Sun, H.; Ang, H. M.; Tadé, M. O. Adsorptive Remediation of Environmental Pollutants Using Novel Graphene-Based Nanomaterials. *Chem. Eng. J.* 2013, 226, 336–347. https://doi.org/10.1016/j.cej.2013.04.070.
[79] Long, R. Q.; Yang, R. T. Carbon Nanotubes as Superior Sorbent for Dioxin Removal. *J. Am. Chem. Soc.* 2001, 123(9), 2058–2059. https://doi.org/10.1021/ja003830l.
[80] Cudahy, J. J.; Helsel, R. W. Removal of Products of Incomplete Combustion with Carbon. *Waste Manag.* 2000, 20(5–6), 339–345. https://doi.org/10.1016/S0956-053X(99)00335-9.
[81] Long, R. Q.; Yang, R. T. Carbon Nanotubes as a Superior Sorbent for Nitrogen Oxides. *Ind. Eng. Chem. Res.* 2001, 40, 4288–4291. https://doi.org/10.1021/ie000976k.
[82] Hsu, S.; Lu, C. Modification of Single-Walled Carbon Nanotubes for Enhancing Isopropyl Alcohol Vapor Adsorption from Air Streams. *Sep. Sci. Technol.* 2007, 42(12), 2751–2766. https://doi.org/10.1080/01496390701515060.

[83] Tanabe, K. K.; Cohen, S. M. Postsynthetic Modification of Metal–Organic Frameworks—a Progress Report. *Chem. Soc. Rev.* 2011, 40, 498–519. https://doi.org/10.1039/c0cs00031k.

[84] Hu, P.; Liang, X.; Yaseen, M.; Sun, X.; Tong, Z.; Zhao, Z.; Zhao, Z. Preparation of Highly-Hydrophobic Novel N-Coordinated UiO-66(Zr) with Dopamine via Fast Mechano-Chemical Method for (CHO-/Cl-)-VOCs Competitive Adsorption in Humid Environment. *Chem. Eng. J.* 2018, 332, 608–618. https://doi.org/10.1016/j.cej.2017.09.115.

[85] Terencio, T.; Di Renzo, F.; Berthomieu, D.; Trens, P. Adsorption of Acetone Vapor by Cu-BTC: An Experimental and Computational Study. *J. Phys. Chem. C* 2013, 117(49), 26156–26165. https://doi.org/10.1021/jp410152p.

[86] Yang, K.; Sun, Q.; Xue, F.; Lin, D. Adsorption of Volatile Organic Compounds by Metal-Organic Frameworks MIL-101: Influence of Molecular Size and Shape. *J. Hazard. Mater.* 2011. https://doi.org/10.1016/j.jhazmat.2011.08.020.

[87] Ma, M.; Noei, H.; Mienert, B.; Niesel, J.; Bill, E.; Muhler, M.; Fischer, R. A.; Wang, Y.; Schatzschneider, U.; Metzler-Nolte, N. Iron Metal-Organic Frameworks MIL-88B and NH2-MIL-88B for the Loading and Delivery of the Gasotransmitter Carbon Monoxide. *Chem. - A Eur. J.* 2013, 19, 6785–6790. https://doi.org/10.1002/chem.201201743.

[88] Böhme, U.; Barth, B.; Paula, C.; Kuhnt, A.; Schwieger, W.; Mundstock, A.; Caro, J.; Hartmann, M. Ethene/Ethane and Propene/Propane Separation via the Olefin and Paraffin Selective Metal-Organic Framework Adsorbents CPO-27 and ZIF-8. *Langmuir* 2013, 29(27), 8592–8600. https://doi.org/10.1021/la401471g.

[89] Liu, J.; Wei, Y.; Li, P.; Zhao, Y.; Zou, R. Selective H_2S/CO_2 Separation by Metal-Organic Frameworks Based on Chemical-Physical Adsorption. *J. Phys. Chem. C* 2017, 121, 13249–13255. https://doi.org/10.1021/acs.jpcc.7b04465.

[90] Yu, K.; Kiesling, K.; Schmidt, J. R. Trace Flue Gas Contaminants Poison Coordinatively Unsaturated Metal-Organic Frameworks: Implications for CO_2 Adsorption and Separation. *J. Phys. Chem. C* 2012, 116, 20480–20488. https://doi.org/10.1021/jp307894e.

[91] DeCoste, J. B.; Browe, M. A.; Wagner, G. W.; Rossin, J. A.; Peterson, G. W. Removal of Chlorine Gas by an Amine Functionalized Metal-Organic Framework via Electrophilic Aromatic Substitution. *Chem. Commun.* 2015, 51, 12474–12477. https://doi.org/10.1039/c5cc03780h.

[92] Wang, Z.; Wang, W.; Jiang, D.; Zhang, L.; Zheng, Y. Diamine-Appended Metal-Organic Frameworks: Enhanced Formaldehyde-Vapor Adsorption Capacity, Superior Recyclability and Water Resistibility. *Dalt. Trans.* 2016, 45, 11306–11311. https://doi.org/10.1039/c6dt01696k.

[93] Britt, D.; Tranchemontagne, D.; Yaghi, O. M. Metal-Organic Frameworks with High Capacity and Selectivity for Harmful Gases. *Proc. Natl. Acad. Sci.* 2008, 105, 11623–11627. https://doi.org/10.1073/pnas.0804900105.

[94] Zhou, R.; Hu, G.; Yu, R.; Pan, C.; Wang, Z. L. Piezotronic Effect Enhanced Detection of Flammable/Toxic Gases by ZnO Micro/Nanowire Sensors. *Nano Energy* 2015, 12, 588–596. https://doi.org/10.1016/j.nanoen.2015.01.036.

[95] Hooker, S. Nanotechnology Advantages Applied to Gas Sensor Development. Proc. 5th Annu. BCC Nanoparticles …, 2002.

[96] Zaporotskova, I. V.; Boroznina, N. P.; Parkhomenko, Y. N.; Kozhitov, L. V. Carbon Nanotubes: Sensor Properties. A Review. *Mod. Electron. Mater.* 2016, 2(4), 95–105. https://doi.org/10.1016/j.moem.2017.02.002.

[97] Bhawana, P.; Fulekar, M. H. Nanotechnology: Remediation Technologies to Clean Up the Environmental Pollutants. *Res. J. Chem. Sci.* 2012, 2(2), 90–96.

[98] Lin, C.; Xu, W.; Yao, Q.; Wang, X. Nanotechnology on Toxic Gas Detection and Treatment. In *Novel Nanomaterials for Biomedical, Environmental and Energy Applications.* 2018, pp. 275–297. https://doi.org/10.1016/B978-0-12-814497-8.00009-6.

[99] Chaudhary, V.; Kaur, A. Enhanced Room Temperature Sulfur Dioxide Sensing Behaviour of in Situ Polymerized Polyaniline-Tungsten Oxide Nanocomposite Possessing Honeycomb Morphology. *RSC Adv.* 2015, 5, 73535–73544. https://doi.org/10.1039/c5ra08275g.

[100] Zhang, D.; Wu, J.; Li, P.; Cao, Y. Room-Temperature SO_2 Gas-Sensing Properties Based on a Metal-Doped MoS2 Nanoflower: An Experimental and Density Functional Theory Investigation. *J. Mater. Chem. A* 2017, 5, 20666–20677. https://doi.org/10.1039/c7ta07001b.

[101] Falla, P. H.; Peres, H. E. M.; Gouvêa, D.; Ramirez-Fernandez, F. J. Doped Tin Oxide Nanometric Films for Environment Monitoring. In *Materials Science Forum.* 2005, pp. 498–499. https://doi.org/10.4028/0-87849-984-9.636.

[102] Wu, G.; Zhang, J.; Wang, X.; Liao, J.; Xia, H.; Akbar, S. A.; Li, J.; Lin, S.; Li, X.; Wang, J. Hierarchical Structured TiO_2 Nano-Tubes for Formaldehyde Sensing. *Ceram. Int.* 2012, 38(8), 6341–6347. https://doi.org/10.1016/j.ceramint.2012.05.004.

[103] Liu, L.; Li, X.; Dutta, P. K.; Wang, J. Room Temperature Impedance Spectroscopy-Based Sensing of Formaldehyde with Porous TiO2 under UV Illumination. *Sensors Actuators, B Chem.* 2013, 185, 1–9. https://doi.org/10.1016/j.snb.2013.04.090.

[104] Vaishnav, V. S.; Patel, S. G.; Panchal, J. N. Development of ITO Thin Film Sensor for the Detection of Formaldehyde at Room Temperature. *Sensors Actuators, B Chem.* 2014, 202, 1002–1009. https://doi.org/10.1016/j.snb.2014.04.090.

[105] Zhang, Y.; Liu, T.; Hao, J.; Lin, L.; Zeng, W.; Peng, X.; Wang, Z. Enhancement of NH_3 Sensing Performance in Flower-like ZnO Nanostructures and Their Growth Mechanism. *Appl. Surf. Sci.* 2015, 357(A), 1263–1269. https://doi.org/10.1016/j.apsusc.2015.08.170.

[106] Azam, M. A.; Alias, F. M.; Tack, L. W.; Seman, R. N. A. R.; Taib, M. F. M. Electronic Properties and Gas Adsorption Behaviour of Pristine, Silicon-, and Boron-Doped (8, 0) Single-Walled Carbon Nanotube: A First Principles Study. *J. Mol. Graph. Model.* 2017, *75*, 85–93. https://doi.org/10.1016/j.jmgm.2017.05.003.

[107] Rajagopalan, V.; Boussaad, S.; Tao, N. J. A Nanocontact Sensor for Heavy Metal Ion Detections. In *Nanotechnology and the Environment: Applications and Implications.* Karn, B., Masciangioli, T., Zhang, W., Colvin, V., Alivisatos, P., (Eds.), 2004, pp. 173–178. https://doi.org/10.1021/bk-2005-0890.ch022.

[108] Liu, G.; Lin, Y. Y.; Wang, J.; Wu, H.; Wai, C. M.; Lin, Y. Disposable Electrochemical Immunosensor Diagnosis Device Based on Nanoparticle Probe and Immunochromatographic Strip. *Anal. Chem.* 2007, 79(20), 7644–7653. https://doi.org/10.1021/ac070691i.

[109] Wang, J.; Liu, G.; Lin, Y. Immunosensors Based on Functional Nanoparticle Labels. *ECS Trans.* 2007, 2(19), 1–7. https://doi.org/10.1149/1.2408982.

[110] Filipponi, L.; Sutherland, D. Chapter 4 Fundamental "Nano - Effects." In *NANOYOU Teachers Training Kit in Nanotechnologies.* 2010.

[111] WHO. Coronavirus disease (COVID-19) pandemic.

[112] Wikipedia.org. Covid-19 pandemic. https://en.wikipedia.org/wiki/COVID-19_pandemic (Last edited, 16 Nov. 2021)

[113] Ruiz-Hitzky, E.; Darder, M.; Wicklein, B.; Ruiz-Garcia, C.; Martín-Sampedro, R.; del Real, G.; Aranda, P. Nanotechnology Responses to COVID-19. *Adv. Healthc. Mater.* 2020, 9(19), 1–26. https://doi.org/10.1002/adhm.202000979.

[114] van Doremalen, N.; Bushmaker, T.; Morris, D. H.; Holbrook, M. G.; Gamble, A.; Williamson, B. N.; Tamin, A.; Harcourt, J. L.; Thornburg, N. J.; Gerber, S. I.; et al. Aerosol and Surface Stability of SARS-CoV-2 as Compared with SARS-CoV-1. *New Engl. J. Med.* 2020, 382, 1564–1567. https://doi.org/10.1056/nejmc2004973.
[115] Bogdan, J.; Zarzyńska, J.; Pławińska-Czarnak, J. Comparison of Infectious Agents Susceptibility to Photocatalytic Effects of Nanosized Titanium and Zinc Oxides: A Practical Approach. *Nanoscale Res. Lett.* 2015, 10(1), 1023. https://doi.org/10.1186/s11671-015-1023-z.
[116] Darnell, M. E. R.; Subbarao, K.; Feinstone, S. M.; Taylor, D. R. Inactivation of the Coronavirus That Induces Severe Acute Respiratory Syndrome, SARS-CoV. *J. Virol. Methods* 2004, 121(1), 85–91. https://doi.org/10.1016/j.jviromet.2004.06.006.
[117] Weiss, C.; Carriere, M.; Fusco, L.; Fusco, L.; Capua, I.; Regla-Nava, J. A.; Pasquali, M.; Pasquali, M.; Pasquali, M.; Scott, J. A.; et al. Toward Nanotechnology-Enabled Approaches against the COVID-19 Pandemic. *ACS Nano* 2020, 14(6), 6383–6406. https://doi.org/10.1021/acsnano.0c03697.
[118] Kampf, G.; Todt, D.; Pfaender, S.; Steinmann, E. Persistence of Coronaviruses on Inanimate Surfaces and Their Inactivation with Biocidal Agents. *J. Hosp. Infect.* 2020, 104(3), 246–251. https://doi.org/10.1016/j.jhin.2020.01.022.
[119] Lu, J.; Gu, J.; Gu, J.; Li, K.; Xu, C.; Su, W.; Lai, Z.; Zhou, D.; Yu, C.; Xu, B.; et al. COVID-19 Outbreak Associated with Air Conditioning in Restaurant, Guangzhou, China, 2020. *Emerg. Infect. Dis.* 2020, 26(7), 1628–1631. https://doi.org/10.3201/eid2607.200764.
[120] Lewis, D. Is the Coronavirus Airborne? Experts Can't Agree. *Nature* 2020, 580(7802), 175. https://doi.org/10.1038/d41586-020-00974-w.
[121] ESPACENET Patent Search. European Patent Office.
[122] Konda, A.; Prakash, A.; Moss, G. A.; Schmoldt, M.; Grant, G. D.; Guha, S. Aerosol Filtration Efficiency of Common Fabrics Used in Respiratory Cloth Masks. *ACS Nano* 2020, 14(5), 6339. https://doi.org/10.1021/acsnano.0c03252.
[123] Joe, Y. H.; Park, D. H.; Hwang, J. Evaluation of Ag Nanoparticle Coated Air Filter against Aerosolized Virus: Anti-Viral Efficiency with Dust Loading. *J. Hazard. Mater.* 2016, 301, 547–553. https://doi.org/10.1016/j.jhazmat.2015.09.017.
[124] Liu, G.; Xiao, M.; Zhang, X.; Gal, C.; Chen, X.; Liu, L.; Pan, S.; Wu, J.; Tang, L.; Clements-Croome, D. A Review of Air Filtration Technologies for Sustainable and Healthy Building Ventilation. *Sustain. Cities Soc.* 2017, *32*, 375–396. https://doi.org/10.1016/j.scs.2017.04.011.
[125] Zhang, S.; Liu, H.; Tang, N.; Ali, N.; Yu, J.; Ding, B. Highly Efficient, Transparent, and Multifunctional Air Filters Using Self-Assembled 2D Nanoarchitectured Fibrous Networks. *ACS Nano* 2019, 13(11), 13501–13512. https://doi.org/10.1021/acsnano.9b07293.
[126] Yang, A.; Cai, L.; Zhang, R.; Wang, J.; Hsu, P. C.; Wang, H.; Zhou, G.; Xu, J.; Cui, Y. Thermal Management in Nanofiber-Based Face Mask. *Nano Lett.* 2017, 17(6), 3506–3510. https://doi.org/10.1021/acs.nanolett.7b00579.
[127] Statnano.com. Coronavirus: Nanotech surface sanitizes Milan with nanomaterials remaining self-sterilized for years.
[128] Sim, W.; Barnard, R. T.; Blaskovich, M. A. T.; Ziora, Z. M. Antimicrobial Silver in Medicinal and Consumer Applications: A Patent Review of the Past Decade (2007–2017). *Antibiotics* 2018, 7(4), 93. https://doi.org/10.3390/antibiotics7040093.
[129] Talebian, S.; Wallace, G. G.; Schroeder, A.; Stellacci, F.; Conde, J. Nanotechnology-Based Disinfectants and Sensors for SARS-CoV-2. *Nat. Nanotechnol.* 2020, 15, 618–624. https://doi.org/10.1038/s41565-020-0751-0.

[130] Sohal, I. S.; O'Fallon, K. S.; Gaines, P.; Demokritou, P.; Bello, D. Ingested Engineered Nanomaterials: State of Science in Nanotoxicity Testing and Future Research Needs. Part. *Fibre Toxicol.* 2018, 15(1), 29. https://doi.org/10.1186/s12989-018-0265-1.

[131] Shimabuku, Q. L.; Ueda-Nakamura, T.; Bergamasco, R.; Fagundes-Klen, M. R. Chick-Watson Kinetics of Virus Inactivation with Granular Activated Carbon Modified with Silver Nanoparticles and/or Copper Oxide. *Process Saf. Environ. Prot.* 2018, *117*, 33–42. https://doi.org/10.1016/j.psep.2018.04.005.

[132] Govorov, A. O.; Richardson, H. H. Generating Heat with Metal Nanoparticles. *Nano Today* 2007, 2(1), 30–38. https://doi.org/10.1016/S1748-0132(07)70017-8.

[133] Loeb, S.; Li, C.; Kim, J. H. Solar Photothermal Disinfection Using Broadband-Light Absorbing Gold Nanoparticles and Carbon Black. *Environ. Sci. Technol.* 2018, 52, 205–213. https://doi.org/10.1021/acs.est.7b04442.

[134] Nazari, M.; Xi, M.; Lerch, S.; Alizadeh, M. H.; Ettinger, C.; Akiyama, H.; Gillespie, C.; Gummuluru, S.; Erramilli, S.; Reinhard, B. M. Plasmonic Enhancement of Selective Photonic Virus Inactivation. *Sci. Rep.* 2017, 7, 11951. https://doi.org/10.1038/s41598-017-12377-5.

[135] Han, J.; Chen, L.; Duan, S. M.; Yang, Q. X.; Yang, M.; Gao, C.; Zhang, B. Y.; He, H.; Dong, X. P. Efficient and Quick Inactivation of SARS Coronavirus and Other Microbes Exposed to the Surfaces of Some Metal Catalysts. *Biomed. Environ. Sci.* 2005, 18(3), 176–180.

[136] Warnes, S. L.; Little, Z. R.; Keevil, C. W. Human Coronavirus 229E Remains Infectious on Common Touch Surface Materials. *MBio* 2015, 6, 01697–15. https://doi.org/10.1128/mBio.01697-15.

[137] Shi, M.; De Mesy Bentley, K. L.; Palui, G.; Mattoussi, H.; Elder, A.; Yang, H. The Roles of Surface Chemistry, Dissolution Rate, and Delivered Dose in the Cytotoxicity of Copper Nanoparticles. *Nanoscale* 2017, 9(14), 4739–4750. https://doi.org/10.1039/c6nr09102d.

[138] Grass, G.; Rensing, C.; Solioz, M. Metallic Copper as an Antimicrobial Surface. *Appl. Environ. Microbiol.* 2011, 77, 1541–1547. https://doi.org/10.1128/AEM.02766-10.

[139] Han, W.; Zhang, P. H.; Cao, W. C.; Yang, D. L.; Taira, S.; Okamoto, Y.; Arai, J. I.; Yan, X. Y. The Inactivation Effect of Photocatalytic Titanium Apatite Filter on SARS Virus. *Prog. Biochem. Biophys.* 2004, 31(11), 982–985.

[140] Sametband, M.; Kalt, I.; Gedanken, A.; Sarid, R. Herpes Simplex Virus Type-1 Attachment Inhibition by Functionalized Graphene Oxide. *ACS Appl. Mater. Interfaces* 2014, 6(2), 1228–1235. https://doi.org/10.1021/am405040z.

12 Nanomaterials in the Environment

Sources, Fate, Transport and Ecotoxicology

Thabitha P. Dasari Shareena,
Asok K Dasmahapatra, and
Paul B Tchounwou
Jackson State University

CONTENTS

DOI: 10.1201/9781003129042-12

12.1 INTRODUCTION

The enhanced production of nanomaterial (NM)-based consumer products has a wide variety of potential applications due to their unique properties. Several factors affect the global market of NMs. These factors include the decrease in the price of NMs due to their abundance in the market, as well as the increase in nanotechnology studies, and the improvement in the quality of NMs that has increased public awareness and support by the government organizations (Inshakova and Inshakov 2017). There has been a significant increase in NM-based products released into the environment due to their unique applications and production, thereby receiving greater attention from the general public and governmental organizations regarding their safety concerns. These exceptional properties of NMs have raised concerns regarding their effects on environmental health and safety, emphasizing the need to assess their potential risks to the environment and human health (Lowry et al. 2012). So far, there have been many toxicological reports on the NMs; however, only a few researchers have investigated their environmental effects.

The research in applied nanotechnology has paved a path for inventions in environmental science and engineering. Hence, many engineered nanomaterials (ENMs) have been produced. However, their development must be focused on a sustainable framework for environmental applications with safety considerations (Mauter and Elimelech 2008). The global economy accounts for the mass production of ENMs in the form of consumer goods such as coatings, paints, pigments, electronics and optics, biomedicines and pharmaceuticals, cosmetics, energy and environmental applications and catalysts (Keller et al. 2013). Moreover, data related to ecotoxicity and the potential underlying mechanisms in the biogeochemical cycle remain insufficient. This chapter focuses on the various NMs in the environment, their sources, fate, transport and ecotoxicity.

12.2 SOURCES

12.2.1 Classes of Nanomaterials

The demand for the classification of NMs is vital to understand the risk assessment that guides the regulatory agencies such as Organization for Economic Cooperation and Development (OECD) and United States Environmental Protection Agency (US EPA). Also, the classification of NMs determines the ability to predict their behaviour in various environmental compartments (land, air and water). The first idea of NM classification was given by Gleiter et al. (2000), followed by Pokropivny and Skorokhod (2007) based on their dimensions. The properties of NMs such as particle morphology (fibrous, nonfibrous and granular), water solubility (high, low and insoluble), chemical reactivity (chemically reactive; chemically non-reactive) and toxic potential (high, low and non-toxic) are also considered for their classification (Arts et al. 2015; Landvik et al. 2018).

In general, according to the International Organization for Standardization (ISO), NMs can be defined as a material with any external dimension or internal

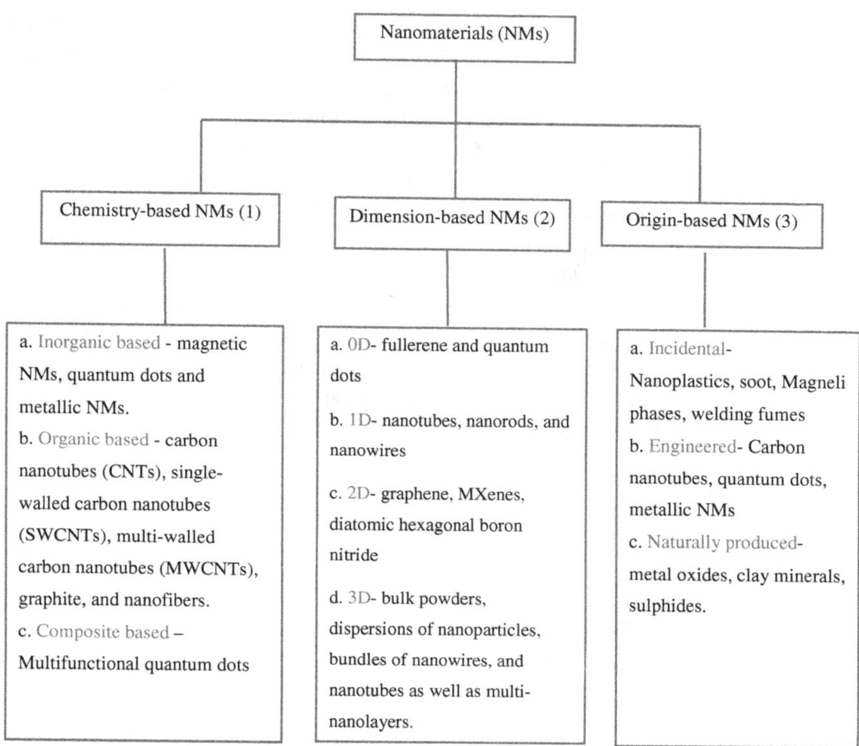

FIGURE 12.1 Classes of nanomaterials grouped into three categories based on the chemistry, dimensions and origin. Nanomaterials (with any external, internal or surface structure, in the nanoscale range of 1–100 nm; Crane et al. 2008) are categorized into three main classes depending on composition (1), size (2) and origin (3). The families of each class are presented in the boxes under each class.

structure or surface structure in the nanoscale range of 1–100 nm (Crane et al. 2008). Naturally occurring NMs have emerged due to natural and anthropogenic processes. The smaller size of NMs provides them with unique properties and reactivity compared to the larger particles, even with the same chemical composition (Hochella et al. 2008; Auffan et al. 2009). Environmental Protection Agency (EPA) confirms that NMs are present in more than 1300 commercial products, including medical equipment, textiles, fuel additives, cosmetics and plastics (EPA Web Resource). Generally, NMs are categorized depending on their composition, size and origin. Figure 12.1 shows three main categories along with the family of NMs.

12.2.2 Nanomaterials in Air, Water and Soil

Natural NMs have been present in the environment for the past billion years, long before the nanotechnology era. Airborne nanocrystals of sea salts, soil colloids, fullerenes, carbon nanotubes and biogenic magnetite are a few examples of natural NMs (Boros and Ostafe 2020). Later on, incidental NMs (anthropogenic NMs) emerged from human activities through an industrial revolution. These anthropogenic NMs can be further divided into two categories: incidental and engineered/manufactured. Incidental NMs are produced unintentionally in man-made processes, for example carbon nanotubes, carbon black and fullerenes and platinum- and rhodium-containing nanoparticle by-products from combustion. ENMs are produced intentionally (Dhawan and Sharma 2010). Recently, ENMs have come into existence for applications to human health and well-being. Figure 12.2 explains how the nanomaterial cycle works in the environment (Hochella et al. 2019). The precursors represent the beginning of a nanomaterial life cycle. These precursors originate from various climates and organisms that evolve into weathering of rocks and biological matter. The reversible chemical and physical pathways generate nanomaterial formation. The next phase of

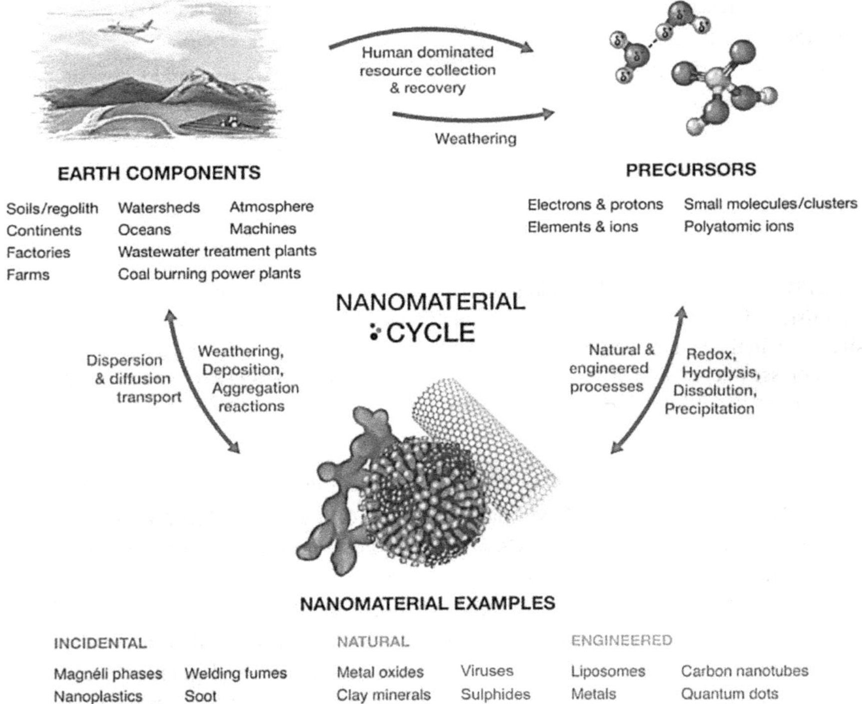

FIGURE 12.2 Nanomaterial cycle in the environment. The precursors originating from various sources represent the beginning of a nanomaterial life cycle. The next phase of the cycle involves a connection between the engineered and natural nanomaterials. These formed nanomaterials are released into various environmental compartments through dispersive and diffusive movements. (From Hochella et al. 2019. Copyright © American Association for the Advancement of Science.)

the cycle involves a connection between the engineered and natural NMs. These formed NMs are released into the environment through dispersive and diffusive movements (Oberdorster Eva 2004; Hochella et al. 2008).

Products such as cosmetics, paints and coatings containing NMs may release NMs into the environment in various ways. There are three possible ways through which NMs might be discharged into the environment. Mainly, NMs are released during the production and utilization of products via emission in indoor air and ambient atmosphere. Then they are finally released into sewage and water streams during the disposal. The stability, aggregation and suspension of NMs in the water are the key factors that determine the transport and fate of NMs (Lin et al. 2010).

NMs in consumer products such as personal care products and health care products could lead to inhalation exposure. It is estimated that 1% from 75% to 95% of ENMs in cosmetics are released into the air. Aerosolization of NMs leads to human exposure via inhalation, either intentionally or unintentionally. Most metal and metal oxide nanoparticles tend to generate inhalation exposures (Abbott and Maynard 2010; Vance et al. 2015; Keller et al. 2013).

Natural NMs exist in rivers, lakes, oceans and groundwater, whereas ENMs are produced for medical and biotechnological applications. Mostly, natural NMs are released into the atmosphere and surface waters via mineral weathering, sea spray, volcanoes and forest fires. Watersheds contain most sources of natural, incidental and ENMs that can enter drinking water treatment plants (Wigginton et al. 2007; Hochella et al. 2008). The aquatic environment plays a crucial role in the entry and dispersion of NMs to the environment (Hochella et al. 2008, 2019).

Moreover, ENMs are released into the aquatic environment via atmospheric release and water infiltration (non-point source) to the soil and by intentional point source discharges (Weinberg et al. 2011). The fate and transport of NMs depend on their physical and chemical properties such as temperature, pH, the concentration of natural organic matter, ionic strength, salinity and water hardness (Selck et al. 2016). However, the potential risks from ENMs are not well understood due to the scarcity of data on NMs' ecotoxicity in air, water and soil. ENMs pose a potential risk to the environment compared to the natural NMs.

NMs are used in various consumer products such as personal and health care products due to their unique properties. After the use, NMs may reach surface waters in run-off (Baker et al. 2014; Mueller and Nowack 2008; Rocha et al. 2017; Rocha et al. 2015) in dissolved or suspended form in water or get adsorbed to eroding soil particles. They may undergo physical, chemical and/or biological transformations to produce other forms of NMs in their respective environmental compartments. However, these NMs end up in soils as they are the waste receiver system for any pollutants. In this process, they may not be completely degraded, but might be transformed into intermediate products that are less safe or more harmful than the parent NMs. In soil compartments, NMs may be immobilized by sorption or partitioning into soil particles such as organic matter and inorganic materials. The fate of NMs is dependent on the long-term stability of the immobilized NMs/soil complexes. Furthermore, the NMs enter the biological membranes of microorganisms, plants and earthworms and affect the organism's physiology, causing various toxicity effects. Moreover, the exposure to NMs in the soils affects microbial biomass and diversity, decreases plant growth and inhibits soil invertebrate growth and reproduction (McKee and Filser 2016).

12.3 RELEASE, TRANSPORT, BEHAVIOUR AND FATE OF NANOMATERIALS IN THE ENVIRONMENT

To better understand NM's environmental fate and behaviour, it is significant to review the scientific information on their entry and dispersion into the environment. Due to the widespread use of consumer products containing ENMs, they are released into the environment during production, use and disposal. Entry into the environment may emerge due to an intentional or non-intentional release of NMs. Mostly, environmental conditions and intrinsic properties of NMs play an essential role in their behaviour and fate in the environment. Hence, a comprehensive understanding of these processes and the quantitative and qualitative estimation of NMs in the environment is essential. In determining the environmental risk assessment, there is a need to know the predicted environmental concentrations (PECs) or measured environmental concentrations of NMs. In recent years, many models and frameworks have emerged to explain the release, fate and distribution of NMs (Baalousha et al. 2016; Isigonis et al. 2019). A great challenge is understanding the physicochemical properties of NMs and their influence and behaviour in the environment during the transformation processes (Batley et al. 2013).

12.3.1 Exposure Pathways of Nanomaterials into the Environment

Due to the unique applications and mechanistic pathways of NMs, their release, transport, behaviour and fate are complex. Exposure evaluation is one crucial step in risk assessment. Also, the quantification of NMs in the environment is a demanding task due to the shortage of well-standardized procedures (von der Kammer et al. 2012). In order to assess the exposure pathways of NMs, it is necessary to understand at what stage of their life cycle they are released either intentionally or unintentionally. NMs could be released during production via mixing and weighing NMs in the form of powders into the air.

Additionally, NMs are released into lakes, rivers, oceans and aquatic sediments and soils based on their solubility and binding to the respective environmental compartments (Batley et al. 2013). After the release of NMs into the water, they tend to aggregate, bind to the organic matter or settle in the sediment layer. The detection of ENMs in the environment is a critical task as there are no standard protocols to estimate the natural NMs that are already present in the environment. Similarly, seasonal variability will affect the quantity and quality of the release of NMs into the environment, which is a major concern (Gottschalk and Nowack 2011 and Figure 12.3).

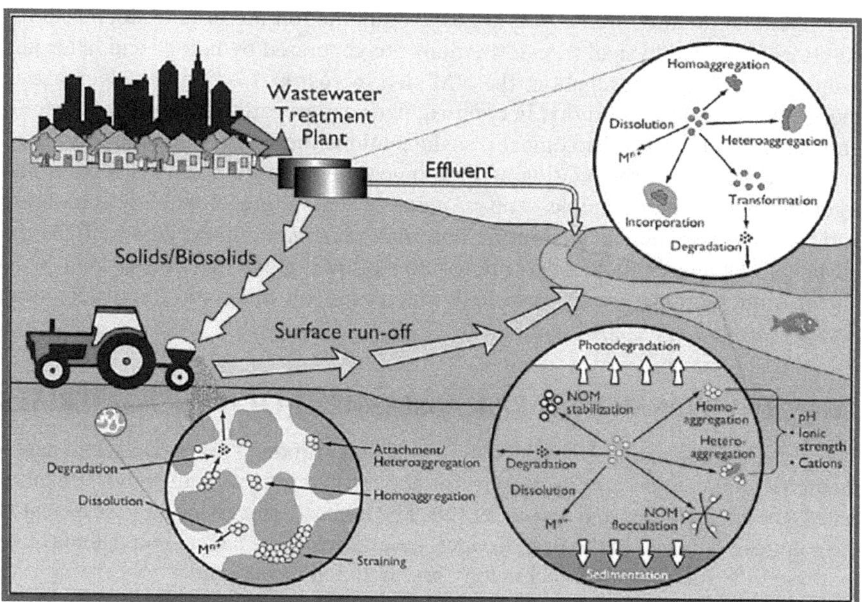

FIGURE 12.3 Possible transformations and pathways of nanomaterials (NMs) released into the environment. NMs could be released during production via mixing and weighing nanomaterials in the form of powders into the air. NMs are released into lakes, rivers, oceans and aquatic sediments and soils based on their solubility and binding to the respective environmental compartments (Batley et al. 2013). After the release of NMs into the water, they tend to aggregate, bind to the organic matter or settle in the sediment layer. (From Batley et al. 2013. Copyright © American Chemical Society.)

12.3.2 Nanomaterials Transformation Processes in the Environment

As the environmental systems are dynamic and unpredictable, the release, transport and fate of NMs are intricate due to the physicochemical changes that may occur. There is limited information on the types and transformations of NMs in the environment. To evaluate NM's environmental and human risks, there is a need to fully understand their transformations and environmental behaviours. The key transformation processes that determine the fate and behaviour of ENMs in the environment are physical and chemical transformations, biologically mediated transformations, and interactions with macromolecules and biomacromolecules (Hegde et al. 2016 and Figure 12.3).

The majority of the toxicity reports focus on the original forms of ENMs rather than the transformed versions of NMs. The toxicity of transformed NMs depends on the conditions of exposure. Transformations are categorized into physical, chemical and biological changes as well as interactions with cellular macromolecules (Figure 12.4). Physical transformations are the processes that influence the physical appearance of ENMs, whereas chemical processes change the chemical composition and speciation of ENMs. Processes such as oxidation and photodegradation, release of ions and sulphidation increase or decrease dissolution of NMs. Physical transformations are generated by homo- and heteroaggregation. In homoaggregation, the NM size increases, leading to a decrease in the surface area and reactivity. In contrast, heteroaggregation affects the transport and reactivity of NMs. Biological transformations include biologically mediated processes such as carboxylation and biological degradation. NM absorption and interaction with biomolecules and proteins affects aggregation, biodistribution and dissolution, whereas adsorption onto natural organic material stabilizes the NMs without aggregation. Adsorption onto natural organic matter assists in NM–bacteria interactions, which eventually decreases ion dissolution and decreases toxicity (Lowry et al. 2012 and Figure 12.4).

12.4 ENVIRONMENTAL RISK ASSESSMENT OF NANOMATERIALS

The wide range of applications of NMs in consumer products, agro-foods and nanomedicine applications has raised concerns about whether NMs cause environmental risk to the environment and human health. The highest priority in risk assessment is the safer use of NMs. In the past decades, several scientific papers and patents have increased the focus on nanotechnology research and development concerning risk assessment and toxicity (Kuhnel and Nickel 2014; Holden et al. 2016). Moreover, public knowledge on the use of these NMs is limited. Risks cannot be generalized due to the diversity of NMs. The behaviour of NMs also differs between humans and the environment. The unique properties of NMs are valuable to the nanotechnology industries. Economic sustainability is necessary to assess nanotechnology's impact to guide the general public and policymakers (Bleeker et al. 2014; Wentworth and Milodowski 2017). According to the National Academy of Sciences (NAS) report, risk assessment is categorized into four different stages (NRC Web Resource). In this section, we briefly discuss the four stages in the context of ENMs.

12.4.1 Hazard Assessment

The hazard assessment emphasizes the evaluation of possible risks of NMs. Hazard assessment is specifically dependent on the type of the particle and its environment. The hazard might be in the form of toxicity or ecotoxicity depending on the class of NMs. The assessment of hazards includes measuring various endpoints by correlating the ecotoxicological adverse effects with the physicochemical properties of NMs with respect to size, shape, surface area, charge and reactivity, and solubility. Apart from the particle-specific properties, the environmental effects play a significant role in atmospheric processes such as soil stability and aquatic organic matter. The most available hazard data are related to the aquatic environment, but there is little information available for the soil and air compartments (Bleeker et al. 2014; Wentworth and Milodowski 2017; NRC Web Resource; Rana and Kalaichelvan 2013). Blinova et al. (2010) emphasized that the ecotoxicological tests are essential tools for the hazard evaluation of ENMs and discussed both the harmful and mitigating effects under standardized experimental conditions.

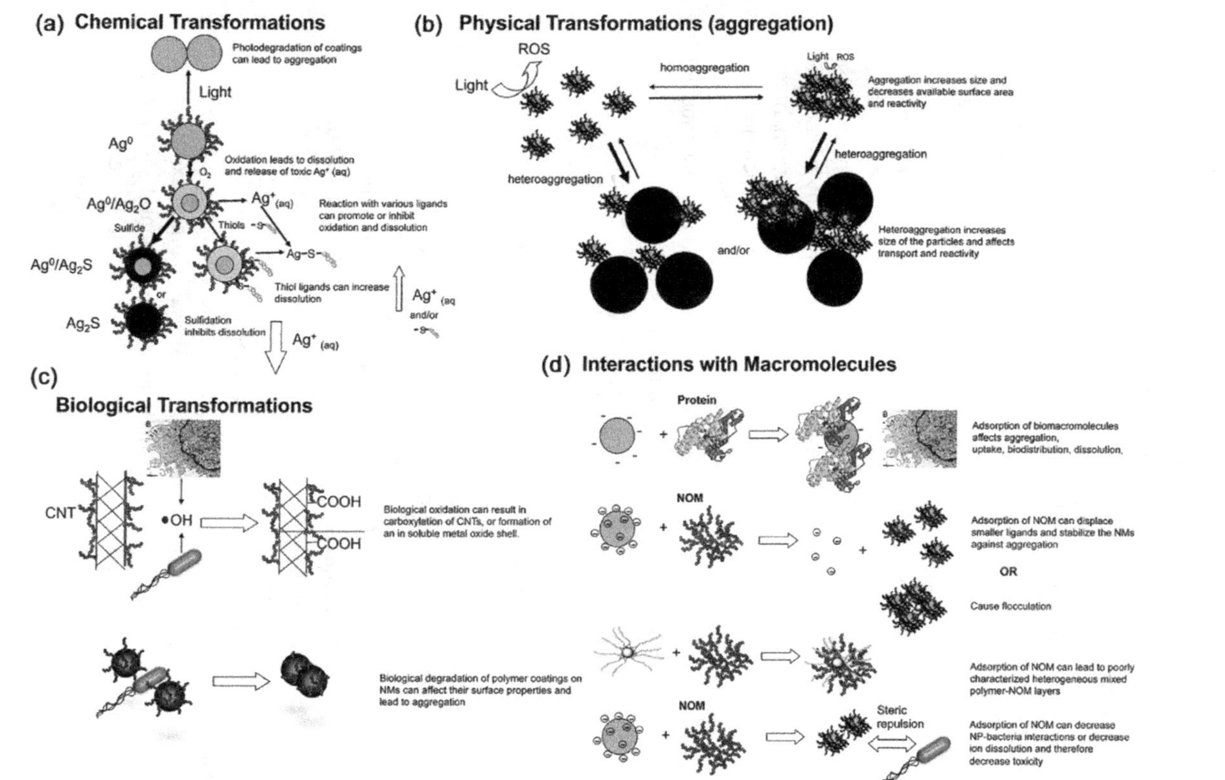

FIGURE 12.4 Nanoparticle transformations in the environment. Transformations are categorized into chemical (a), physical (b) and biological changes (c) as well as interactions with cellular macromolecules (d). Physical transformations influence the physical appearance of ENMs; chemical processes change the chemical composition and speciation of ENMs; biological transformations include biologically mediated processes such as carboxylation and biological degradation. (From Lowry et al. 2012. Copyright © American Chemical Society.)

12.4.2 Dose–Response Assessment

Dose–response assessment determines how a specific chemical or NM can adversely affect human health at various doses or the ecosystems at various concentrations. The estimation of dose/concentration of ENMs in different environmental compartments is vital to determine the exposure levels of humans and ecosystems to these NMs (Keller et al. 2013). The evaluation of dose–response correlation might involve conducting laboratory experiments and mathematical or computational models to predict the toxicity and environmental impacts of NMs before mass production. Several studies have reported on the dose–response assessment to develop a theoretical framework for quantifying the release of ENMs from specific applications (Puzyn et al. 2011).

The studies on epidemiologic data are needed for risk assessment, but there are no reports available on the NMs' quantitative exposure data. Henceforth, experimental data from animals, microbes and aquatic species are needed to determine the dose–response relationships that could be extrapolated to humans especially targeting the organs specificity (i.e. absorption, distribution, metabolism and elimination). In toxicology study designs, quantitative risk assessment considers BMD (benchmark dose method) data. A BMD is defined as a risk-associated dose estimation by model curve fitting to the dose–response data. BMD estimates have been used in cancer and non-cancer risk assessments and even in ENMs (Kuempel and Castranova 2016).

12.4.3 Exposure Assessment

Nanoparticles are released from a variety of sources and deposited in various environmental compartments. In the aquatic environment, they might be adsorbed onto organic or inorganic materials based on their physical and chemical properties. After physical and/or chemical transformations, they might be transported based on run-off and affect native biota. Risk assessment methodology should be developed with a detailed perspective on persistence, bioaccumulation and toxicity. Also, the methodology should include particle, concentration, distribution and decomposition products. Empirical data are essential in predicting the persistence and mobility of NMs in air, soil and water.

Additionally, existing standardized assessment methods on exposure may not be sufficient to evaluate the fate of NMs in the environment. Exposure assessment assesses the life cycle of NMs from synthesis to disposal. Human exposure may occur at workplaces, as well as at other public and environmental settings. The vast development in nanotechnological applications may cause increased intentional or unintentional release of NMs into the environment (Tsuji et al. 2006; SCENIHR 2007). The transformation of NMs in soil may be affected by individual or a combination of abiotic or biotic pathways. The extensive characterization of soil matrices requires advanced analytical and spectroscopic methods to obtain comprehensive knowledge of the interaction between sorption phenomena and nanoparticles (Theng and Yuan 2008). NMs may be aerosolized during manufacturing, production, disposal and recycling. Based on their physicochemical characteristics, they may disperse/aggregate or sorb onto dust particles. NMs that are released into the air may sorb onto dust particles or onto the food. The route of exposure via inhalation or ingestion might depend on the NMs settling out in the air or food. The NMs released into the air require proper characterization techniques to quantify the risk assessment. Cost-effective quantification and analysis methods should be evaluated at predictable environmental concentrations of NMs to understand the exposure (Tsuji et al. 2006; EPA Web Resource).

12.4.4 Risk Characterization

Risk characterization takes into account all the above three steps to evaluate the risk to the exposed population or ecosystems. This risk assessment data might be essential to regulate and assist the safer use of NMs in consumer-based products. The risk is often assessed by comparing the exposure concentration with an exposure level with no effects (Kuempel and Castranova 2016; Joner et al. 2008). Even though mass may be the final concentration metric in risk characterization, it is essential to include physicochemical descriptors such as mass, the number of particles, surface area and surface charge of NMs. The environmental risk characterization is mainly based on the association between the PEC and the predicted no-effect concentration (PNEC) (Scott-Fordsmand et al. 2017). Further research is needed to understand the knowledge gaps to determine the parameters the fate, exposure, transport and toxicity of these NMs. These uncertainties may be mitigated with the right strategic approaches. Figure 12.5 illustrates the essential questions and knowledge gaps to perform environmental risk assessment (Bleeker et al. 2014). These questions might address the difficulties in understanding the problem in the long run clearly.

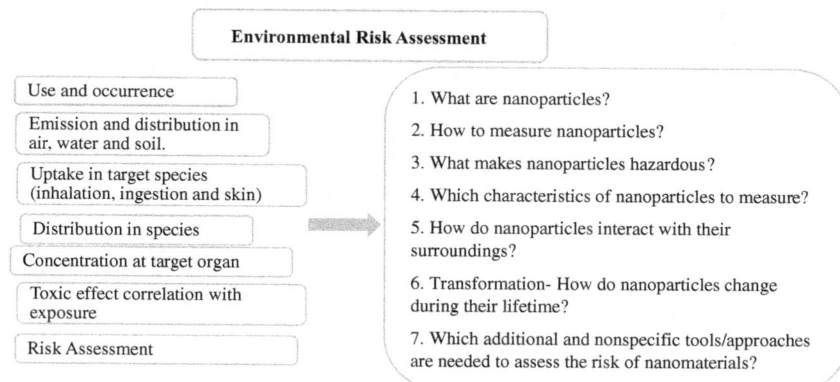

Environmental Risk Assessment
Use and occurrence
Emission and distribution in air, water and soil.
Uptake in target species (inhalation, ingestion and skin)
Distribution in species
Concentration at target organ
Toxic effect correlation with exposure
Risk Assessment
1. What are nanoparticles?
2. How to measure nanoparticles?
3. What makes nanoparticles hazardous?
4. Which characteristics of nanoparticles to measure?
5. How do nanoparticles interact with their surroundings?
6. Transformation- How do nanoparticles change during their lifetime?
7. Which additional and nonspecific tools/approaches are needed to assess the risk of nanomaterials?

FIGURE 12.5 Essential questions and knowledge gaps to perform environmental risk assessment. The risk assessment data are essential to regulate and assist the safer use of NMs in consumer-based products. The environmental risk characterization is mainly based on the association between the predicted environmental concentration (PEC) and the predicted no-effect concentration (PNEC). (From Bleeker et al. 2015. Copyright ©RIVM 2015.)

12.5 ECOTOXICITY

The adverse effects of NMs could be evaluated by the factors mobility, ecotoxicity and modification. So far, in the previous sections, we have discussed the mobility and modification of NMs. Toxicity relates to the toxic properties of NMs for safer use by human beings, whereas ecotoxicity focuses on the trophic organism levels to protect populations and ecosystems. Mainly, ecotoxicity focuses on the study of how chemicals interact with organisms in the environment. It incorporates natural uptake mechanisms and the effect of environmental factors on the bioavailability of NMs (Rana and Kalaichelvan 2013). This section focuses on the ecotoxic effects and the uptake mechanisms of NMs on microorganisms, fresh and marine water organisms, and plants. Moreover, there are very few reports available on the ecotoxicity of NMs, and our knowledge on the mechanisms is still insufficient. Also, it is difficult to compare the results due to controversial findings in published reports. Most of the studies on the ecotoxicity assessments on different trophic levels, including microorganisms, plants, invertebrates, vertebrates and test systems, have systematically been standardized for organisms and various exposure conditions. It has now become a common practice to include the ecotoxicity tests for ENMs, as part of the safe-by-design concept of creation, fabrication, utilization and disposal. The assessment of ecotoxicological effects of previously untested substances, as in the case of ENMs, is a challenging task. Acute assays are frequently used to evaluate the survival of the organisms. The results from chronic tests and acute tests combined with large safety factors suggest that there may be risks of these NMs to the environment. These assays are helpful to evaluate the sublethal effects on organism growth or reproduction (Crane et al. 2008). Classical ecotoxicological studies have assessed the toxic effects of various NMs through both a simplified and realistic experimental approach. It is required to have a comprehensive evaluation process to predict the behaviour of ENMs in the environment to microbes, invertebrates and vertebrates. Several concerns, including agglomeration, dissolution, association with dissolved chemical species, sedimentation, chemical transformation, or adhesion of ENMs, could interfere with the testing methods. Handy et al. (2008) illustrated the key challenges and knowledge gaps in NMs' ecotoxicology. Table 12.1 shows the ecotoxicological studies of select NMs on microbes, amphibians and crustaceans.

12.5.1 Ecotoxicity of Nanomaterials on Microbes, Amphibians and Crustaceans

Microbes are considered excellent ecological indicators for evaluating the ecotoxicity of nanoparticles. Also, bacteria play an essential role as decomposers in natural ecosystems; the by-products of a xenobiotic substance may produce harmful effects on the environment. They have the key environmental relevance and are involved in nutrient cycling and waste decomposition (Hu et al. 2009; Neal Andrew 2008). Mainly, soil microbial communities play an essential role in ecology for the balance of the ecosystem. Hedge et al. (2016) summarized the impact

TABLE 12.1
Physiological, Biochemical and Molecular Effects of Nanomaterials on Various Organisms

Nanomaterials	Microbes/Animals	Ecotoxicity	Refs
1. AgNPs (silver nanoparticles)	Natural bacterial assemblages in estuarine sediments	Negligible impact on bacterial diversity in estuarine sediments at (25–1000 μg/L)	Bradford et al. (2009)
2. AgNPs	Natural bacterial community	Negative effects on total abundance of bacteria at 0.01–2.0 mg-Ag/L	Das et al. (2012)
3. AgNPs (5 μM)	Natural aquatic bacterial assemblages	Light-induced toxicity at 5 μM	Dasari and Hwang (2010)
3. CuO (copper oxide) and ZnO (zinc oxide)	Daphnia magna, *Thamnocephalus platyurus* and protozoan *Tetrahymena thermophila*	Decreased the toxicity of nano-CuO ((90–224 mg Cu/L) Increased the toxicity of nano-ZnO (1.1–16 mg Zn/L)	Blinova et al. (2010)
4. C_{60} (fullerene)	*Escherichia coli* and *Bacillus subtilis*	Lack of growth at 0.4 ppm and decreased aerobic respiration rates at 4 ppm	Fortner et al. (2005)
5. C_{60}	River biofilms	0.30 and 3 μg/L	Freixa et al. (2018)
6. C_{60}	*Bacillus subtilis*	Exhibited strong antibacterial activity at an MIC of 0.09 ± 0.35 mg/L	Lyon et al. (2006)
7. Single-walled carbon nanotubes	*E. coli* K12	5 to ~300 mg/L increased inhibition of both total cell growth and biofilm formation	Rodrigues et al. (2010)
8. nC_{60}	*Scenedesmus obliquus*	At sublethal concentration of 0.09 mg/L, decreased algal Mg^{2+} and inhibited the activity of Mg^{2+} ATPase	Tao et al. (2015)
9. Cu and CuO, MgO (magnesium oxide), fullerenol and graphene oxide	*Photobacterium phosphoreum* and recombinant *Escherichia coli* strains	Toxicity ranked as Cu > (MgO, CuO) > fullerenol, graphene oxide based on EC_{50}	Deryabin et al. (2016)
10. ZnO, NiO (nickel oxide) and FeO (iron oxide)	*P. phosphoreum*	EC_{50} values ranged from 0.34 to 3747 mg/L	Wang et al. (2014)

(*Continued*)

TABLE 12.1 (*Continued*)
Physiological, Biochemical and Molecular Effects of Nanomaterials on Various Organisms

Nanomaterials	Microbes/Animals	Ecotoxicity	Refs
11. nZnO	*Vibrio fischeri*	EC_{50} ranged from 1.17 to 319.24 mg/L	Chen et al. (2020)
12. Nanosized titanium dioxide (TiO_2), silicon dioxide (SiO_2) and zinc oxide (ZnO)	*Bacillus subtilis* and *Escherichia coli*	Exhibited antibacterial activity towards *B. subtilis* and to a lesser extent to *E. coli*	Adams et al. (2006)
13. AgNPs (PVA-coated)	*Nitrifying bacteria* and *Escherichia coli*	AgNPs inhibited respiration and nitrification; the effect varied depending on the size and bioavailability of the NPs	Hansch and Emmerling (2010)
14. AgNPs	Heterotrophic bacterial community	At 0.0032–0.32 mg/kg dosage, AgNPs caused a decrease in carbon microbial biomass and an increase in the metabolic quotient	Grün and Emmerling (2018)
15. AgNPs	Soil microbial community	At 0.01 mg/kg dosage, negative effects on soil microbial biomass and bacterial ammonia oxidizers observed	Grün and Emmerling (2018)
		At 0.01–1 mg AgNPs/kg, soil caused disadvantages for the autotrophic ammonia oxidation, organic carbon transformation and chitin degradation by causing harmful effects on the liable bacterial phyla	Grün et al. (2018)
16. AgNPs and AgMPs (microparticles)	Bacterial and fungal assemblages	At 0.066% and 6.6% dosage, AgNPs are more toxic (decrease in levels of respiration, decrease in bacterial fatty acids and changes in bacterial and fungal DNA sequence)	Kumar et al. (2014)
		Disrupted the natural seasonal progression of tundra assemblages	
17. TiO_2 or ZnO, Ag ENMs added	*Eisenia fetida*	Decreased the earthworm survival and reproduction rates at higher concentrations	Lahive et al. (2017)

(*Continued*)

TABLE 12.1 (*Continued*)
Physiological, Biochemical and Molecular Effects of Nanomaterials on Various Organisms

Nanomaterials	Microbes/Animals	Ecotoxicity	Refs
18. AgNPs	Soil invertebrates *Eisenia andrei* and *Folsomia candida*; plants *Elymus lanceolatus* and *Trifolium pratense*	9–833 mg/kg (*E. andrei*) 10–833 mg/kg (*F. candida*) AgNPs are more toxic than $AgNO_3$ to *E. andrei*, but there was no difference for *F. candida* Reproduction was mainly affected	Handy et al. (2018)
19. AgNPs and CoNPs (cobalt oxide)	*Lumbricus rubellus*	At 500 mg/kg, Ag caused accumulation of Ag in organisms (more with AgNPs than with $AgNO_3$) and a decrease in the unsaturation degree of fatty acids Accumulation of CoNPs and Co^{2+} in earthworm tissues was two and three times greater than that of AgNPs	Antisari et al. (2016)
20. Carbon nanotubes	*Ambystoma mexicanum*	Carbon nano tubes are neither toxic nor genotoxic (larvae) at concentrations 1–1000 mg/L	Mouchet et al. (2008)
21. ZnO NPs	*Pelophylax ridibundus*	Physiological and cytological disorders	Falfushynska et al. (2017)
22. Au NPs	*Lithobates sylvaticus*	Decrease in metamorphosis time	Fong et al. (2016)
23. $TiSiO_4$-NPs (titanium silicon oxide)	*Pelophylax perezi*	Alterations in intestinal mucosa by elevation of gastrointestinal acidity	Salvaterra et al. (2013)
24. CuO NPs	*Xenopus laevis*	Increase in mortality and decrease in metamorphosis rate	Nations et al. (2011)
25. QDs (quantum dots)	*Xenopus laevis* and *Daphnia magna*	Dysfunction of the gills, lungs and intestines	Galdiero et al. (2017)
26. IO NPs (iron oxide nanoparticles)	*Xenopus laevis*	Defects in embryo body shape, such as bent spine or enlarged ventral fin	Marín-Barba et al. (2018)
27. CeO_2 NPs	*Xenopus laevis*	Increase in mortality; decrease in growth; no genotoxicity at concentrations of 0.1, 1.0 and 10 mg/L	Bour et al. (2015)
28. CeO_2 NPs (triammonium citrate)	*Pleurodeles waltl*	Decrease in growth; increase in DNA damage at 0.1, 1 and 10 mg/L	Bour et al. (2015)
29. CeO_2 NPs (citrate)	*Pleurodeles waltl*	Genotoxicity was observed with bare and coated NPs; increase in MN: naked NPs > N-coated NPs	Bour et al. (2015)
30. CeO_2 NPs	*Pleurodeles waltl*	Toxicity observed in *Pleurodeles* in mesocosm may be indirect, due to microorganism's interaction with CeO_2 NPs	Bour et al. (2017)

of NMs on microbial community. Rocha et al (2015) reviewed the ecotoxicological impacts of ENMs on bivalve molluscs. do Amaral et al. (2019) reviewed nano-ecotoxicological studies on 11 amphibian species, of which eight of them were from the order Anura and three were from the order Caudata. Most of these studies mainly used inorganic NMs compared to organic NMs. They concluded that nano-ecotoxicity is dependent upon the behaviour and transformation of NMs in the environment. Amphibians are suitable models for ecotoxicity studies due to their two-phase life cycle (aquatic larval phase and adult terrestrial phase). The ecotoxic effects of various NM classes on microbes, amphibians, crustaceans, and others are summarized in Table 12.1. These effects include no damage to moderate damage to cellular processes to cell mortality. There should be comprehensive data on the mobility, transfer and uptake that affect environmental matrices to precisely determine the risks of NMs to microorganisms (He et al. 2014; Joner et al. 2008).

12.5.2 Potential Mechanisms of Ecotoxicity

The ecotoxicity mechanisms have not been elucidated for all the NMs. The mechanism of ecotoxicity mainly depends on the fate of NMs in the environment. The NMs might undergo various transformation reactions such as aggregation, agglomeration, dissolution, sedimentation, sorption and surface reactions. The toxicity mechanisms are the direct or indirect impact of NMs on microbes and animals (Hedge et al. 2016). Even though the ecotoxicity mechanisms of NMs have not yet been elucidated, the possible toxicity mechanisms are categorized into direct and indirect mechanisms. The possible mechanisms include disruption of membranes or membrane potential, oxidation of proteins, genotoxicity, interruption of energy transduction, formation of reactive oxygen species (ROS) and release of toxic metal ions (Klaine et al. 2008; do Amaral et al. 2019). Figure 12.6 shows a schematic diagram of the possible ecotoxicity mechanisms of NMs. Generally, all these mechanisms may not occur alone, but a combination of these mechanisms may involve in generating ecotoxicity.

12.5.3 Environmental Regulations/Legislation for Nanomaterials

The availability of exposure and hazard data is very limited for government agencies' strict legislation in many countries. However, there has been progress in nanotechnology regulations in developed countries, Europe and the USA. In the USA, NMs are considered as chemical substances under the Toxic Substances Control Act (TSCA). Agencies and acts such as EPA, FDA, TSCA and FIFRA (Federal Insecticide, Fungicide, and Rodenticide Act) also regulate NMs. So far, EPA has reviewed 160 more new chemicals under TSCA on nanoscale materials. The EPA and TSCA gather more comprehensive information on NMs from manufacturers for one-time reporting and record-keeping prior to manufacturing. EPA takes action to effectively control the NMs that cause risk to human health and environment. There are only few reports available on the research of NMs exposure; it is a concern that the population that work with NMs are at higher risk (EPA Web Resource).

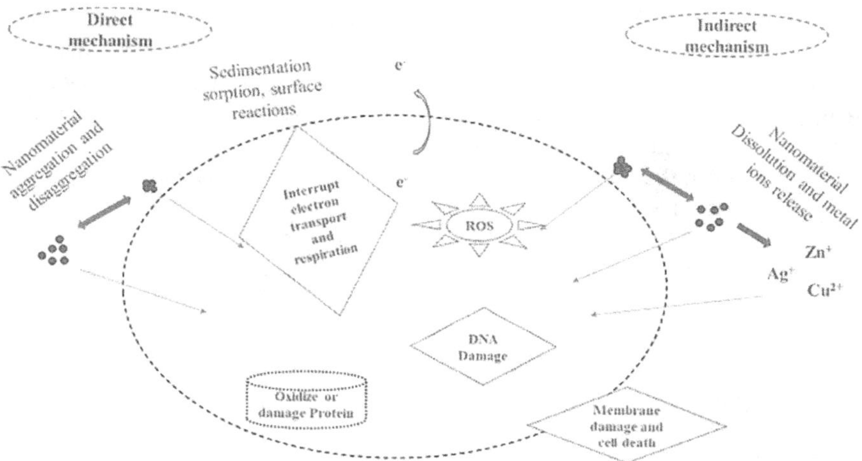

FIGURE 12.6 Schematic diagram of possible ecotoxicity mechanisms of NMs. The possible mechanisms include disruption of membranes or membrane potential, oxidation of proteins, genotoxicity, interruption of energy transduction, formation of reactive oxygen species (ROS) and release of toxic metal ions. All these mechanisms may not occur alone, but in combination induce ecotoxicity. (Data adapted from do Amaral et al. (2019), Klaine et al. (2012) and Hedge et al. (2016)).

Other agencies such as OECD published several test guidelines on NMs that provide data to support the knowledge on the risks of NMs. Agencies such as National Institute for Occupational Safety and Health (NIOSH), Engineered Nanoparticles Risk Assessment (ENPRA), European Agency for Safety and Health at Work (EU-OSHA), European Commission – Registration, Evaluation, Authorization and Restriction of Chemicals (REACH), National Institute of Environmental and Health Sciences – Nanotechnology Consortium (NIEHS) and National Research Council (NRC) (EPA Web Resource). EPA also contributed significantly to the risk assessment of NMs disposed in the environment. Handy et al. (2008) reviewed the key challenges and knowledge gaps in NMs' ecotoxicology and developed a flow chart as shown in Figure 12.1 of their article.

12.6 CONCLUSIONS

The unique and tunable properties of NMs enable new developments in applied nanotechnology research. These unique properties of NMs include size, shape, surface area, sorption properties, molecular interactions and optical and thermal properties. Despite the natural NMs present on the earth, the incidental and ENMs might positively or negatively impact the environment. There are uncertainties concerning the potential risks of NMs released into the air, water and soil due to limited toxicity data. The classification of NMs is essential for regulatory purposes particularly to define the safe use of NMs. Generally, NMs are classified based on their chemistry, dimensions and origin. Particle size, shape, chemical composition and surface area of NMs play a significant role in the environment. Despite the origin of NMs, they may enter air, water and soil via various exposure routes. The biological and physicochemical transformations alter their fate, behaviour and toxicity of NMs in their respective environmental compartments. However, the information on the release of NMs into various environmental compartments is limited. The stability, aggregation and suspension of NMs in the water are the key factors determining the transport and fate of NMs.

The highest priority in risk assessment is the safer use of NMs. The risk assessment of NMs is comprised of four components including hazard identification, dose–response assessment, exposure assessment and risk characterization. The environmental risk assessment data are essential to regulate and assist the safer use of NMs in consumer-based products. Regulations for risk assessment of NMs must include the case-by-case scenario approach.

Research on the application and ecotoxicity studies on NMs increased significantly during 2010–2019. The current ecological tests on models use wide range of organisms, including algae and invertebrates, as simple organisms and higher plants and vertebrates are the more complex organisms. The tests on algae, duckweed, amphipod, daphnia and chironomids in the aquatic environment, on terrestrial plants, and in nematodes and earthworms are suitable for the terrestrial environments. (Boros and Ostafe 2020). Based on the EC_{50} values, algae tests are found to be more sensitive and duckweed tests are less sensitive in aquatic models; in terrestrial models, the most sensitive test was on the nematodes and the less sensitive test was on the earthworms. Although there is difficulty in assessing and testing NMs in the environment, further research could resolve these issues.

As the wide range of NMs demand increases, the regulatory agencies face challenges to promote the safety and reduce or eliminate the health and ecological risks associated with their release in various environmental compartments. Another challenge is related to the uncertainty associated with the interpretation and translation of laboratory-based experimental data to determine the risks to human health and the environment. It is necessary to standardize, test and classify NMs so that the risks may be minimized or evaluated efficiently (Drasler et al. 2017; Boverhof et al. 2015). Also, green nanotechnology should be incorporated in the design of safer ENMs to reduce the risk of NMs' environmental and human health effects (Mauter and Elimelech 2008). There should be demand for implementing a combined strategy for using experimental, theoretical and computational approaches to develop risk assessment strategies to handle the NMs' ecotoxicological problems (Hedge et al. 2016). It is essential to apply a sustainable framework to develop environmentally friendly NMs to minimize the risk of NMs. Green nanotechnology should be implemented in product design and application with minimal dispersal of NMs into the environment.

Most importantly, there should be a multidisciplinary collaboration between chemists, environmental scientists and toxicologists (Mauter and Elimelech 2008). Further studies are considered necessary to evaluate NMs' fate, transport and ecotoxicological risks in various environmental compartments. Finally, our ultimate goal must be to help protect the environment and human health. The knowledge gaps concerning NMs' fate, behaviour and transport, need to be addressed to generate new scientific data. Moreover, it is recommended to integrate data on the regulation, sustainable nanotechnology and risk assessment of NMs.

ACKNOWLEDGEMENTS

This research was supported by NIH/NIMHD grant # G12MD007581 (RCMI Center for Environmental Health), NIH/NIMHD grant #1U54MD015929 (RCMI Center for Health Disparities Research) and NSF grant #HRD 1547754 (CREST Center for Nanotoxicity Studies) at Jackson State University, Jackson, Mississippi, USA.

REFERENCES

Abbott, L.C., & Maynard, A.D. (2010). Exposure assessment approaches for engineered nanomaterials. *Risk Analysis.* 30: 1634–1644. DOI: 10.1111/j.1539-6924.2010.01446.x.

Adams, L.K. Lyon, D.Y., & Alvarez, P.J.J. (2006). Comparative eco-toxicity of nanoscale TiO_2, SiO_2, and ZnO water suspensions. *Water Research.* 40: 3527–3532. DOI: 10.1016/j.watres.2006.08.004.

Antisari, L.V., Carbone, S. Gatti, A. Ferrando, S. Naucchi, M. Pascalis, F.D. Gambardella, C. Badalucco, L., & Laudicina, V.A. (2016). Effect of cobalt and silver nanoparticles and ions on Lumbricus rubellus health and on microbial community of earthworm faeces and soil. *Applied Soil Ecology.* 108: 62–71. DOI: 10.1016/j.apsoil.2016.07.019.

Arts, J.H.E., Hadi, M. Irfan. M.A. Keene, A.M. Kreiling, R. Lyon, D. Maier, M. Michel, K. Petry, T. Sauer, U.G. Warheit, D. Wiench, K. Wohlleben, W., & Landsiedel, R. (2015). A decision-making framework for grouping and testing of nanomaterials (DF4nanoGrouping). *Regulatory Toxicology and Pharmacology.* 71: S1–S27. https://doi.org/10.1016/j.yrtph.2015.03.007.

Auffan, M. Rose, J. Bottero, J.Y. Lowry, G.V. Jolivet, J.P., & Wiesner, M.R. (2009). Towards a definition of inorganic nanoparticles from an environmental, health and safety perspective. *Nature Nanotechnology.* 4: 634–641. DOI: 10.1038/nnano.2009.242.

Baalousha, M. Cornelis, G. Kuhlbusch, T.A.J. Lynch, I. Nickel, C. Peijnenburg. W., & van den Brink, N.W. (2016). Modeling nanomaterial fate and uptake in the environment: current knowledge and future trend. *Environmental Science: Nano.* 3: 323–345. DOI: 10.1039/C5EN00207A.

Baker, T.J. Tyler, C.R., & Galloway, T.S. (2014). Impacts of metal and metal oxide nanoparticles on marine organisms. *Environmental Pollution.* 186: 257–271. DOI: 10.1016/j.envpol.2013.11.014.

Batley, G.E. Kirby, J.K., & McLaughlin, M.J. (2013). Fate and risks of nanomaterials in aquatic and terrestrial environments. *Accounts of Chemical Research.* 46: 854–862. DOI: 10.1021/ar2003368.

Bleeker, E.A.J., Evertz, S. Geerstma, R.E. Peijnenburg, W.J.G.M. Westra, J., & Winjnhoven, S.W.P. (2014). Assessing health and environmental risks of nanomaterials, current state of affairs in policy, science and areas of application. The Netherlands, National institute of public health and Environment. RIVM Report 2014–0157. https://www.rivm.nl/bibliotheek/rapporten/2014-0157.pdf.

Blinova, I. Ivask, A. Heinlaan, M. Mortimer, M., & Kahru, A. (2010). Ecotoxicity of nanoparticles of CuO and ZnO in natural water. *Environmental Pollution*. 158: 41–47. https://doi.org/10.1016/j.envpol.2009.08.017.

Boros, B.V., & Ostafe, V. (2020). Evaluation of ecotoxicology assessment methods of nanomaterials and their effects. *Nanomaterials*. 10(4):610. DOI: 10.3390/nano10040610.

Bour, A. Mouchet, F. Cadarsi, S. Silvestre, J. Baqué, D. Gauthier, L., & Pinelli, E. (2017). CeO_2 nanoparticle fate in environmental conditions and toxicity on a freshwater predator species: a microcosm study. *Environmental Science and Pollution Research*. 24: 17081–17089. https://doi.org/10.1007/s11356-017-9346-1.

Bour, A. Mouchet, F. Verneuil, L. Evariste, L. Silvestre, J. Pinelli, E., & Gauthier, L. (2015). Toxicity of CeO_2 nanoparticles at different trophic levels – effects on diatoms, chironomids and amphibians. *Chemosphere*. 120: 230–236. DOI: 10.1016/j.chemosphere.2014.07.012.

Boverhof, D.R. Bramante, C.M. Butala, J.H. Clancy, S.F., Lafranconi, M., West, J., & Gordon, S.C. (2015). Comparative assessment of nanomaterial definitions and safety evaluation considerations. *Regulatory Toxicology and Pharmacology*. 73: 137–150. https://doi.org/10.1016/j.yrtph.2015.06.001.

Bradford, A. Handy, R.D. Readman, J.W. Atfield, A., & Muhling, M. (2009). Impact of silver nanoparticle contamination on the genetic diversity of natural bacterial assemblages in estuarine sediments, *Environ Science and Technology*. 43: 4530–4536. DOI: 10.1021/es9001949.

Chen, F. Wu, L. Xiao, X. Rong, L. Li, M., & Zou, X. (2020). Mixture toxicity of zinc oxide nanoparticle and chemicals with different mode of action upon Vibrio fischeri. *Environmental Sciences. Europe*. 32: 41. DOI: 10.1186/s12302-020-00320-x.

Choi, O., & Hu, Z. (2008). Size dependent and reactive oxygen species related nanosilver toxicity to nitrifying bacteria. *Environmental Science and Technology*. 42: 4583–4588. DOI: 10.1021/es703238h.

Crane, M. Handy, R.D. Garrod, J., & Owen, R. (2008). Ecotoxicity test methods and environmental hazard assessment for engineered nanoparticles. *Ecotoxicology*. 17: 421–437. DOI: 10.1007/s10646-008-0215-z.

Das, P. Williams, C.J. Fulthorpe, R.R. Hoque, M.E. Metcalfe, C.D., & Xenopoulos, M.A. (2012). Changes in bacterial community structure after exposure to silver nanoparticles in natural waters. *Environmental Science and Technology*. 46: 9120–9128. DOI: 10.1021/es3019918.

Dasari, T.P., & Hwang, H.M. (2010). The effect of humic acids on the cytotoxicity of silver nanoparticles to a natural aquatic bacterial assemblage. *The Science of the Total Environment*. 408: 5817–5823. DOI: 10.1016/j.scitotenv.2010.08.030.

Deryabin, D.G. Efremova, L.V. Karimov, I.F. Manukhov, I.V. Gnuchikh, E.Y., & Miroshnikov, S.A. (2016). Comparative sensitivity of the luminescent *Photobacterium phosphoreum*, *Escherichia coli*, and *Bacillus subtilis* strains to toxic effects of carbon-based nanomaterials and metal nanoparticles. *Microbiology*. 85: 198–206. DOI: 10.1134/S0026261716020053.

Dhawan, A., & Sharma, V. (2010). Toxicity assessment of nanomaterials: methods and challenges. *Analytical Bioanalytical Chemistry*. 398: 589–605. DOI: 10.1007/s00216-010-3996-x.

do Amaral, D.F. Guerra, V. Motta, A.G.C. de Melo E Silva, D., & Rocha, T.L. (2019). Ecotoxicity of nanomaterials in amphibians: a critical review, *Science of the Total Environment*. 686: 332–344. DOI: 10.1016/j.scitotenv.2019.05.487.

Drasler, B, Sayre, P. Steinhäuser, K.G. Petri-Fink, A., & Rothen-Rutishauser, B. (2017). In vitro approaches to assess the hazard of nanomaterials. *NanoImpact*. 8: 99–116. https://doi.org/10.1016/j.impact.2017.08.002.

EPA Web Resource. https://www.epa.gov/expobox/exposure-assessment-tools-chemical-classes-nanomaterials, Accessed on December 13, 2020.

EPA Web Resource. https://www.epa.gov/reviewing-new-chemicals-under-toxic-substances-control-act-tsca/control-nanoscale-materials-under, Accessed December 13, 2020.

Falfushynska, H. Gnatyshyna, L. Horyn, O. Sokolova, I., & Stoliar, O. (2017). Endocrine and cellular stress effects of zinc oxide nanoparticles and nifedipine in marsh frogs *Pelophylax* ridibundus. *Aquatic Toxicology.* 185: 171–182. DOI: 10.1016/j.aquatox.2017.02.009.

Fong, P.P. Thompson, L.B. Carfagno, G.L.F., & Sitton, A.J. (2016). Long-term exposure to gold nanoparticles accelerates larval metamorphosis without affecting mass in wood frogs (Lithobates sylvaticus) at environmentally relevant concentrations. *Environmental Toxicology and Chemistry.* 35: 2304–2310. DOI: 10.1002/etc.3396.

Fortner, J.D. Lyon, D.Y. Sayes, C.M. Boyd, A.M. Falkner, J.C. Hotze, E.M. Alemany, L.B. Tao, W. Guo, W. Ausman, K.D. Colvin, V.L., & Hughes, J.B. (2005). C60 in water: nano crystal formation and microbial response. *Environmental Science and Technology.* 39: 4307–4316. http:/doi.org/10.1021/es048099n.

Freixa, A. Acuna, V. Gutierrez, M. Sanchís, J. Lucia, Santos, H.M.L.M. Rodriguez-Mozaz, S. Farre, M. Barcelo, D., & Sabater, S. (2018). Fullerenes influence the toxicity of organic micro-contaminants to river biofilms, *Frontiers in Microbiology.* 9: 1426. DOI: 10.3389/fmicb.2018.01426.

Galdiero, E. Falanga, A. Siciliano, A. Maselli, V. Guida, M, Carotenuto, R. Tussellino, M. Lombardi, L. Benvenuto, G., & Galdiero, S. (2017). Daphnia magna and Xenopus laevis as in vivo models to probe toxicity and uptake of quantum dots functionalized with gH625. *International Journal of Nanomedicine* 12: 2717–2731. DOI: 10.2147/IJN.S127226.

Gleiter, H. (2000). Nanostructured materials: basic concepts and microstructure. *Acta Materialia.* 48: 1–29. DOI: 10.1016/S1359–6454(99)00285-2.

Gottschalk, F., & Nowack, B. (2011). The release of engineered nanomaterials to the environment. *Journal of Environmental Monitoring*, 13: 1145–1155. DOI: 10.1039/C0EM00547A.

Grün, A.-L. Straskraba, S. Schulz, S. Schloter, M., & Emmerling, C. (2018). Long-term effects of environmentally relevant concentrations of silver nanoparticles on microbial biomass, enzyme activity, and functional genes involved in the nitrogen cycle of loamy soil. *Journal of Environmental Sciences.* 69: 12–22. DOI: 10.1016/j.jes.2018.04.013.

Grün, A.-L., & Emmerling, C. (2018). Long-term effects of environmentally relevant concentrations of silver nanoparticles on major soil bacterial phyla of a loamy soil. *Environmental Sciences. Europe.* 30: 31. DOI: 10.1186/s12302-018-0160-2.

Handy, R.D. Owen, R., & Valsami-Jones, E. (2008). The ecotoxicology of nanoparticles and nanomaterials: current status, knowledge gaps, challenges, and future needs. *Ecotoxicology.* 17: 315–325. DOI: 10.1007/s10646-008-0206-0.

Hansch, M., & Emmerling, C. (2010). Effects of silver nanoparticles on the microbiota and enzyme activity in soil. *Journal of Plant Nutrition and Soil Science.* 173: 554–558. DOI: 10.1002/jpln.200900358.

Hartmann, N.I.B. Skjolding, L.M. Hansen, S.F. Baun, A. Kjølholt, J., & Gottschalk, F. (2014). Environmental fate and behaviour of nanomaterials: New knowledge on important transformation processes. Danish Environmental Protection Agency. Environmental Project, No. 1594. https://orbit.dtu.dk/files/106171114/MST_rapport_Environmental_and_behaviour_of_nanomaterials.pdf.pdf.

He, X. Aker, W.G. Leszczynski, J., & Hwang, H.-M. (2014). Using a holistic approach to assess the impact of engineered nanomaterials inducing toxicity in aquatic systems. *Journal of Food and Drug Analysis*. 22: 128–146. DOI: 10.1016/j.jfda.2014.01.011.

Hegde, K. Brar, S.K. Verma, M., & Surampalli, R.Y. (2016). Current understandings of toxicity, risks and regulations of engineered nanoparticles with respect to environmental microorganisms. *Nanotechnology for Environmental Engineering*. 1: 5. DOI: 10.1007/s41204-016-0005-4.

Hochella, M.F. Jr. Lower, S.K. Maurice, P.A. Penn, L.R. Sahai, N. Sparks, D.L., & Twining B.S. (2008). Nanominerals, mineral nanoparticles, and Earth systems. *Science*. 319: 1631–1635. DOI: 10.1126/science.1141134.

Hochella, M.F. Jr. Mogk, D.W. Ranville, J. Allen, I.C. Luther, G.W. Marr, L.C. McGrail, B.P. Murayama, Q. Mitsu, M. Qafoku, N.P. Ross, K.M. Sahai, N. Schroeder, P.A. Vikesland, P. Westerhoff, P., & Yang, Y. (2019). Natural, incidental, and engineered nanomaterials and their impacts on the Earth system. *Science*. 363: 1–12. DOI: 10.1126/science.aau8299.

Holden, P.A. Gardea-Torresdey, J.L. Klaessig, F. Turco, R.F. Mortimer, M. Hund-Rinke, K. Hubal, E.A.C. Avery, D. Barceló, D. Behra, R. Cohen, Y. Deydier-Stephan, L. Ferguson, P.L. Fernandes, T.F. Harthorn, B.H. Henderson, W.M. Hoke, R.A. Hristozov, D, Johnston, J.M. Kane, A.B. Kapustka, L. Keller, A.A. Lenihan, H.S. Lovell, W. Murphy, C.J. Nisbet, R.M. Petersen, E.J. Salinas, E.R. Scheringer, M. Sharma, M. Speed, D.E. Sultan. Y. Westerhoff, P. White, J.C. Wiesner, M.R. Wong, E.M. Xing, B. Horan, M.S. Godwin, H.A., & Nel, A.E. (2016). Considerations of environmental relevant test conditions for improved evaluation of ecological hazards of engineered nanomaterials. *Environmental Science and Technology*. 50: 6124–6145. DOI: 10.1021/acs.est.6b00608.

Hu, X. Cook, S. Wang, P., & Hwang, H.M. (2009). In vitro evaluation of cytotoxicity of engineered metal oxide nanoparticles. *The Science of the Total Environment*. 407: 3070–3072. DOI: 10.1016/j.scitotenv.2009.01.033.

Inshakova, E., & Inshakov, O. (2017). World market for nanomaterials: structure and trends, *MATEC Web of Conferences. EDP Sciences*, 129: 1–5. DOI: 10.1051/matecconf/201712902013.

Isigonis, P. Hristozov, D. Benighaus, C. Giubilato, E. Grieger, K. Pizzol, L. Semenzin, E. Linkov, I. Zabeo, A., & Marcomini, A. (2019). Risk governance of nanomaterials: review of criteria and tools for risk communication, evaluation, and mitigation. *Nanomaterials*. 9. DOI: 10.3390/nano9050696.

Joner, E.J. Hartnik, T., & Amundsen, C.E. (2008). Environmental fate and ecotoxicity of engineered nanoparticles Norwegian Pollution Control Authority Report no. TA 2304/2007. As, Norway, Bioforsk, pp. 1–64.

Keller, A.A. McFerran, S. Lazareva, A., & Suh, S. (2013). Global life cycle releases of engineered nanomaterials. *Journal of Nanoparticle Research*. 15: 1692 DOI: 10.1007/s11051-013-1692-4.

Klaine, S.J. Alvarez, P.J. Batley, G.E. Fernandes, T.F. Handy, R.D. Lyon, D.Y. Mahendra, S. McLaughlin, M.J., & Lead, J.R. (2008). Nanomaterials in the environment: behavior, fate, bioavailability, and effects. *Environmental Toxicology and Chemistry*. 27: 1825–1851 DOI: 10.1897/08-090.1.

Kuempel, E.D., & Castranova, V. (2016). *Hazard and Risk Assessment of Workplace Exposure to Engineered Nanoparticles: Methods, Issues, and Carbon Nanotube Case Study, Assessing Nanoparticle Risks to Human Health*, Second Edition, William Andrew Applied Science Publishers. pp. 45–82. DOI: 10.1016/B978-0-323-35323-6.00003-7.

Kuhnel, D., & Nickel, C. (2014). The OECD expert meeting on ecotoxicology and environmental fate- towards the development of improved OECD guidelines for testing nanomaterials. *Science of the Total Environment*. 472:347–353.

Kumar, N. Palmer, G.R. Shah, V., & Walker, V.K. (2014). The effect of silver nanoparticles on seasonal change in arctic tundra bacterial and fungal assemblages. *PLOS ONE*. 9: e99953. DOI: 10.1371/journal.pone.0099953.

Lahive, E. Matzke, M. Durenkamp, M. Lawlor, A.J. Thacker, S.A. Pereira, M.G. Spurgeon, D.J. Unrine, J.M. Svendsen, C., & Lofts, S. (2017). Sewage sludge treated with metal nanomaterials inhibits earthworm reproduction more strongly than sludge treated with metal metals in bulk/salt forms. *Environmental Science: Nano*: 4: 78–88. DOI: 10.1039/C6EN00280C.

Landvik, N.E. Skaug, V. Mohr, B. Verbeek, J., & Zienolddiny, S. (2018). Criteria for grouping of manufactured nanomaterials to facilitate hazard and risk assessment, a systemic review of expert opinions. *Regulatory Toxicology and Pharmacology*. 95: 270–279. DOI: 10.1016/j.yrtph.2018.03.027.

Lin, D. Tian, X. Wu, F., & Xing, B. (2010). Fate and transport of engineered nanomaterials in the environment. *Journal of Environmental Quality*. 39: 1896–908. DOI: 10.2134/jeq2009.0423. PMID: 21284287.

Lowry, G.V. Gregory, K.B. Apte, S.C., & Lead, J.R. (2012). Transformations of Nanomaterials in the Environment. *Environmental Science and Technology*. 46: 6893–6899. https://doi.org/10.1021/es300839e.

Lyon, D.Y. Adams, L.K. Falkner, J.C., & Alvarez, P.J.J. (2006). Antibacterial activity of fullerene water suspensions: effects of preparation method and particle size. *Environmental Science and Technology*. 40: 4360–4366. DOI: 10.1021/es0603655.

Marín-Barba, M. Gavilán, H. Gutiérrez, L. Lozano-Velasco, E. Rodríguez-Ramiro, I. Wheeler, G.N. Morris, C.J. Morales, M.P., & Ruiz, A. (2018). Unravelling the mechanisms that determine the uptake and metabolism of magnetic single and multicore nanoparticles in a Xenopus laevis model. *Nanoscale*. 10: 690–704. DOI: 10.1039/c7nr06020c.

Mauter, M.S., & Elimelech, M. (2008). Environmental applications of carbon-based nanomaterials. *Environmental Science and Technology* 42: 5843–5859. DOI: 10.1021/es8006904.

McKee, M.S., & Filser, J. (2016). Impacts of metal-based engineered nanomaterials on soil communities. *Environmental Science: Nano*. 3: 506–533. DOI: 10.1039/C6EN00007J.

Mouchet, F. Landois, P. Sarremejean, E. Bernard, G. Puech, P. Pinelli, E. Flahaut, E., & Gauthier, L. (2008). Characterisation and in vivo ecotoxicity evaluation of double-wall carbon nanotubes in larvae of the amphibian *Xenopus laevis*. *Aquatic Toxicology*. 87: 27–137. DOI: 10.1016/j.aquatox.2008.01.011.

Mueller, N.C., & Nowack, B. (2008). Exposure modeling of engineered nanoparticles in the environment. *Environmental Science and Technology*. 42: 4447–4453. DOI: 10.1021/es7029637.

National Research Council, 1983. Risk Assessment in the Federal Government. Managing the Process. National Academy Press, Washington, DC. https://doi.org/10.17226/366.

Nations, S. Wages, M. Cañas, J.E. Maul, J. Theodorakis, C., & Cobb, G.P. (2011). Acute effects of Fe_2O_3, TiO_2, ZnO and CuO nanomaterials on Xenopus laevis. *Chemosphere* 83: 1053–1061. DOI: 10.1016/j.chemosphere.2011.01.061.

Neal, A.L. (2008). What can be inferred from bacterium–nanoparticle interactions about the potential consequences of environmental exposure to nanoparticles? *Ecotoxicology*. 17: 362–371. DOI: 10.1007/s10646-008-0217-x.

Oberdorster E. (2004). Manufactured nanomaterials (fullerenes, C_{60}) induce oxidative stress in the brain of juvenile largemouth bass. *Environ Health Perspectives.* 112: 1058–1062. DOI: 10.1289/ehp.7021.

Pokropivny, V.V., & Skorokhod, V.V. (2007). Classification of nanostructures by dimensionality and concept of surface forms engineering in nanomaterial science. *Materials Science and Engineering: C.* 27: 990–993. DOI: 10.1016/j.msec.2006.09.023.

Puzyn, T. Rasulev, B. Gajewicz, A. Hu, X. Dasari, T.P. Michalkova, A. Hwang, H.-M. Toropov, A, Leszczynska, D., & Leszczynski, J. (2011). Using nano-QSAR to predict the cytotoxicity of metal oxide nanoparticles. *Nature Nanotechnology.* 6: 175–178. DOI: 10.1038/nnano.2011.10.

Rana, S., & Kalaichelvan, P.T. (2013). "Ecotoxicity of nanoparticles." *International Scholarly Research Notices.* DOI: 10.1155/2013/574648.

Rocha, T.L. Gomes, T. Sousa, V.S. Mestre, N.C., & Bebianno, M.J. (2015). Ecotoxicological impact of engineered nanomaterials in bivalve molluscs: an overview. *Marine Environmental Research.* 111: 74–88. DOI: 10.1016/j.marenvres.2015.06.013.

Rocha, T.L. Mestre, N.C. Sabóia-Morais, S.M.T., & Bebianno, M.J. (2017). Environmental behaviour and ecotoxicity of quantum dots at various trophic levels: a review. *Environment International.* 98: 1–17. DOI: 10.1016/j.envint.2016.09.021.

Rodrigues, D.F., & Elimelech, M. (2010). Toxic effects of single-walled carbon nanotubes in the development of E. coli biofilm. *Environmental Science and Technology.* 44: 4583–4589. DOI: 10.1021/es1005785.

Salvaterra, T. Alves, M.G. Domingues, I. Pereira, R. Rasteiro, M.G. Carvalho, R.A. Soares, A.M.V.M., & Lopes, I. (2013). Biochemical and metabolic effects of a short-term exposure to nanoparticles of titanium silicate in tadpoles of *Pelophylax perezi* (Seoane). *Aquatic Toxicology.* 128: 190–192. DOI: 10.1016/j.aquatox.2012.12.014.

SCENIHR (Scientific Committee on Emerging and Newly-Identified Health Risks). (2007). The appropriateness of the risk assessment methodology in accordance with the Technical Guidance Documents for new and existing substances for assessing the risks of nanomaterials, 21–22 June. https://ec.europa.eu/health/ph_risk/committees/04_scenihr/docs/scenihr_o_010.pdf.

Scott-Fordsmand, J.J. Peijnenburg, W. Semenzin, E. Nowack, B. Hunt, N. Hristozov, D. Marcomini, A. Irfan, M.A. Jiménez, A.S. Landsiedel, R. Tran, L. Oomen, A.G. Bos, P., & Hund-Rinke, K. (2017). Environmental risk assessment strategy for nanomaterials. *International Journal of Environmental Research and Public Health.* 14: 1251. DOI: 10.3390/ijerph14101251.

Selck, H. Handy, R.D. Fernandes, T.F. Klaine, S.J., & Petersen, E.J. (2016). Nanomaterials in the aquatic environment: A European Union–United States perspective on the status of ecotoxicity testing, research priorities, and challenges ahead, *Environmental Toxicology and Chemistry*, 35: 1055–1067. DOI: 10.1002/etc.3385.

Tao, X. Yu, Y. Fortner, J.D. He, Y. Chen, Y., & Hughes, J.B. (2015). Effects of aqueous stable fullerene nanocrystal (nC_{60}) on Scenedesmus obliquus: evaluation of the sub-lethal photosynthetic responses and inhibition mechanism. *Chemosphere.* 122: 162–167. DOI: 10.1016/j.chemosphere.2014.11.035.

Theng, B.K.G., & Yuan, G. (2008). Nanoparticles in the soil environment. *Elements.* 4: 395–399. DOI: 10.2113/gselements.4.6.395.

Tsuji, J. S. Maynard, A.D. Howard, P.C. James, J. T. Lam, C-W. Warheit, D.B., & Santamaria, A.B. (2006). Research strategies for safety evaluation of nanomaterials, part iv: risk assessment of nanoparticles. *Toxicological Sciences.* 89: 42–50. DOI: 10.1093/toxsci/kfi339.

Vance, M.E. Kuiken, T. Vejerano, E.P. McGinnis, S.P. Hochella, M.F. Jr. Rejeski, D., & Hull, M.S. (2015). Nanotechnology in the real world: Redeveloping the nanomaterial consumer products inventory. *Beilstein Journal of Nanotechnology.* 6: 1769–1780 DOI: 10.3762/bjnano.6.181.

von der Kammer, F. Ferguson, P.L. Holden, P.A. Masion, A. Rogers, K.R. Klaine, S.J. Koelmans, A.A. Horne, N., & Unrine, J.M. (2012). Analysis of engineered nanomaterials in complex matrices (environment and biota): General considerations and conceptual case studies. *Environmental Toxicology and Chemistry.* 31: 32–49. DOI: 10.1002/etc.723.

Wang, D. Gao, Y. Lin, Z. Yao, Z., & Zhang, W. (2014). The joint effects on Photobacterium phosphoreum of metal oxide nanoparticles and their most likely coexisting chemicals in the environment. *Aquatic Toxicology.* 154: 200–206. DOI: 10.1016/j.aquatox.2014.05.023.

Weinberg, H. Galyean, A., & Leopold, M. (2011). Evaluating engineered nanoparticles in natural waters. *Trends in Analytical Chemistry.* 30: 72–83. DOI: 10.1016/j.trac.2010.09.006.

Wentworth, J., & Milodowski, R. (2017). POSTNOTE. Risk Assessment of Nanomaterials 562. https://post.parliament.uk/research-briefings/post-pn–0562/

Wiesner, M.R. Lowry, G.V. Alvarez, P. Dionysiou, D., & Biswas, P. (2006). Assessing the risks of manufactured nanomaterials. *Environmental Science and Technology.* 40: 4336–4345. https://doi.org/10.1021/es062726m.

Wigginton, N.S. Haus, K.L., & Hochella, M.F. (2007). Aquatic environmental nanoparticles. *Journal of Environmental Monitoring.* 9: 306–1316. DOI: 10.1039/B712709J.

Index

Note: **Bold** page numbers refer to tables and *italic* page numbers refer to figures.

For Product Safety Concerns and Information please contact our EU
representative GPSR@taylorandfrancis.com
Taylor & Francis Verlag GmbH, Kaufingerstraße 24, 80331 München, Germany

www.ingramcontent.com/pod-product-compliance
Lightning Source LLC
Chambersburg PA
CBHW060759220326
41598CB00022B/2496

* 9 7 8 0 3 6 7 6 5 3 4 8 4 *